Predicting Disasters

CRITICAL STUDIES IN RISK AND DISASTER

Series editors: Kim Fortun, Scott Gabriel Knowles, and Jacob A. C. Remes

Critical Studies in Risk and Disaster explores how environmental, technological, and health risks are created, managed, and analyzed in different contexts. Global in scope and drawing on perspectives from multiple disciplines, volumes in the series examine the ways that planning, science, and technology are implicated in disasters. The series also engages public policy formation, including analysis of science, technology, and environmental policy, as well as welfare, conflict resolution, and economic policy developments where relevant.

A complete list of books in the series is available from the publisher.

PREDICTING DISASTERS

Earthquakes, Scientists, and Uncertainty in Modern Japan

Kerry Smith

PENN

UNIVERSITY OF PENNSYLVANIA PRESS

PHILADELPHIA

Published by
University of Pennsylvania Press
Philadelphia, Pennsylvania 19104-4112
www.pennpress.org

Printed in the United States of America on acid-free paper
10 9 8 7 6 5 4 3 2 1

Hardcover ISBN: 978-1-5128-2537-4
eBook ISBN: 978-1-5128-2536-7

A catalogue record for this book is available from the Library of Congress.

For Ian

CONTENTS

Introduction

Japan is a place where powerful earthquakes have occurred more frequently and caused more harm in the modern era than they have in all but a handful of other locations on the planet. Some of these seismic events—the 1891 Nōbi earthquake (which left more than 7,000 people dead), the 1896 Sanriku earthquake (more than 27,000 deaths), the Great Kantō Earthquake of 1923 (105,000 killed), the Great Hanshin-Awaji Earthquake in 1995 (6,500 killed), and the Great East Japan Earthquake of 2011 (20,000 dead or missing)—have been so profoundly destructive that they are as much ruptures in Japan's national narrative as they are of the lithosphere below. The intervals between those singular catastrophes, meanwhile, were filled with a much larger number of earthquake disasters with relatively smaller historical footprints. Since 1900, earthquakes responsible for at least one fatality have occurred every other year, on average; a dozen were responsible for a hundred or more deaths (in addition to those in 1923 and 1995).[1] Earthquake disasters took almost as many lives in the twentieth century alone as they had in all of Japan's recorded history up to that point.[2]

Seismicity's impacts on modern Japan have taken many forms. The human suffering and harms to the built environment that earthquakes frequently left behind are in a category all their own in that regard, of course, and are the most legible metrics for understanding any single event's effects. A host of efforts to reduce that suffering and those harms, meanwhile, have over time changed what the built landscape looks like in Japan, and to some degree the structures and routines of daily life as well. Ferro-concrete and reinforced masonry began to displace wood as the construction material of choice in Tokyo and elsewhere in Japan as architects and engineers applied what their fields had learned from the effects of the Great Kantō Earthquake, for example.[3] Subsequent earthquake disasters triggered further adaptations to

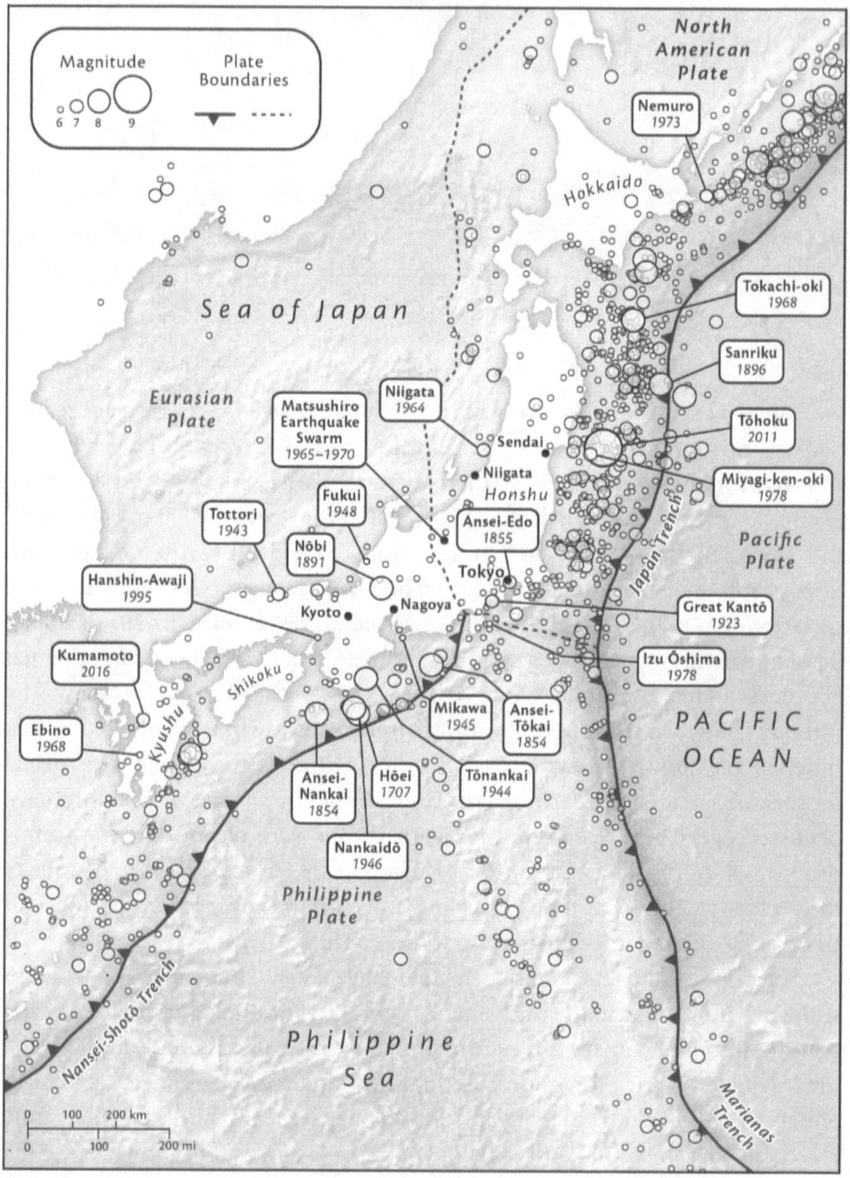

Figure 1. Earthquakes, M6 and above. Includes pre-1900 events mentioned in the text; all others are 1900–2022. Map by Erin Greb Cartography.

building codes and urban planning practices, altering not just how structures looked and functioned but how public space itself was arranged.[4] On a different register, every September 1 since 1960, Japan's "Disaster Preparedness Day" (Bōsai no hi) has brought together community groups, school children, first responders, and other agencies into carefully choreographed performances of the actions they would presumably take after the next big earthquake.[5] Mock evacuations, elaborate search-and-rescue drills, and displays of firefighting equipment and their capabilities are all common exercises. The events and the planning that make them possible are at once unique to the communities they serve and reflections of how commonplace and sophisticated the anticipation of major earthquake disasters has become in Japan.

Seismicity's impacts are also evident in the work scientists in Japan did to try to understand it. Modern seismology's global history essentially begins in late nineteenth-century Japan. One of the first things that the new Meiji government did to prepare the country to compete with the West after it took power in 1868 was to hire engineers, geologists, and other experts from abroad so that they could share what they knew with the relevant officials and the country's growing cohort of students. When a handful of these foreign experts took up the systematic study of earthquakes in the 1880s, they did so out of curiosity, and quite apart from their original purposes for coming to Japan. John Milne, James Alfred Ewing, and Thomas Gray founded the Seismological Society of Japan in 1880, designed and built the first modern seismometers, and helped train the Japanese scientists who led the field for many years thereafter.[6] Aided by a growing network of reporting stations and increasingly sensitive instrumentation, scientists in Japan were singularly well-positioned to record seismic activity and to disseminate their findings to global and local audiences. Japan led the late nineteenth- and early twentieth-century world in the production of knowledge about earthquakes and in the development of seismology as a scientific discipline.

The work of trying to understand how and why earthquakes happen was shaped from the start in Japan by repeated encounters with evidence of just how vulnerable the nation was to them. The 1891 Nōbi disaster, for example, prompted the Meiji government to appoint prominent researchers to an Imperial Earthquake Investigation Committee (IEIC) in an explicit effort to link scientific labor to the production of answers to questions about how to protect the country from seismicity's hazards.[7] Other interventions with similar aims followed, some government-mandated and others prompted by the scientists themselves; as with the creation of the IEIC, they usually took shape in the aftermath of earthquake disasters and in full view of the media and the public. The Earthquake Research Institute (ERI hereafter, or Jishin kenkyūjo) at Tokyo Imperial University opened its doors two years after the Great Kantō Earthquake, in the wake of a post-disaster debate over how seismology and the earth sciences could best help reduce the harm the nation's next catastrophes might inflict. More debates along those same lines and further moves to reshape research practices and priorities in seismology surfaced repeatedly after that: following the 1946 Nankaidō earthquake, in the late 1960s after the Matsushiro earthquake cluster, after the 1995 Hanshin-Awaji disaster, and following the tragedies in Fukushima and elsewhere on March 11, 2011 (known as 3.11).

While the focus of seismological research in Japan was never just about the search for ways to protect the nation from earthquake disasters, for some scientists and at some points in the field's history, that goal was paramount or very close to it. In the broadest possible sense, what all of those experts were after were insights that would allow them to reduce the uncertainty associated with the timing and location of potentially destructive earthquakes. Seismologists in Japan started looking for ways to reveal when and where destructive earthquakes would occur before they actually did—to forecast or predict them—not long after the Nōbi earthquake and, in one form or another, have continued to do so since. It isn't difficult to understand why. Many of the hazards associated with powerful earthquakes could in theory be mitigated to some degree if their locations and when they were most likely to occur were known in advance. Early warning that an earthquake was imminent would give people time they wouldn't otherwise have to evacuate unsafe buildings, flee to higher ground to escape a tsunami, secure flammable materials and potential sources of ignition to prevent all-too-common post-earthquake conflagrations, and so on.

Nor is it difficult to understand why many scientists from the late nineteenth century to the present have viewed earthquake forecasting and prediction as problems they could solve. Seismologists in Japan often drew attention to advances in fields like meteorology as a model for their own aspirations, and to make the case that many natural phenomena that had once defied understanding no longer did, thanks to the hard work of pioneering scientists.[8] The weather forecasts that were already beginning to appear in Japanese newspapers in the late 1880s were one sign of science's rapid advances, each subsequent improvement in their accuracy and scope further evidence that even the most complex natural systems functioned in predictable ways.

The search for ways to reveal where and when destructive earthquakes would occur has included the production of forecasts that, together with probabilistic models, were intended to identify periods of heightened risk for a particular locale that were months, years, or even decades in the future. The pursuit of short-term prediction, in which scientists hoped to rely on the detection of unusual seismic activity, deformations in the earth's crust, or signs of other anomalous phenomenon in order to warn that an earthquake was minutes, hours, or a few days away, has also been part of the search. These strategies have played important roles in shaping both the Japanese public's understanding of the risks they faced from seismicity and the government's disaster-mitigation efforts.

Forecasts come in many different forms. They can be the products of fairly straightforward calculations of how much time has passed between major events at the same site to estimate when the next one was likely to occur. Seismologist Imamura Akitsune's 1905 warning about Tokyo is a case in point; he noted then that since Tokyo seemed to suffer destructive earthquakes every one hundred years or so and, since the last such event was in 1855, it made sense for the city's residents to start preparing for the next one sooner rather than later. Kawasumi Hiroshi's announcement in the 1960s that he had discovered a sixty-nine-year cycle of destructive earthquakes in the vicinity of Tokyo and Yokohama, and that the district was therefore about to enter a period when another one was quite likely, is another, later expression of a similar argument. More recently, the government has turned to more sophisticated probabilistic calculations of future risk, the results of which are shared with the Japanese public in the form of detailed national seismic hazard maps.[9]

Attempts to put short-term earthquake prediction into practice, meanwhile, started earlier and have persisted longer in Japan than in any of the other settings where that approach might once have seemed promising. A nationwide effort to coordinate and promote scientific research into short-term prediction that began in the 1960s was still underway in the 2020s. Many of the country's most prominent earth scientists and research institutions have been involved in the project over the years, as have an array of government agencies. While the pursuit of earthquake prediction was never the only problem that seismologists in Japan were working on, there really hasn't been a time since the 1960s when it wasn't one of them. At its peak in the 1970s, policy makers' confidence in prediction's potential was so strong that they made it the cornerstone of the country's plans for minimizing the effects of the long-awaited and potentially catastrophic Tōkai earthquake. That faith has waned in recent years but the legislative and institutional frameworks that support the pursuit of short-term prediction as part of earthquake disaster-mitigation planning have remained largely in place.

Neither short-term prediction nor longer-term forecasting produced the results scientists had hoped they would. Claims that careful scrutiny would eventually reveal the precursory signs needed for short-term prediction did not pan out. No such signs were found. Researchers in Japan did announce on more than one occasion that they had recorded unusual measurements in parts of the country shortly before an earthquake struck there, but these "predictions" were always offered retroactively, after the event itself and after scientists pored over the monitoring data in search of clues. None of the post-hoc discoveries proved useful when it came to spotting earthquakes that hadn't happened yet. In 2017, after forty years of insisting that short-term earthquake prediction's methods would provide a warning before the long-feared Tōkai earthquake, government officials publicly acknowledged that they no longer believed this to be true.

Forecasting efforts have had issues of their own. Both Kawasumi's warning in the 1960s about the rising risk to Tokyo and another the following decade about a significant threat to the Tōkai district prompted the government to hurriedly introduce disruptive and costly anti-earthquake disaster measures. Neither of those earthquakes followed the schedule that the public had expected them to abide by. The more recently employed probabilistic methods and the maps that they help create are also problematic, critics have said, on more than one register.[10] The specificity of the information the maps conveyed implied that some parts of the country were significantly safer than others, a conclusion that critics said was simply not valid. They have repeatedly assigned low probabilities of a destructive earthquake happening to parts of Japan that subsequently experienced one. This was the case for Kumamoto before it was struck by deadly earthquakes in 2016. Similarly, prior to the March 11, 2011 disaster, government maps had associated the area off Japan's northeast coast with a very high earthquake risk, which was not incorrect, but also significantly and tragically underestimated how powerful that earthquake could be, which was. Finally, these probabilistic methods for assessing future earthquake risks were also called into question by earth scientists who challenged some of their basic premises, including the assumption that past patterns of seismic activity repeat.[11] If the past was not a reliable guide when it came to assessing the likelihood of future earthquakes or their intensity, then neither were probabilistic seismic hazard maps. Some of those experts have called on the government to admit that it cannot be sure when or where the next major earthquake will occur—to fully acknowledge the uncertainty around seismicity—and to invest instead in measures with a proven track record of reducing harm. It has been easier for scientists and policy makers to concede the necessity of the second step than it has been to take the first.

In the early 2020s, the Japanese government was still using hazard maps and the assumptions on which they were based to convey its grasp of the risks facing the nation. Scientists expressed particular concern about three future events that they said posed special hazards. The *shuto chokkagata*, or "directly beneath the capital district" earthquake, is one of them; an earthquake of this kind would follow a rupture (or ruptures) within one or more of the several tectonic plates under Tokyo and its environs. A Sagami Trough earthquake is another; when it happens, its mechanism will be similar to the one that destroyed much of Tokyo and Yokohama in 1923. The likelihood of this earthquake occurring in the near term is said to be low. Both it and a third anticipated event, this one along the Nankai Trough, are expected to be megathrust earthquakes, in which sectors of one tectonic plate "slip" beneath those of another. (Not all megathrust earthquakes are massive, but the planet's most powerful earthquakes are almost invariably megathrust events.) As of 2020, the Japanese government's position was that the likelihood of a Nankai Trough earthquake occurring over the next 30 years was somewhere between 70 and 80 percent.

Were any of these scenarios to come to pass, studies tell us that the devastation could be immense. Researchers estimate that in a worst-case scenario, a *chokkagata* earthquake would leave some 23,000 people dead. A repeat of the earthquake like the one that struck the capital in 1923 might kill almost three times that number.[12] The public has been told that the collapsed buildings, fires, and tsunami that would follow a Nankai Trough earthquake could cause as many as 320,000 deaths up and down Japan's Pacific Coast. Needless to say, the economic impact of any one of these catastrophes would also be far reaching.[13]

Disasters

Predicting Disasters is a history of scientists' efforts to reduce the uncertainty around the timing and location of powerful earthquakes in modern Japan. Earthquake forecasting and prediction have been a focus of scientific inquiry in Japan since the late nineteenth century, and still had the attention of the earth sciences community in the early twenty-first. This book's exploration of that project and the researchers caught up in it is in part an intervention in the typical narratives historians have relied on to describe the challenges modern Japan has faced and overcome. One of its goals is to show how scientists struggled to make large-scale earthquake disasters legible to the public and to policy makers as significant threats to Japan's future and as phenomena that could be understood and prepared for. Another is to make clear why those struggles matter; I argue that disasters, scientists, and uncertainty belong alongside more familiar subjects of analysis in modern Japanese history, such as economic growth and its impacts, political crises and popular protest, and even the legacies of the war, for the work they do in helping us better understand how the past has influenced beliefs about Japan's possible futures, and how beliefs about the future shaped the present. Finally, this study also attempts to make relevant elements of Japan's past more accessible to anyone interested in histories of disaster, scientific communities, and uncertainty more generally.

The idea that disasters have their own histories is a relatively recent one. Historians were latecomers to the study of modern disasters as something other than "natural" events—that is, that they were instead products of a particular cultural and political moment and the rhetorical strategies that defined what they meant. (Sociologists got there first; their studies began as Cold War–era efforts to head off wide-scale panic and disruption in the newly precarious future by understanding how communities had responded under extreme circumstances in the past.)[14] A turn by historians toward thinking more critically about the role disasters in general and earthquake disasters in particular might play in modern society was well underway by the end of the twentieth century. Part of what that shift did was show how the public's understanding of specific catastrophic events and of their implications for the future were shaped—often intentionally—by political and economic agendas that didn't always put the public's best interests first.[15] That work has shown us that the meanings assigned to disasters had never been objective. They didn't emerge fully formed once the dust had settled, but were the products of sometimes conflicting arguments over whose expertise, privilege, or suffering mattered most.

The earliest products of that analytical shift focused mostly on the US experience, but studies of Japan's were not far behind.[16] A wave of scholarship in the first two decades of this century dove deep into the histories of the 1855 Ansei-Edo and 1891 Nōbi earthquakes, for example, and focused particular attention on the Great Kantō Earthquake.[17] We learned from these works how each of these disasters shaped knowledge production across a variety of disciplines, from architecture to seismology and beyond. Each challenged educators, artists, writers, and eventually filmmakers to capture the devastation all around them, and to reflect on its meanings for the nation.[18] In each instance, scholars also showed how efforts to assign meaning to these events highlighted tensions running through Japanese society at the time. In Japan, as elsewhere, how societies prepared for and responded to each disaster was very much a function of their engagement with some of the modern era's most persistent and widespread challenges. Whether the problems were rising economic inequality, unsustainable industrialization and urbanization, or the appeal of political extremism in times of crisis, disasters could be both a lens that brought them into focus for all to see and a mechanism that allowed the state to assert (or, more rarely, subvert) calls for a return to normalcy.[19]

This study differs from that earlier research in several ways. It is the first work in English to focus on disaster and Japan in the latter half of the twentieth century. Recent scholarship on events like the Ansei, Nōbi, and Great Kantō earthquakes have highlighted the importance of their legacies, but it is fair to say that disaster was not viewed as all that relevant to historians documenting Japan's postwar transformation. In most accounts of Japan's long postwar era, disasters were at best ancillary elements of narratives focused on the country's economic recovery, the high-speed growth that followed, the rise of the middle class, and the war's legacies.[20] As a category of analysis, the disasters of speculative fiction and film for many years attracted more attention from scholars than its real ones.[21] After Japan's defeat, *Gojira*'s 1954 debut marked the start not just of the eponymous *kaijū*'s film career but of a new cinematic genre in which mass destruction—usually of Tokyo but nowhere in Japan was truly safe—was a central and inevitable theme. Depictions of dystopian and post-apocalyptic futures for Japan became commonplace elements in *anime* and other media not long afterwards. Novels, movies, and the occasional television show in which catastrophic "natural" disasters are central to the plot began to make their appearances in the 1970s. Komatsu Sakyō's bestselling 1973 novel *Nihon Chinbotsu* (*Japan Sinks*), for example, borrowed some of its ideas from newly emergent plate tectonics theory to construct a scenario in which massive earthquakes, tsunamis, and volcanoes were precursors to the country's total annihilation.[22] Scholars often turned to *Gojira*, films like it, and other examples of what Susan Napier has described as cultural "panic sites" to better understand how the trauma of the atomic bombings, of defeat, and of occupation were processed in postwar Japan.[23]

The war and its legacies loom large in critical histories of the postwar, and for good reason, but Japan's encounters with deadly earthquake disasters and with the prospects of more ahead have not. The work that has gone into understanding the experiences of wartime, as useful as it is, does not shed much light on the role that other disasters have played in shaping Japan's history since the middle of the twentieth century. Average citizens, policy makers, and scientists seldom invoked the experiences and legacies of wartime mass civilian trauma in their assessments of postwar disasters, including those that had yet to occur. I say this with some confidence because I had assumed that references to those experiences would surface in the postwar era as awareness of the potential for new disasters grew. I kept a close eye out in popular journals, scientific literature, works of fiction, and other sources for any evidence that would support that assumption. To my surprise, the war's effects seldom came up in analyses of or speculation about Japan's postwar disasters. I suspect that one reason they didn't is not that the horrors associated with them were forgotten, but rather that they are remembered as singular events, products of circumstances that Japan would hopefully never again confront. Their utility as reminders of what future earthquake disasters might do was limited. What I discovered instead, and engage with throughout the book, is that it was accounts and memories of encounters with previous "natural" disasters that provided the shared vocabulary for descriptions of their postwar counterparts. There are, of course, important exceptions. Because it shelters the remains of victims of both the Great Kantō Earthquake and the firebombing of March 1945, the Tokyo Metropolitan Memorial Hall is itself an argument that nothing separated one disaster from the other. Yoshino Yatarō, who built his own private disaster shelter in the Kōtō delta in the 1970s, cited his encounters with both the 1923 earthquake and the 1945 firebombing as the sources of his fear about what would happen if another earthquake struck the city. After 3.11, it was not uncommon for observers to mention the aftermaths of the atomic bombings of Hiroshima and Nagasaki when discussing conditions in Fukushima.

The earthquake disasters this book focuses on include several that hadn't happened yet but that experts and average citizens alike feared were about to. Sometimes the anxiety they invited was short-lived, as was the case when earthquakes that scientists had said were imminent failed to materialize; sometimes the anticipated disasters remained just over the horizon for quite a long time. The earthquake threats to Tokyo and to the Tōkai district that were first described in detail in the 1960s and 1970s are two important examples of events in the latter category. The processes that created these potential catastrophes began with the discovery of previously unrecognized seismic threats and continued in the form of efforts to describe the potential effects of such an event. Plans for how best to prepare for the coming disasters soon followed. In each of these two instances, the scientific understanding of the underlying seismic mechanisms and ideas about how to defend against them changed over time, but did so without significantly altering the generally grim assessments of what would happen when those earthquakes struck. If anything, expectations about the severity of the potential disasters tied to the earthquakes that were believed to be in Tokyo's and the Tōkai district's futures became more dire, not less, over the years.

The work that went into the creation of the anticipated earthquake disasters of the 1960s and 1970s and the choices that have kept them relevant since then were well documented at almost every step of the way by the scientists involved and by journalists, among other observers. This study draws on those sources to show that the advent of potential catastrophes like the ones associated with Tokyo and the Tōkai district marked an inflection point in the history of the relationships between disaster and society in modern Japan. The warnings that began in the 1960s, along with the descriptions of the earthquakes to come and the harm they would do, all argued that Japan's future was going to have complex, far-reaching earthquake disasters in it, and probably always would. That idea obtained widespread acceptance fairly quickly and has been part of common perceptions of what lies ahead for Japan ever since. If the status quo before this shift took place was one in which earthquake catastrophes were unpredictable events that would have to be suffered through only if they actually happened, the one that replaced it insisted that constant anticipation of a dangerous future was the new normal.

The changes in the relationships between disaster and society that this book analyzes includes some that were documented in legislative initiatives and public opinion surveys and others that took more concrete forms, sometimes quite literally. This book tells the story of how Tokyo's leaders tried to protect the capital from the earthquake believed to be threatening it by building a network of massive shelters in the city's Kōtō delta, for example. That project began in the late 1960s and was followed by other interventions for the entire metropolis. We will examine the circumstances that led to the passage of a sweeping new national disaster prevention law in the early

1960s and of a package of policies in the late 1970s that enlisted scientists and their work on prediction into the government's struggle against the next Tōkai earthquake.[24] *Predicting Disasters* also explores changes over time in the imagery and discursive shortcuts used by experts and non-expects alike to describe future earthquake catastrophes and to explain their significance in the present.

The significance of this shift toward anticipated earthquakes and their effects is also in part a function of how long it has continued to shape attitudes toward disaster in Japan. One of the goals of the book is to show how and explain why many of the beliefs about earthquake disasters that were new in the 1960s and 1970s—and the interventions they triggered—were still in place decades later. This was despite a growing body of evidence challenging the premises behind the choices policy makers and some scientists made early on in that process. Those tensions were most evident in the aftermaths of Japan's major postwar earthquake disasters that were not in Tokyo or the Tōkai district, which as of 2023 was still all of them. This study looks to the debates that followed the 1995 Hanshin-Awaji earthquake and the catastrophes of March 2011 to understand how those events called into question long-standing assumptions about what parts of Japan were at greatest risk, and what to do about it. I'm able to show that while both tragedies did alter Japan's relationships with disaster to some degree, those changes did not come at the expense of the status quo that had been in place since the 1970s. Rather, they simply added new layers of complexity to decades-old frameworks for thinking about and preparing for potential earthquake disasters.

Finally, this study argues that the Japanese public experienced and learned about earthquake disasters in the modern era in multiple temporalities—there were those that had happened a long time ago, those that had happened only recently, and those that hadn't happened yet—and that these encounters with past, present and future destructive seismic events were often untethered from any single incident as it unfolded in the present. Seen in that light, disaster in Japan is best thought of as an ongoing phenomenon, as the product of what historians Andy Horowitz and Jacob Remes have described as "the processes that create them as ideas, cause them as material facts, and define them as human experiences."[25] The time between disasters, I argue, is when scientists, policy makers, and the public did the work not just of recovery and reconstruction, as important as those steps were, but also took on the burden of imagining the next catastrophe in light of the one just past.

Scientists

Science and scientists, not unlike disasters, have been something of a blindspot in our understanding of the processes that have shaped Japan's history since the mid-twentieth century.[26] Their absence is particularly striking when considered alongside the relative abundance of analyses of scientists, their work, and their communities in the contemporaneous United States, the Soviet Union, and Western Europe. To the extent that it was the arms race or other elements of the Cold War that provided the analytical and narrative frameworks for studies of scientists in those settings—as was often the case—the gaps where Japanese scientists would have been are understandable. Like their counterparts in the West and the Soviet Union, Japanese researchers studying the effects of exposure to radiation or the threats posed by nuclear fallout were interested in problems that were novel products of that era's tensions.[27] Japan's status as a non-nuclear "pacifist" power, however, its dependence on the United States for its security, and its government's relentless pursuit of economic growth arguably produced different relationships between scientists and the state and scientists and society more broadly than the ones the analytical frameworks associated with "Cold War science" were designed to illuminate.[28] This study argues that one way to begin seeing scientists' interventions in Japan's postwar history is by exploring the evolution of those relationships at the intersections of knowledge production about disasters, earthquakes, and uncertainty.

When I first started work on this project, my focus was on seismologist Imamura Akitsune (1870–1948) and his involvement in the post-1923 debates over earthquake prediction and whether it should be part of his chosen field's research priorities. One of the qualities that made Imamura so interesting (and challenging) to research and write about was that he had been a prolific author with wide-ranging professional and personal interests, and a fondness for being in the public eye. He left quite a paper trail.[29] The publication in 1905 of his new textbook on seismology, for example, was a first for the field in Japan and an early sign of Imamura's desire to engage with the broadest possible audience.[30] Unfortunately for him, Imamura's speculation about the timing of Tokyo's next earthquake in the book and in public triggered a professional backlash that for many years limited his opportunities to write for a general audience. After the Great Kantō Earthquake, Imamura's ambitions were rekindled and he not only maintained a steady stream of publications within his field but also found time to write for the broader readership he had long sought. A significant share of the considerable content Imamura produced after 1923 and up until his death in 1948 had to do with his efforts to forecast and perhaps even predict earthquakes.

For most of the first half of the twentieth century, Imamura was one of the few seismologists in Japan to write about earthquakes and uncertainty at length, and perhaps the only one who tried to put his theories on the topic into practice.[31] (At one point, he funded his own monitoring network in the hopes of catching precursory signs of the next Nankaidō earthquake.) The cohort of scientists who inherited the debates over earthquakes and uncertainty that Imamura had helped launch was considerably larger, and it is that group that *Predicting Disasters* is primarily concerned with.

"Larger" in this context is relative; the number of scientists whose careers and public profiles were closely tied to efforts to radically reduce the uncertainty around earthquakes in Japan since the 1950s was smaller than one might think. Among the key scientists in that cohort were Kawasumi Hiroshi (1904–1972) and Hagiwara Takahiro (1908–1999), both of whom started in the field early enough that their professional lives overlapped with Imamura's at Tokyo Imperial University, if only briefly, as well as with the launch of the earthquake prediction project in the 1960s. Like Imamura, both Kawasumi and Hagiwara witnessed the horrors of the Great Kantō Earthquake firsthand. Kawasumi was a high school student in Tokyo at the time and, in later years, often referred to his experiences then to explain why he was so worried about the capital's vulnerabilities to future earthquakes.[32] The younger Hagiwara was still in middle school in 1923 and living in Sumida ward. He had been adopted by a family there after the death of his mother; many of his relatives by birth lived in nearby Honjo ward. Hagiwara seldom discussed what he witnessed in the aftermath of the quake but among the memories he did share later in his life were ones of him watching from Shirahige Bridge as a massive fire storm erupted over the former Army Clothing Depot in Honjo and of accompanying his father days later as they attempted to locate the many members of his birth family—grandparents, aunts and uncles, cousins—who had taken shelter at the Depot grounds on September 1. All they found were a few of the family's belongings among the piles of clothing and household articles that had been recovered from the more than 30,000 who died there that day.[33]

Kawasumi's "sixty-nine-year theory" about the frequency of powerful earthquakes in the capital district set the stage for the city's long anticipation of the next disaster in its future. Hagiwara, whose early research focused on ground-motion measurement, was instrumental in the launch of Japan's earthquake prediction research project in the late 1960s. As chair of the national Coordinating Committee on Earthquake Prediction (Jishin Yochi Renrakukai, or Yochiren for short) from 1969 to 1981 and also of the Earthquake Assessment Committee (a group of scientists responsible for warning the prime minister if they thought the Tōkai earthquake was imminent), the avuncular Hagiwara was the project's public face and its primary spokesman for many years.

Ishibashi Katsuhiko (b. 1944) also belongs in that cohort. Ishibashi was just finishing his graduate training at the University of Tokyo in 1976 when he uncovered evidence to suggest that the already-anticipated Tōkai earthquake was going to do much more damage than previously thought. The debates over what to do about Ishibashi's discovery culminated in the passage of the Large-Scale Earthquake Countermeasures Act two years later, with all that it entailed for the Tōkai district and for the earthquake prediction project. Ishibashi was then out of the public eye for many years but when he returned to it in the 1990s it was to once again share concerns about future earthquake disasters. He argued that Japan still hadn't done enough to prepare for them and that the government's focus on prediction as an anti-disaster measure had been misguided. It was Ishibashi who first raised the alarm about Japan's vulnerability to "nuclear power plant earthquake disasters" (*genpatsu shinsai*), more than a decade before 2011's catastrophes.[34] His warnings about a potential Nankai Trough earthquake, meanwhile, reached a wide audience after 3.11.[35]

Finally, there was seismologist Robert J. Geller, Japan's earthquake prediction and forecasting efforts' most outspoken and passionate critic. Geller finished his doctoral work at the California Institute of Technology in 1977 and was at Stanford for several years before joining the geophysics faculty at the University of Tokyo in 1984, the first foreigner with a permanent faculty appointment at that institution. Geller's primary research interests involved the use of seismic waveform data for mapping the interior structures of the Earth, but he has also pursued something like a second simultaneous career challenging the premises behind Japan's earthquake prediction project, and calling on the scientific community to acknowledge its mistakes and change course. His detailed critiques have appeared frequently in print and other media in English and Japanese. *Nature*, *Science*, and other professional journals have provided one variety of platform for Geller's arguments and he has also made them available to colleagues and to a wider readership in Japan. Geller published two trade books for the Japanese market. *The Truth about*

Earthquake Prediction: What the Japanese Don't Know (*Nihonjin wa shiranai 'jishin yochi' no shōtai*) came out in August 2011; *Hey Japan, Mr. Geller Has a Few Things to Say to You* (*Gerā san Nippon ni mono mōsu*), which includes sections on prediction (but touches on other topics as well), reached bookstore shelves in 2018.[36]

This is not meant to be a full accounting of the backgrounds or experiences of the researchers who appear in the book, but aspects of these particular scientists' histories are illustrative of themes that run through it. Almost all of the seismologists and other scientists mentioned in this book, for example, were products of the same educational institution, which is not something one would be likely to encounter in histories of research initiatives set elsewhere. Most of the scientists in *Predicting Disasters* received their degrees from the University of Tokyo, Japan's premier public university. (Those who graduated before the end of the Second World War received theirs from Tokyo Imperial University; the name was changed in 1947.) The University of Tokyo—Tōdai—was the first and for many years the only institution of higher education in Japan to offer advanced training in seismology, which is one of the reasons that so many of the scientists who appear in this book are its graduates. Another is that even after other top-tier universities began to add seismology faculty and eventually departments of their own in the postwar era, Tōdai retained many of the advantages it had started out with and that continued to set it apart from other schools. Its location in the capital was one of them, as was the fact that, for most of the twentieth century, the lion's share of top-ranking officials in the ministries that managed the education system and research funding were Tōdai graduates.

Another factor that helped it maintain its central role in seismology in general and in the work on prediction and the management of uncertainty in particular was its association with the prestigious Earthquake Research Institute (ERI), which has been based at the university's Hongo campus since the 1920s. The ERI was set up in the wake of the Great Kantō Earthquake in an attempt to establish new research priorities for seismology in Japan. Its scientists maintained a high profile in the field, aided by the institute's access to facilities and funding and by the name recognition that is another product of its long history.[37]

It is useful to know that so many of the scientists involved in the efforts to reduce uncertainty around earthquakes in Japan were trained or employed (or both) by Tōdai and the ERI in part because it speaks to perceptions of their credibility. Kawasumi was serving as director of the ERI and nearing the end of his career when he introduced his "sixty-nine-year theory" to legislators and the public; Hagiwara was in almost identical professional circumstances right before he took over leadership of the new, national Coordinating Committee on Earthquake Prediction (CCEP) in the late 1960s. The next four chairs of the CCEP after Hagiwara were either former Tōdai professors themselves or had at least trained under one.[38] It wasn't until 2012, more than forty years after its establishment, that the CCEP had its first chairperson whose career hadn't originated at Tōdai.[39] The fact that someone hailed from Tōdai was no guarantee of their brilliance or veracity, but it almost certainly made a difference that the warnings about future earthquake disasters in Tokyo and the Tōkai district and proposals for a long-term earthquake prediction research project came from University of Tokyo scientists first. At the very least, such credentials made it unlikely that policy makers or the press could dismiss their ideas and proposals out of hand.

That said, another point to make about scientists and their roles as advocates for or against plans to pursue earthquake prediction research or other uncertainty-reducing projects is that it was never a given that their arguments would elicit the results they might have hoped for. Being from Tōdai mattered, yes, but that alone was almost never enough to convince legislators, bureaucrats, or the public to act on the arguments and information provided by the scientists, regardless of how dire and obvious their implications were to those in the latter group. The national earthquake prediction research project that finally won support from legislators in the 1960s was successful only after two earlier, similar proposals that were not. The outcome the third time around was still in doubt when the Niigata earthquake of June 1964 helped convince legislators to back the most recent initiative. Similar contingencies were frequently decisive factors in determining whether or not warnings about potential disasters ahead were acted on and what form those actions took. Ishibashi's announcement in 1976 that a future Tōkai earthquake would probably do much more damage than previously thought attracted a lot of attention when he first made it, but the government's response was slow to materialize and still of limited scope as of early 1978. Much as had happened in 1964, it was only after a deadly earthquake struck—this one offshore of the Izu peninsula on January 14, 1978—that policy makers acted.

Predicting Disasters examines the processes of translating scientists' expert knowledge relating to future hazards and uncertainty into interventions that might reduce either or both with several questions in mind. One asks how scientists like Kawasumi, Hagiwara, Ishibashi, and others conveyed their understanding of the potential threats ahead in language and imagery that made sense to their different audiences. What sorts of narratives did they rely on to describe how they had reached the conclusions they had? How did they distinguish between the findings they had confidence in and those they might be less sure about? For Ishibashi and especially Geller, those dialogues were even more complicated because their goals often involved persuading colleagues, policy makers, and the public that other scientists' arguments were flawed, in contexts where the consequences of having been wrong would be significant. In the debates over earthquake prediction's validity, what rhetorical strategies did different groups turn to in order to persuade other experts and the public that they were right and their opponents wrong?

Finally, an observation about what I came to see as my responsibilities to the scientists in *Predicting Disasters* from the perspective of someone writing about them in the early twenty-first century. The book's narrative arc begins with scientists' efforts to reduce the uncertainty around earthquakes, follows the subsequent translation of their theories into institutions and policy, and then documents the growing tensions between those theories, other scientists' critiques of them, and Japan's actual experiences with seismicity. As of 2023, there had been no successful examples of short-term earthquake prediction of the kind researchers in Japan had been pursuing since the 1960s. Similarly, as of 2023, the earthquakes anticipated for Tokyo and the Tōkai district had not occurred, but others that the public had not been warned about had, at great cost. Many of the institutional arrangements and policies developed in the 1960s and 1970s to facilitate the pursuit of prediction or prepare for those specific earthquake disasters were still in place, reminders of what had once seemed possible but apparently no longer was. Even their days were beginning to feel numbered.

One of my responsibilities to the scientists in this book, given those developments, was to make sure that readers had the chance to consider the arguments behind theories like Kawasumi's and the roles played by Hagiwara, Ishibashi, and the others, in something like the contexts in which they had been encountered by other experts and by the public at the time. I wanted readers to be able to appreciate both scientists' anxieties about future earthquake disasters and their excitement when opportunities to potentially reduce the uncertainties associated with them arose. Absent access to those contexts, from a contemporary perspective, the scientists working on prediction or raising alarms about future earthquakes could easily have come across to readers as deceptive or perhaps just naive. Assigning those qualities to some or all of the experts discussed in the book would in fact have made it relatively simple to explain the origins and persistence of their claims that it was possible to know when and where powerful earthquakes would occur. Robert Geller's critiques of the prediction project sometimes took that approach; greed and self-interest are easy motives to parse.[40]

Predicting Disasters take a less cynical perspective. Having spent time in the company of these scientists (so to speak) while researching and writing this book I came away convinced of the sincerity of their fears about the earthquake disasters in Japan's future and, more often than not, with respect for the methods with which they chose to address them. Could a significant number of the scientists involved in prediction over the past fifty years have been acting mostly in self-interest and bad faith? Yes, of course, they could have. Did the historical sources—the paper trail—support that argument? Not really, in my assessment. What I've tried to offer is a history that takes scientists and their ideas seriously, and leaves room for critical analysis of projects and promises that fell short of their stated goals.

Prediction and Uncertainty

Prediction didn't take on an important role as one of the defining elements of scientific practice until relatively recently. The idea that anticipating the outcomes of geological processes was something the earth sciences could do on a regular basis is even newer. Naomi Oreskes dates the latter development to the 1960s and attributes it in part to plate tectonics' emergence as "the new paradigm of the earth sciences."[41] "Both scientists and philosophers of science," she writes, "declared that what separated this theory from the older theory of continental drift, debated in the 1920s, was that *this* one made predictions."[42] As Oreskes points out, however, the development of a new relationship between the earth sciences and prediction can't be attributed solely to plate tectonics' novelty. One of the broader contexts for that change was the twentieth-century transformation of people's beliefs about scientific knowledge and their understanding of how it was produced. "It became conventional wisdom," Oreskes observes, "that science works by testing theories through their logical predictions, and therefore that the goal of science *is* to test theories by comparing their predictions to observations."[43] The move by the earth sciences to make prediction part of what they did was thus in keeping with developments in other fields and in line with "conventional wisdom."

As Oreske's observations imply, the Japanese weren't alone in their pursuit of earthquake prediction. The search for reliable forms of earthquake forecasting and prediction was never confined to a single country. Governments devoted resources and scientists their expertise to prediction research in China, Greece, the United States, the USSR, and elsewhere at different points throughout the twentieth century.[44] In the United States, scientists in the 1970s modeled some of their plans for a prediction program on what the Japanese were already doing and eventually launched an ambitious study focused on Parkfield, California, that was designed to test theories about the timing of the next earthquake there and monitor the area where it was expected to occur for precursory signs.[45] In China, officials claimed in 1975 that their successful prediction of the powerful Haicheng Earthquake had both saved countless lives and demonstrated the sophistication of China's research initiatives and its citizen scientists alike.[46] Research teams in Greece and the USSR announced discoveries of their own that they said opened up promising new opportunities to put prediction into practice. For a while in the 1970s and 1980s, Japan's project was just one among many prediction-related initiatives underway with government support and high hopes for the future.

The goal of reducing the uncertainty of future phenomena in the modern world was one that seismologists in Japan shared not just with other scientists but also with experts working in a wide array of other settings. One of the defining features of the late nineteenth and early twentieth centuries was the emergence of new forecasting methods targeting crop yields, financial markets, the climate, and the weather, among many other phenomena. Historians have shown that the cultural, economic, and social effects of those practices were far reaching.[47] The rise of Cold War–era futurism and forecasting projects later in the century was also a global phenomenon, accompanied on occasion by commodified expressions of a collective faith in the future (the 1964 New York World's Fair, for example), as well as by the production of studies warning of dire environmental or social consequences ahead.[48] Works like Rachel Carson's *Silent Spring* (1962) and Paul Ehrlich's *The Population Bomb* (1968) both put scientific expertise to work to describe bleak but potentially avoidable futures.[49]

In Japan, too, where economic and weather forecasting were already familiar tools, pollution's long-term impacts attracted the attention of experts and was another area in which predictions became commonplace. Newspapers and other media began reporting on pollution as a national problem in earnest in the early 1960s and expanded their coverage as more and more cases came to light.[50] When environmental scientist Shōji Hikaru and his economist co-author Miyamoto Kenichi released *Fearsome Pollution* (*Osorubeki kōgai*) in April 1964, they captured the scale and severity of the crisis in a format that found a large and appreciative audience. "Inside its covers," Simon Avenell observes, "readers encountered a shocking compendium of images and statistics. Along with photos of smoggy industrial cities, polluted lakes, and distraught victims, Miyamoto and Shōji provided a 'pollution map' of Japan identifying cases in almost every prefecture."[51] The book's impact on the Japanese public is often equated with the effects that Carson's *Silent Spring* had on American readers in terms of how it raised the alarm about pollution's long-term effects and in gathering support for a resolute response to it.[52]

Kawasumi Hiroshi's announcement forecasting a major seismic event in Tokyo's future came just a few months after the publication of *Fearsome Pollution*. His warning and the descriptions that appeared in the media soon afterwards of what that disaster would look like were remarkable on many levels. Those developments in 1964 helped mark the start of a process of normalizing work on earthquake forecasting and prediction in the Japanese earth sciences and thrust the scientists involved in it into new roles as public figures and adjuncts to policy making. From that point onward, Tokyo's future had an earthquake disaster in it and other parts of Japan soon learned that theirs did, too. The country's relationships with disasters, prediction, and uncertainty changed radically in the 1960s and 1970s. At the same time, the underlying belief that experts could produce reliable, useful knowledge about the future and reduce the uncertainty associated with it was by then not novel at all. The prospect of forecasting or even predicting earthquakes in Japan and elsewhere echoed the anticipation that had preceded breakthroughs in knowledge production about the future in so many other fields over the course of the twentieth century.

What scientists gradually discovered, however, was that their pursuit of short-term earthquake prediction was not producing the breakthroughs they (and the public) had hoped for. The success at Haicheng was not replicated elsewhere. The Parkfield experiment came to a disappointing end and researchers in other settings also had little to show for their efforts. As a result, by the 1990s, scientists and governments that had once supported work on short-term earthquake prediction began to back away from it. In California and elsewhere, researchers and policy makers shifted their focus toward the development of probabilistic seismic hazard maps and the forecasting work they required. Disaster prevention policies based on those maps and those forecasts followed. Interest in short-term prediction didn't entirely disappear, of course, but by the end of the 1990s the skepticism that had once dominated discussions about prediction before the 1960s and 1970s was doing so again.[53]

Scientists in Japan, however, continued to pursue earthquake prediction after most other research communities stopped taking it seriously. Japanese scientists were no more successful than their colleagues overseas at identifying earthquake precursors or finding other methods that would yield reliable short-term predictions, but they responded differently to those circumstances. Funding for prediction-related research was not cut. The Large-Scale Earthquake Countermeasures Act remained in effect, its presumptions that short-term earthquake prediction was possible and that it would protect the country from the Tōkai earthquake intact. The Earthquake Assessment Committee remained in place, with all that its persistence implied, while the Yochiren continued reviewing prediction-related research and screening reports of potential precursory signs. There was a brief period after the 1995 Hanshin-Awaji earthquake when the prediction project did appear to be in jeopardy. The government launched a new nationwide probabilistic seismic hazard mapping program that year that some speculated signaled the end of support for work on prediction. They were mistaken; the prediction project's supporters rallied behind it and kept it going. (Funding actually increased.) It wasn't until 2017 that the Japanese government acknowledged that short-term prediction's potential was far more limited than scientists had once hoped it would be and that predicting the Tōkai earthquake would in fact be difficult, if not impossible.

This book explores the tensions between the choices made by seismologists in Japan and the general trend away from prediction elsewhere from three angles. The first focuses on contingencies specific to the institutions and structures involved in the prediction project in Japan at those moments when doubts about its future were at their height. One of the conclusions to emerge from that approach is that, by the time other research communities were moving away from prediction, the project in Japan was already so spread out across different institutions, agencies, and researchers that altering its forward momentum would have required immense, coordinated effort. Ken Hewitt warned in 1983 that making predictability and uncertainty the problems to be solved in order for disasters to be less of a threat to society would have significant, largely unwelcome consequences. Hewitt argued that the most worrisome potential outcomes of such an approach would be a reliance on "centrally controlled systems to protect those at risk," a "colossal commitment to means of improved forecasting," and entanglements "with the fashionable spin-off of the enormous statistics and probability industry," each of which took root in Japan after the 1970s.[54]

Another point about institutions and contingencies is that definitions of what researchers were looking for in order to solve the short-term prediction problem, and the methods they used to do the looking, were not static. The project's public profile changed over time, sometimes dramatically. For its critics, those changes were more cosmetic than substantive, since the overall goal remained the same whether researchers were focused on dilatancy or slow-slip events in the search for earthquake precursors. Its supporters, however, could point to changes in focus and methods as signs of progress, and as evidence that scientists' understanding of the problem was always evolving.

A second perspective on prediction's longevity in Japan comes from approaching the debates over it as one case among many in which controversies in science have moved toward some form of closure. In this instance, the first well-organized critiques of the science behind the earthquake prediction project date back to the mid-1990s. That the project's supporters were still advocating on its behalf almost three decades later was challenging for those observers anticipating what Bruce Lewenstein once described as "the idealized scientific outcome," or "closure through sound argument."[55] In that scenario, "rational arguments, based on well-established data, convince everyone of the outcome," and the field can move on.[56] That was not how the debates over prediction played out in Japan, however. As we'll see, the project's critics have sometimes interpreted their inability to achieve closure on those terms as proof that one would have to look somewhere other than the science to explain why researchers in Japan continued to pursue short-term prediction. This double critique—of prediction's premise and of the motives of its supporters—has played the important role of framing the academic discussions and to some extent the public ones, too, around the choices seismologists in Japan have made over the years with regard to prediction. There are other ways besides "closure through sound argument" that controversies in science get resolved, however, and I argue that some of these are helpful for thinking about how seismologists in Japan have engaged with prediction over the years.

Finally, *Predicting Disasters* also explores the roles uncertainty played in shaping what scientists and the public believed was knowable about Japan's future seismicity. As noted earlier, historians and other scholars have traced the evolution of forecasting and prediction as methods for reducing uncertainty in daily life, markets, and elsewhere and shown how important the ability to calculate risk was to the emergence of everything from the habits of daily life to the larger structures of the modern economy.[57] The prediction project's efforts to determine the timing and location of powerful earthquakes that hadn't happened yet were likewise all about reducing uncertainty. For most of the project's history the scientists involved in it were confident they understood not just how, but why it was possible for them to turn uncertainty into risk, even if they didn't as yet have much to show for their efforts. One of the key arguments against the pursuit of earthquake prediction, however, was that the project's scientists had been wrong from the start about the type of uncertainty they were confronting. The uncertainty surrounding seismogenesis, critics argued, was permanent and irreducible.

Engineers and scientists concerned with understanding large-scale physical processes in the early twenty-first century—such as climate change or seismogenesis, for example—would likely use terms like "aleatory" and "epistemic" to differentiate between the categories of uncertainty they confronted in their research. The former refers to uncertainty associated with the random, stochastic nature of a given phenomenon and the latter to uncertainty that is "attributable to incomplete knowledge about a phenomenon that affects our ability to model it."[58] Standard examples of where aleatory uncertainty comes into play include the outcome of a coin flip, or of the throw of a dice. It may be possible to calculate the probabilities that a certain outcome will be obtained—that the coin will land heads up rather than tails—but what one cannot do is know for certain ahead of time what the result of any given toss will be.[59] Aleatory uncertainty cannot be reduced; epistemic uncertainty can.

The prediction project's effort to determine the timing and location of powerful earthquakes that hadn't happened yet is in the broadest sense a history of scientists trying to reduce epistemic uncertainty. An important premise of that undertaking, for example, was that the primary barrier to successful predictions was not the nature of the seismogenic process itself but rather scientists' ability to observe it and gather data about it. The project focused, as we know, on acquiring data and developing models that would allow experts to eventually describe the seismogenic process with enough precision to make earthquake predictions possible. Every improvement in instrumentation and every refinement of the theories that helped explain how earthquakes happened was another step toward that goal. So long as there were scientists who believed that the uncertainty they were struggling against could be reduced, there was no reason to stop trying to do so. It was always possible that some new discovery or closely observed event would solve the puzzle.

The basis for many of the arguments against the continued pursuit of prediction in Japan, on the other hand, were beliefs that earthquakes were stochastic events and that the seismogenic processes that produced them were wrapped in irreducible aleatory uncertainty. More data and more sensitive instrumentation were beside the point, at least so far as prediction was concerned. It wasn't that scientists didn't know enough and that was why they couldn't predict earthquakes; the problem was that the forces acting on tectonic plate boundaries and faults of all sizes were so many and their interactions with each other so complex that no new algorithm or model could possibly account for them all.[60] There was nothing that could be done to change that fact. Scientists who believed that the processes responsible for earthquakes belonged at the aleatory end of the uncertainty spectrum were likely to view characterizations of them as anything other than utterly unpredictable events as a disservice to the public, and any further investments in earthquake prediction research as misguided.

This book shows how differing beliefs about uncertainty have shaped prediction's trajectory in Japan and defined key aspects of the debate among seismologists over whether it should continue or not. It also makes clear that, in their encounters with legislators and the public, scientists often struggled to explain how uncertainty figured in their analyses of Japan's future seismicity. Part of the problem was that uncertainty can be a difficult concept to explain, especially for specialists who normally would not have been called on to do so for audiences of nonspecialists. Seismologists in Japan watched closely as prosecutors in the Italian city of L'Aquila charged and later convicted half a dozen scientists and one official with manslaughter in the aftermath of the April 2009 earthquake that left more than 300 of the city's residents dead. The scientists had been members of a committee tasked with assessing the risks posed by increased seismic activity that had been observed in the L'Aquila area. Public statements by a Civil Protection Agency official attributed to the committee appear to have downplayed the risk of a major event; the earthquake struck a week later.[61]

The charges reflected the prosecutor's belief, which was also held by many of the earthquake's survivors, that had scientists shared all that they knew about the risks to which city residents were exposed, fewer lives would have been lost. The issue was not so much that science had failed to predict the earthquake—as one resident pointed out, "We all know that the earthquake could not be predicted and that evacuation was not an option."[62] It was, instead, that scientists and officialdom had failed to adequately explain what the science meant and the parameters of the uncertainty associated with it.

As we'll see, seismologists in Japan struggled with how to explain uncertainty's role in earthquake forecasting and prediction—what the science meant—since Imamura's day. The criminal case in Italy attracted as much attention among Japanese earth scientists as it did in part because the tensions between what experts understood about the limits on their abilities to forecast future seismicity and what the public had come to expect were even more problematic in the early twenty-first century than they had been in the early twentieth. Although all six of the Italian scientists had their convictions overturned and were acquitted of the original charges against them in 2015, by then many seismologists in Japan knew what it was like to be accused of having failed the public.[63]

Narrative Arrangements

This book follows a timeline that begins with the 1905 publication of Imamura Akitsune's warning to Tokyo's residents that an earthquake was coming and continues through the early twenty-first-century debates over the efficacy of prediction and probabilistic approaches to talking about earthquake risk. The chapters proceed chronologically, for the most part, starting with Chapter 1's focus on Imamura's early warnings and his efforts to pursue prediction after the Great Kantō Earthquake. Chapter 2 looks at seismology in Japan after World War II and at Imamura's ongoing attempts to detect the signs that he was sure would precede the earthquake disaster he worried would strike next, this one in the Nankaidō district. It also shows how, as the American Occupation was underway, a group of scientists who previously had shown little interest in earthquake prediction began to either practice versions of it themselves or admit that it might be worth researching, after all. While neither the warnings nor the expressions of interest came to much, in the short run, I show how the process of decoupling earthquake prediction from its previous association with Imamura and pre-surrender Japanese science set the stage for its re-emergence later. Chapter 3's attention is on the emergence of a new set of assumptions about Japan's future disasters and how to prepare for them; it also examines the first steps toward a national research project in earthquake prediction.

Chapter 4 analyzes the origins of and responses to Kawasumi Hiroshi's mid-1960s assertion that the odds of a powerful earthquake striking Tokyo were about to go up, dramatically. It shows how the Niigata disaster was an inflection point in postwar Japan's understanding of its vulnerabilities to earthquake hazards and how, in that context, Kawasumi's theory about the rising risk to Tokyo was taken very seriously. Chapter 5 explores the impact of the Matsushiro earthquake cluster (1965–1970) on the public's interest in earthquake prediction and on scientists' willingness to accommodate their concerns.

Chapters 6 and 7 describe the steps that local and national policy makers took in response to scientists' warnings about the threats to Tokyo and the Tōkai district, respectively. In Tokyo, the city's preferred defense was a huge project to build half a dozen massive disaster-shelter complexes at different locations in the Kōtō delta. Chapter 6 tracks the history of the project from its planning stages in the 1960s through the start of construction in the 1970s and the completion of the functioning complex at Shirahige-East. Chapter 7 shows how another earthquake that hadn't happened yet, this one offshore of the Tōkai district, took shape in the public imagination in the mid-1970s, and Chapter 8 takes up questions about how the unfounded belief that scientists would be able to warn the public that a Tōkai earthquake was imminent ended up as the central premise of the government's disaster preparedness plans. The 1978 Large-Scale Earthquake Countermeasures Act gave the government a range of interventions it could take to mitigate the harm the Tōkai earthquake might do, but they were all designed to go into effect only after scientists were sure that the quake was coming.

Chapter 9 explores the growing tension between the persistence of policies that took for granted that prediction worked and the mounting evidence that it did not. This tension arose from several sources: one was the growing recognition in the 1990s that scientists outside of Japan who had once shared their Japanese colleagues' enthusiasm for short-term earthquake prediction for the most part no longer did. Another was the disconnect between disaster preparedness policies that focused on threats to Tokyo and the Tōkai district and the harm being done by earthquakes elsewhere, such as the one that struck Kobe in 1995.

The Epilogue assesses the effects of the multiple disasters of March 2011 on the prediction project and suggests several reasons why the pursuit of prediction has held on in Japan much longer than it has elsewhere. Finally, the Epilogue returns to where the book began, to scientists' efforts to make the earthquake disasters in Tokyo's future legible and survivable.

Conventions

Because this is a book about disasters, earthquakes, and scientists who study them, it makes frequent use of the terms and conventions that the latter rely on to describe the former. Here are several that appear throughout the book along with brief definitions to help clarify what they're meant to convey.

We're likely all familiar with references to an earthquake's *magnitude* and with shorthand descriptions of one as an "M6.0 earthquake," for example. Scientists have taken a number of different approaches to measuring earthquake "size" over the years and relied on different magnitude scales to represent the results of their calculations. In 1931, Japanese seismologist Wadati Kiyō showed how it might be possible to use seismographic data to generate a standard measurement of an earthquake's size. In 1935, Charles Richter took Wadati's work one step further and described his own method for calculating an earthquake's size from seismographic records on a ten-stage logarithmic magnitude scale.[64] The "Richter scale" and variations on it were in use for the next several decades.

Since the 1980s, however, scientists have turned to the "moment-magnitude scale" as their preferred metric. "Seismic moment" is a measure of how much energy is released by the faulting action and dislocation that occurs during an earthquake; one of the reasons experts prefer the moment-magnitude scale is that it does a much better job than its predecessors of reflecting the relative sizes of very large earthquakes.[65] When I refer in the text to an earthquake's magnitude (or represent it as M8.0, for example), it is safe to assume that it's moment-magnitude we're talking about.

The moment-magnitude scale is logarithmic, which can be confusing when it comes to understanding the information it is designed to convey. What might be helpful would be to remember that a single step (up or down) on the moment-magnitude scale represents a thirty-two-fold difference in the amount of energy release it describes. An M6.0 earthquake, for example, is thirty-two times as powerful as an M5.0 event, and roughly one thousand times as powerful as an M4.0.

Seismic intensity is another frequently invoked metric for describing an earthquake and, more specifically, for describing its effects at a particular location. While there is a correlation between an earthquake's magnitude and how much energy ultimately reaches the surface in the form of potentially destructive seismic waves, their effects at the surface will vary widely depending on a number of variables, including how far away that location is from the earthquake's epicenter, the characteristics of the soil at the surface, and so on. Seismic intensity scales start out with an idea of the range of the different types of effects that an earthquake might produce at the surface, whether on the built landscape or on humans experiencing the shaking first-hand, from the inconsequential to the catastrophic. The scale's designers must then decide how many subdivisions to assign and which effects and at what levels of severity to associate with each.

As with magnitude, standards for the measurement of intensity have varied considerably over time and from place to place. Scientists recognized early on that approaches developed to describe the effects of earthquakes on late nineteenth-century European landscapes and buildings did not adequately reflect their effects on Japanese ones and, therefore, developed one that did.[66] In 1884, Japan's first such scale encouraged observers to describe an earthquake's effects as either faint, weak, strong, or violent. Three additional tiers were added in 1898 and numbers from 0–6 assigned to each of the now seven levels of seismic intensity, or *shindo*.[67] After the 1948 Fukui earthquake, government scientists introduced a new maximum tier of Shindo 7, "Severe Earthquake" (after it was determined that what had happened in Fukui was much more serious than Shindo 6, "Disastrous," could convey). In 1996, the Japan Meteorological Agency (JMA) refined the scale again by somewhat confusingly subdividing intensity levels 5 and 6 into "weak" and "strong" tiers, thus bringing the total number of *shindo* to 10.[68] Before 1996, JMA technicians determined the seismic intensity levels for a given earthquake and location based on their own observations; after 1996, intensity levels were determined automatically based on real-time data from seismometers installed throughout Japan.

Finally, as is standard practice in most English-language scholarship on Japan, I will refer to Japanese people by their surnames first.

CHAPTER 1

Scientists, Disasters, and Prediction Before the War

For most of the first half of the twentieth century, Imamura Akitsune was the only scientist to claim in public, repeatedly, that it was possible to significantly reduce the uncertainty around when and where powerful earthquakes were likely to strike in some parts of Japan. He did this first in 1905, when he explained why he believed that Tokyo could soon suffer a major event, and again in the 1930s, when he described the rising risk of a large and potentially devastating earthquake in the Nankaidō district and explained his plans to monitor the area for signs that one was imminent.[1] Imamura was also the only scientist to repeatedly propose that seismologists in Japan focus their resources and research on the pursuit of earthquake prediction. This he did first a few years before the Great Kantō Earthquake and then again and more ambitiously shortly after that disaster.

What Imamura was trying to do matters to us in part because there are echoes of his ideas and proposals in the commitments Japanese seismologists eventually made to earthquake prediction research in the late 1940s and then more wholeheartedly in the 1960s. Teasing out Imamura's contributions to those initiatives clarifies some of the reasoning behind the choices that younger generation of scientists made, especially in the 1960s. At one level, those contributions are fairly easy to document and include elements of the methods and equipment that Imamura argued were most likely to reveal to scientists when and where powerful earthquakes would occur, in either the short term (as predictions) or the long term (as forecasts), for example. Imamura's approaches to communicating with the public about earthquake threats and about what could be done to prepare for them were also there for other scientists to consider.

Figure 2. Imamura Akitsune, 1924. (The formula on the blackboard is a standard result for wave propagation.) Photo by Mainichi Newspaper/AFLO.

Imamura's younger colleagues and students who went on to play important roles in postwar Japanese seismology were also familiar with aspects of his career and public persona as a scientist and raiser-of-alarms that might have given them pause. They would have known, for example, that Imamura's colleagues at the time had repeatedly dismissed his warnings to the public as overblown and opposed his calls for the field to commit itself to earthquake prediction research. In 1905, Imamura's comments about a potential earthquake threat to Tokyo were rejected as unscientific and unsound by Ōmori Fusakichi, the most prominent seismologist in the country. After 1923, Imamura's call to invest in earthquake prediction research was rejected in favor of a very different set of priorities for Japanese seismology. His later attempts to call attention to what he said were signs of an imminent earthquake in the Nankaidō district were largely ignored by other researchers. Imamura's track record when it came to enlisting the support of other scientists for his work with forecasting and prediction-adjacent activities was a singularly dismal one.

One other consequential facet of Imamura's efforts to put science to work reducing the uncertainty around earthquakes in early twentieth-century Japan is made up of the narratives about that work that he himself helped construct.

It was Imamura's analysis of the Great Kantō Earthquake that shaped much of the public's understanding of that event and the disaster that followed. Newspaper reporters sought him out, as did the editors of popular magazines, all of them anxious to share what he had to say. In those articles, Imamura provided what were presumably objective, scientific analyses of the earthquake as a seismological process. He also, however, frequently wrote about himself and made his own actions and experiences an important part of the disaster's context for readers. Imamura focused in particular on how he had tried to draw attention to the seismic risks to the capital in 1905 and had called on the city to prepare for the earthquake disaster ahead, only to have his work undermined by his senior colleague (Ōmori) and his warnings ignored. The consequences of not heeding his advice, he wrote, were there in the ruins of the city for anyone to see. After the December 1946 Nankaidō earthquake, Imamura made essentially the same argument, just in the wake of a different disaster.

On the one hand, Imamura—or "Professor Earthquake" as the press took to calling him in 1923—was making a powerful case that science in general and seismology in particular had vital roles to play in protecting the country from disasters like the one that had just devastated Tokyo. On the other, his portrayal of how science actually worked didn't inspire a lot of confidence, given the way he and his ideas had been treated. Readers could take some comfort after 1923 knowing that Imamura had finally achieved a degree of notoriety and the status to say what he thought, but it wasn't at all clear where that left the rest of the field or what criteria experts relied on when assessing which ideas had merit and which didn't. By entangling the twists and turns of his own career with earthquake prediction's trajectory in Japan, Imamura was constructing a complicated legacy for other scientists to deal with.

Tokyo's Next Earthquake

The history of the dispute between Ōmori Fusakichi and Imamura Akitsune over the timing of Tokyo's next catastrophic earthquake has long been a template for understanding how scientists in early twentieth-century Japan went about trying to reduce the uncertainty around earthquakes. It's easy to see why; the rift between the two was as much personal as professional, the clash went on for years, and its resolution was dramatic and poignant. It also matters that Imamura had the last word and wrote the first draft of his own history in the aftermath of the Great Kantō Earthquake.

An abridged version of the story of the disagreement between the two seismologists goes something like this: Imamura was just a year-and-a-half younger than Ōmori and started his training in the field not long after Ōmori did. That gap made all the difference in the world to how their careers unfolded. Ōmori was appointed to Tokyo Imperial University's sole professorship in seismology in 1896. He was only twenty-eight when he took up the post and mandatory retirement wasn't until sixty. His financial and professional security were assured, whereas Imamura had to teach mathematics to Imperial Japanese Army officer candidates to make ends meet, while continuing his work in the field as an unpaid adjunct—under Ōmori's supervision—at the university.

In late 1905, Imamura was preparing for the publication of his first monograph, *Jishingaku* (*Seismology*). The book would be the first major work in the field in Japanese that was not a translation or reworking of a Western publication. In the lead up to its release, Imamura put together a short article for the popular general readership magazine *Taiyō*, in which he summarized some of the monograph's more interesting conclusions.[2] In the article and in *Jishingaku*, Imamura described the faulting processes associated with the great earthquakes of Tokyo's past, drew readers' attention to the patterns those earthquakes had exhibited, and wrote in specific terms about the steps he and others in the field had taken to apply their growing body of knowledge to the earthquake problem.[3] These included making recommendations that the city improve its fire-fighting capacity and take other steps that would slow the spread of fires that would inevitably ignite after a powerful earthquake.

The problematic parts of Imamura's summary were the ones dealing with his references to the timing of Tokyo's next major earthquake. It wasn't unlikely, he wrote, that the city would see a major seismic event within the next fifty years. His reasoning was this: the three most recent major earthquakes in the city's history (1650, 1704, and 1855) occurred on average once every 100 years. "In light of the fact that 50 years have passed since the last quake, in 1855," Imamura wrote, "even if one allows that ample time still remains before the next quake, given the occurrence of the Genroku [1704] quake 54 years after the Keian [1650] event, disaster preparedness must not be delayed for even one day."[4]

Were the city to be struck by a quake more or less equivalent to the one that hit Edo in 1855, he pointed out, Tokyo would suffer tremendous damage. The tremors themselves would be responsible for some of it, but it was the fires in the quake's aftermath that would cause the most harm. "Tokyo's major earthquakes tend to happen at night," he wrote.

> Fires from the oil lamps will come from all directions, and there will be absolutely no way to deal with them. Should there be even a slight wind, it is possible that the entire city will be reduced to ashes. Even if there isn't much wind, more than half the city could well burn to the ground. The number of deaths will increase with the scope of the fires. Property losses will be in the hundreds of millions of yen, and deaths could well exceed 100,000. One can't help but be horrified just thinking about it.[5]

The *Taiyō* article attracted little attention when it appeared in September but that soon changed. Over the next few months, Tokyo newspapers began to carry reports highlighting the parts of the article that discussed the timing of the next quake and the damage it would do, while leaving out the central point of Imamura's argument, which was that there was plenty the city could do to prepare so that such a disaster never came to pass.[6] Signs that the sensationalization was having an impact surfaced after Tokyo was struck by a series of mild earthquakes, the first in late January 1906 and then two more in quick succession, on February 23 and 24. These were fairly run-of-the-mill earthquakes for Tokyo but as reports surfaced of city residents on the verge of panic, some opting to sleep on the street in the cold rather than risk returning to their homes, it became apparent that the public's reactions were out of character for the capital. People were truly alarmed.

Imamura and Ōmori both decided that the arguments in the junior scientist's article were somehow connected to the post-earthquake unrest. Imamura, who blamed the press for taking his ideas out of context and for emphasizing only the most alarmist parts of his tract, wrote to the newspapers responsible, demanding that they retract their exaggerations and replace them with coverage of his real message about the importance of disaster prevention and planning. Some, like the *Hōchi shinbun*, began a series of articles reporting on common experiences and recommended behaviors during an earthquake. While this may have been just the sort of discussion Imamura would have welcomed under other circumstances, it appears to have done little to calm public fears.[7] Other publications took a different approach and began to feature more, not fewer, articles and commentary on earthquakes. It didn't help matters that the publisher of Imamura's new book chose that moment to begin promoting the book in the nation's major newspapers. "Read this book and get prepared for the big earthquake!" read the advertisements.[8]

Ōmori's strategy was different. He turned to newspapers and magazines to address the public's fears, as Imamura had, but, rather than use those opportunities to clarify his colleague's arguments, Ōmori attacked them instead. He started with a series of short articles in the *Kokumin* newspaper and followed those with a longer piece published in the March 1906 issue of *Taiyō*.[9] "The Groundless Rumor of a Great Tokyo Earthquake" (*Tokyo to daijishin no fusetsu*) was directed not just at general misconceptions about earthquakes but at Imamura's arguments in particular. Ōmori asserted that Imamura's analysis of the historical statistics relating to earthquake frequency in the capital district was flawed; his own calculations had led him to believe that a one-hundred-year interval was much more likely than the shorter one Imamura had highlighted. (The *Taiyō* article was followed by a similar one in the flagship journal for earthquake researchers in which the same criticisms of Imamura's methods were offered.)[10] Moreover, he argued, as serious as the Ansei quake had been, there was no reason to assume that a similar event in contemporary Tokyo would be all that devastating but might well result in only limited damage.[11]

If Ōmori's reaction seems out of proportion to the relatively low levels of distress Imamura's observations had supposedly provoked, the professor's choices that fall and winter make a little more sense in light of contemporaneous events in the capital that had nothing to do with earthquakes. Imamura's original article in *Taiyō* appeared in September 1905 and the drama around its contents over the next few months unfolded alongside far more incendiary developments. The Japanese government's announcement of the terms of the Treaty of Portsmouth on September 5 both signaled an end to a bloody and hugely expensive war with Russia and marked the beginning of an "era of popular violence" at home. What began that day in Tokyo's Hibiya Park not far from the grounds of the Imperial Palace as a peaceful protest against the government for having agreed to a treaty that did not include the payment of an indemnity by Russia had, by nightfall, turned violent and spread throughout the city. Over the next two days, protestors set fire to police boxes, streetcars, newspaper offices, and other symbols of authority; seventeen people died and thousands were arrested before the police and the military retook control of the capital. Martial law remained in place until late November.[12]

Ōmori didn't cite the Hibiya Riot or concerns over wider political and social unrest to explain his harsh critique of Imamura and it is possible that he would have taken the same position toward him regardless of what was happening in the streets of the capital. At the same time, it's hard to imagine someone in his position simply dismissing reports of panicky activity prompted by fears of an earthquake while memories of the riots were still fresh. The record suggests instead that Ōmori's anxiety ratcheted up with each new report of people taking Imamura's warnings seriously and with each of his subordinate's failures to address the problem himself.

Given his circumstances, there was little Imamura could do to rebut Ōmori's assessment and he was left having to accept what amounted to a public humiliation at the hands of his senior colleague. From that point on, Imamura avoided discussing the timing of Tokyo's next earthquake in public almost entirely, with one notable exception. In late November 1915, while Ōmori was in Kyoto attending enthronement ceremonies for the Taishō Emperor, Tokyo again experienced a series of minor tremors. In circumstances not unlike those of a decade earlier, anxious city dwellers reminded themselves that the sixtieth anniversary of the Ansei quake was near and feared the worst. Rumors that a powerful earthquake was imminent quickly spread. Sought out by reporters, Imamura provided a cautious statement in which he downplayed the likelihood that the current cluster of smallish earthquakes was leading up to a single catastrophic event. But he did not rule out that possibility altogether and, in his statement, advised citizens to not let their guard down and to take extra care when dealing with fires, just in case. Ōmori was furious when he learned of what Imamura had said, again because of its potential to trigger unrest among city dwellers, and returned to Tokyo ahead of schedule to chastise him, again, and to reassure those unnerved by his junior colleague's comments.[13] The relationship between the two scientists fell to a new low.

The friction between Ōmori and Imamura was real but the differences between the two did not reflect fundamental clashes over basic scientific principles. Ōmori shared his younger colleague's concerns about Tokyo's firefighting capabilities and the risks associated with gas lamps, for example. Neither man questioned whether predicting earthquakes was something seismologists should pursue. Their disagreements were never about whether forecasting was possible and, in fact, the timing of Ōmori's rebuttals made it clear that his concerns had more to do with how the public was reacting to Imamura's warnings than with the science they were based on. Ōmori didn't doubt that prediction in some form was theoretically possible and believed—as did Imamura—that history offered essential clues in their efforts to understand when and where the next major earthquakes would occur.[14]

The Great Kantō Earthquake

The long-running Ōmori-Imamura dispute came to a dramatic end in 1923. We understand it to have been dramatic in part because Imamura's accounts of events portray it as such but, even without his narration to rely on, there is no mistaking the fact that the two scientists' fortunes changed abruptly and radically after the Great Kantō Earthquake. Ōmori was in Australia for a conference when the earthquake struck.[15] He was reportedly not feeling well even before hastening back to Japan and his condition worsened during the long voyage home. (Doctors later determined that he was suffering from a brain tumor.) When his ship arrived in Yokohama in early October, medical staff were waiting dockside to transport him directly to Tokyo Imperial University Hospital.[16] He died there about a month later.

Much of what we know about the interactions between Imamura and Ōmori after the earthquake come from the younger scientist's accounts, which emphasize two points. One is that Ōmori said he felt responsible for what had happened to Tokyo; the other is that the two men reconciled at the end, as was reflected in some of Ōmori's last administrative acts. Ōmori reportedly lobbied university officials on Imamura's behalf from his hospital bed, essentially designating him as his successor. In December, at long last and as Ōmori had wished, Imamura was promoted to the rank of Professor. Imamura was 53 and had only a few years remaining before mandatory retirement from the university but his new status afforded him the authority and public voice he had long been denied. Ōmori's actions on his behalf, Imamura wrote, "did away with any lingering doubts I may have had about the Professor's intentions. The dispute that had gone on for so many years between us was over."[17]

In 1923, the story of the "Ōmori-Imamura Dispute" would have been familiar enough to other scientists in their orbit at the university but to few outside of that community. Imamura's warnings over the years about the threat to the capital district had never stayed in the public eye for very long and, for that matter, neither had Imamura. Ōmori's efforts to challenge statements that might disturb the public peace certainly account for some of the silence that settled around Imamura's occasional outbursts of concern for the future of the city but it is also true that, after 1905, Imamura himself seldom wrote about earthquake forecasting or about what he feared might happen to Japan's urban centers. His research publications tended toward highly detailed reports on recent earthquakes, along with analyses that drew on the rich catalog of data he himself had helped compile about the seismic events of the early modern era. He had not gone out of his way to focus the public's attention, or that of his colleagues, on his issues with Ōmori. After so many years of not talking about his concerns or his run-ins with Ōmori, at least in print, Imamura's post–September 1 embrace of that part of his past and of the controversy associated with his warnings marks a sharp change in direction for him. The impact of this new discourse on seismology and on science's relationships with the Japanese public would be considerable.

The disaster came at a moment in which the mass media was growing in leaps and bounds, in search of new audiences, and deploying new mediums to win them.[18] Japanese publishers in the late 1910s and early 1920s had, for example, helped foster the nation's appetite for magazines and books devoted to "popular science." As Hiromi Mizuno and others have suggested, this era marked a turning point in the relationship between the general public and science, one which saw the introduction of new ways for average citizens to "consume" science in the form of books, magazines, lectures, and hobbies developed especially with them in mind.[19] The first issues of *Kagaku chishiki* (*Scientific Knowledge*) appeared in 1921; it was joined by *Kagaku gahō* (*Science Illustrated*) in 1923, and by *Kodomo no kagaku* (*Children's Science*) in 1924.[20] "All three magazines," writes Mizuno, "presented science as a variety of knowledge that readers could and should acquire through their pages."[21]

Despite the damage done to the printing industry by the earthquake and fires, publishers recovered quickly. Many of the best-known journal titles were back on sale in October with editions devoted to coverage of the earthquake and were joined on bookstore shelves and magazine racks by one-off omnibus publications that also focused solely on the disaster.[22] Imamura's articles and commentary were fixtures of the initial wave of post-quake magazines and omnibus collections of photographs, first-person accounts, and descriptive essays.[23] He provided the first estimates of the earthquake's location and strength, and was the first to describe the event in print as "The worst catastrophe in the recorded history of the world."[24]

Imamura's post-quake contributions to the public discourse on disasters were numerous and prediction played multiple important roles in the narrative he helped create about the disaster's meanings for the nation. One such role bordered on the vindictive; "reminding" readers of his past warnings about the risks Tokyo faced was as much an attack on the credibility of those who had tried to silence Imamura as it was an assertion that recent tragic events had proven him right. It is clear from the language and tone that Imamura strikes in much of what he wrote after September 1923 that he felt he had been wrong to remain silent about the risks to the city but that others were even more to blame by repeatedly ignoring Imamura's warnings and continuing to insist that the capital was safe. Imamura sometimes identified Ōmori by name, pointing out the many instances in which his former colleague had dismissed his predictions as nonsense. Writing in the first post-disaster edition of *Scientific Knowledge*, Imamura recapped both his arguments from the 1905 *Taiyō* article and Ōmori's critical response before concluding that, "My prediction about the nature of the earthquake itself was fully realized; the damages to the city, however, were even worse than I had imagined."[25] As Gregory Clancey notes, Imamura likely had a hand in arranging to have a piece written by Ōmori in early 1923 published posthumously, in 1924. It reads, in part: "Tokyo may be assumed to be free in future from the visitation of a violent earthquake like that of 1855, as the later shock originated right under the city itself, and as destructive earthquakes do not repeat from one and the same origin, at least not in the course of 1,000 or 1,500 years."[26] That Ōmori had been wrong and Imamura right was a point that few readers of these and similar accounts would miss. One thread that runs through the post-quake analytical and descriptive narratives is very much concerned with settling old grudges.

Making a history of prediction a focal point in his coverage of the disaster allowed Imamura to do more than just complain about the past. As noted above, warnings ahead of time and efforts to mitigate against future harm were tightly paired in Imamura's published analyses and public discussions of the disaster. Being able to take the necessary steps early—whether encouraging citizens to be especially mindful of fires or convincing local authorities to make the switch from gas to electric lighting—was what made forecasting and prediction worthwhile undertakings. By contrasting his description of the damage done to Tokyo and Yokohama by fire with what might have been, had steps been taken ahead of time to address well-understood hazards, Imamura again and again insisted that much of what had happened on September 1 had been unnecessary.

Perhaps more importantly, given what had already transpired, Imamura suggested it was still within the nation's power to mitigate the effects of future earthquakes. The list of what still needed doing was little changed from the one Imamura (and others, Ōmori among them) had developed decades earlier, in which better fire prevention and suppression, improved construction practices, and more outreach and education programs at the local level figured prominently. Applicable to almost any community in Japan, these were steps that could significantly reduce the harms associated with earthquakes.

As obvious and necessary as these steps seemed to Imamura, he faced several dilemmas in the aftermath of the Kantō earthquake. Because the risk to the capital district of a second major quake so soon after the one in September was low and because Imamura had gone on the record to reassure residents that it would likely be many years before another damaging earthquake occurred, there were few incentives for citizens and officials in those already devastated communities to pay much attention to his recommendations on preparedness, planning, and education, at least in the short term. The most receptive audiences and, arguably, those most likely to benefit from the sort of advice Imamura had to offer, would be those least affected by what had happened on September 1.

Deciding where in Japan was most at risk from a powerful earthquake in the near term was another problem. The combination of historical data and local knowledge that decades prior had led Imamura to conclude that the Tokyo area had been at risk seemed, on the face of it, to shed little light on future risks to other parts of the country. In other words, it wasn't immediately evident that his "prediction" of the September 1 Kantō earthquake was based on principles and practices that could be applied—successfully—elsewhere, ever again.

Imamura, of course, was singularly well suited to speak to these concerns and it is clear from his diary and other accounts that the work of mapping out a national post-disaster research and preparedness agenda began very soon after September 1. While his was by no means the only voice weighing in on issues related to protecting the nation from earthquakes, Imamura's frequent presence in public forums, his academic credentials, and prolific publication rate did mean that his perspective was among the most visible and most direct in its claims on the public's attention.

In the short term, he drew attention to the possibility that Osaka might be the next victim of a major earthquake. He offered no timeline to go along with this warning and only limited analytical support for his findings but, for a few months after the Kantō disaster, at least, he encouraged local leaders in the Kansai area to be proactive in their preparations for a disaster closer to home.[27] Speaking to audiences elsewhere, Imamura was no less blunt about the likelihood that the nation would suffer another disaster soon. He often urged listeners worried about earthquakes to come to Tokyo, where he assured them they would be safe. Comments along those lines were frequently followed by the droll observation that anyone interested in experiencing for themselves what had happened in Tokyo on September 1, 1923, should, on the other hand, move to Osaka.[28]

Imamura's willingness to exploit the tension between the public's growing awareness of the risks they faced and the considerable uncertainty that remained about when or if that risk would manifest itself—a tension he himself had helped create and sustain—meant that earthquake prediction and disaster preparedness remained topics of ongoing interest to the press for many months after September 1. Asserting the relevance of forecasting and prediction to the post-disaster nation was at some level self-serving, as this was an issue with which Imamura himself was closely associated, and he certainly appeared to relish the attention his expertise attracted. At the same time, however, it was also evident that the disaster and Ōmori's death had together created an opportunity, a necessity even, to rethink how seismology "worked" in the aftermath of the earthquake.

Planning for Prediction

The possibility of reforming the existing institutional arrangements for the study of earthquakes had occurred to Imamura before. In 1918, he had put forward a series of proposals to do just that to several academic societies and, in 1921, presented one to the newly formed National Research Council (Gakujutsu kenkyū kaigi), of which he was a member. The proposal, which the council eventually endorsed and sent on to the Diet for further action (where none was taken), called for the establishment of temporary observation facilities in earthquake regions.[29] It argued, in part, that seismologists in Japan had by and large identified those locations most at risk from earthquakes. What remained for the field to determine was when those events would occur; progress toward that next goal called for the construction of large numbers of observation sites, where sensitive instruments would gather data that might help scientists identify changes in the earth's crust or other phenomena that might be precursors to a seismic event.

From Imamura's perspective, the problems identified in these early proposals were just as pressing in the aftermath of the September 1 earthquake as they had been prior to it. His visit to Osaka had included conversations with local officials about setting up observation facilities there and a number of his published articles made similar proposals for ongoing study. Over the next several months, however, the center of the debate over how seismology might be organized, and the place of prediction in that new structure, shifted to Tokyo, where Imamura, his colleagues in the IEIC, and other Tokyo Imperial University faculty started work on detailed plans for the future of the field. Their intent was to collaborate on a draft proposal that would be submitted to the Diet in June, although the precise framework of that proposal was not well defined at the start.

The discussions among the key actors in the proposal-writing process are not well documented. The few details that can be teased out from later accounts agree on at least two things: one is that the plan presented to the Diet in mid-July was Imamura's from start to finish. As the IEIC's new chair, Imamura was well-positioned to control that body's deliberations and he appears to have taken full advantage of that opportunity. The other observation that can be safely made about how the proposal went forward is that many of Imamura's colleagues were displeased with his tactics and with his plans for seismology's trajectory.

In many respects, the 1924 *Proposition Regarding the Establishment of Special Facilities for Earthquake Research* was a bigger and more ambitious version of the plan Imamura had submitted to the National Research Council several years prior.[30] The newer proposition called for the creation of an Earthquake Research Institute based in Tokyo, with nine remote facilities for observation and data collection (four in the Kantō district, four in Kansai, and one in the Shinano region).[31] Staffing these facilities would require no fewer than 140 personnel. The estimated start-up cost for the institute and the local facilities came to more than a million yen; the annual operating budget was more than 400,000.[32]

Imamura's proposal did its work in part by reminding legislators that earthquakes posed a very real and ongoing threat to the nation. The point was not that listeners would have forgotten the horrors of Tokyo and Yokohama so much as it was to get them to imagine what it would be like if Japan's other cities and towns were to suffer a similar fate. The historical record of the nation's earthquakes that the IEIC had labored for so many years to produce, listeners were told, revealed how geographically dispersed earthquakes were in Japan. It didn't take much to realize that the same record also revealed how frequently they happened. Just as there were no safe places, there were no safe moments; one could never be sure when the next major event might occur.

If these reminders of the threats earthquakes posed to the nation were the bad news, Imamura's arguments for legislators turned next to the good, namely the prospect that seismologists were on the cusp of being able to offer timely warnings of impending quakes. Scientists had recently discovered evidence suggesting that the ground in parts of the Bōsō Peninsula had risen significantly over the course of the three years leading up to the Kantō earthquake, policy makers were told.[33] Such changes in the relative height of segments of the earth's crust were thought to be precursors, or signals, of an impending earthquake. The reason these changes had not been noticed before September 1, said Imamura and others, was that there were too few observation facilities and too few scientists trained to analyze the data to which they did have access for the connection to have been made in time. Had the equipment been in place, the proposition's sponsors argued, it was not impossible that warnings could have gone out days, if not weeks, prior to the earthquake.

The testimony offered in both houses of the Diet painted a clear picture of the way forward for seismology in Japan. The ongoing threat to Japan from earthquakes was significant, Imamura argued, but the combination of research into prediction and the pursuit of other measures to prepare the nation to defend itself against an array of potential harms were rational responses to that challenge. While no one involved in presenting the proposal to the Diet promised that earthquake prediction would be possible right away, the optimistic tone of the testimony, the reference to the recently discovered data, and hoped-for improvements in instruments to measure strain in the earth's crust all boded well for the initiative.

Implicit in this package of reforms was an acknowledgement that the study of earthquakes could not continue as it had. The IEIC's accomplishments to date had been significant, as each of the legislators dutifully noted, but its structures and practices were not up to the tasks the field now faced. Scale was one problem—the IEIC had too few scientists available to it and lacked the resources to provide much support to those who did contribute to its research and reporting functions. Equipment was another—neither the IEIC nor any of the handful of university-based seismology programs had access to the sort of precision measuring devices that prediction was going to require. And, finally, there was the issue of the role that prediction research and its connection to public policy ought to play in the field's future. Imamura's argument, and the one put forward in the Diet testimony, was that prediction was the future of the field and the new earthquake research facility should be organized and equipped in accord with that agenda.

Legislators and Ministry of Education officials both expressed polite interest in Imamura's proposal but ultimately supported an alternative way forward that had almost nothing to do with prediction. Criticism of Imamura's plan had come from two directions. At the time of Imamura's proposal, the IEIC's annual budget was only 30,000 yen, which meant that the new initiative would have required a more than ten-fold increase in spending, year-to-year. Several of Imamura's colleagues, including some within the membership of the committee itself, argued that the new proposal was just too grandiose and expensive and that it would never be approved by the Diet. A second and perhaps more telling set of issues challenged the underlying premises of what Imamura wanted to do. These critics suggested that the new plan merely replicated the weaknesses of the existing system as they had developed under Ōmori. Imamura's insistence that scientists continue to focus on instrumentation, documentation, and prediction, critics complained, did little more than continue the practices and methods that had dominated the field since the early years of seismology in Japan. Physicists and others in Western Europe doing innovative work in geophysics and basic science had usurped Japan's once-prominent position in seismology and opponents of Imamura's plan pushed hard to develop an alternative that would allow them to pursue similar research agendas.[34]

Imamura refused to give ground when others raised these concerns and stubbornly refused to scale back the scope of his proposal or to tolerate any revisions in its focus. As a result, a competing plan took shape under the guidance of Tokyo Imperial University professors Suehiro Kyoji and Professor of Physics Terada Torahiko.[35] Both were well-respected and well-connected scholars; neither had been much involved in seismology prior to 1923, a quality that may have worked in their favor, at that point. The two advocated for the creation of an institute that would replace the IEIC and which would coordinate basic research in earth sciences and seismology by a new generation of researchers trained in physics, mechanical engineering, and similar disciplines. As their staffing plans suggested, the new institute would maintain some elements of the IEIC's focus on earthquake resistant construction and materials development, even as it supported work in new fields.

In the end, and over Imamura's objections, their vision won. The Diet quietly abandoned Imamura's plan in favor of the much more modest proposal brought forward by Terada and Suehiro. The funds the Diet agreed to commit to the new Earthquake Research Institute—while more money than the IEIC had ever had access to—was a small fraction of what Imamura's proposal had called for.[36]

Earthquake prediction and forecasting were not high on the new institute's agenda, which instead encouraged its scientists to study basic problems in geophysics and, in so doing, initiated a shift in the field away from the observational, often speculative work that had characterized Ōmori and Imamura's approach to research. (Terada made no secret of his own skepticism about earthquake prediction's prospects at the time.)[37] This shift was also evident when commentators had occasion to assess seismology's accomplishments to date, and its future course in Japan. As historian Kim Boumsoung points out, over the course of the 1930s, in publications and presentations intended for both domestic and international audiences, Ōmori's "style" of scientific inquiry was repeatedly critiqued and a sharp line drawn between the old ways of doing things and the new, more rigorous approaches.[38] If Ōmori and Imamura's work could be described as having sought to arrive at an understanding of earthquakes through observation, the emergent research agenda represented a turn to theory and to a more fundamental understanding of the forces and processes at work in the earth's crust. Imamura's conception of what earthquake prediction research would involve—very much an artifact of that earlier, increasingly discredited regime—had little traction in this new environment.

Imamura and the Nankaidō Earthquake

Despite these setbacks, Imamura remained one of the most visible spokesmen for seismology and for science more generally long after the priorities of the Earthquake Research Institute had been decided. He accepted an appointment in the new institute, where he served until his retirement in 1931. Imamura made a number of appearances on the radio and was involved in the production of at least one "earthquake" film; Victor Records in 1926 released an album of lectures by the famous "Doctor of Earthquakes." Before his death in 1948, Imamura published another three volumes of work on earthquakes, each intended for general readers. *Earthquake Lectures* (*Jishin kōwa*) appeared in 1924, *Overcoming Earthquakes* (*Jishin no seifuku*) in 1926, and *The Catfish's Pranks* (*Namazu no zaregoto*) in 1941.[39] A fourth book, coauthored with Kyoto University astronomy professor Yamamoto Issei, was written especially for children; *The Stars and the Sky, Volcanoes and Earthquakes* (*Hoshi to sora kazan to jishin*) came out in 1930.[40] Just a year earlier, Imamura had helped launch both the Jishin Gakkai (Seismological Society) and its new journal, *Jishin* (*Earthquakes*). He hoped the magazine would make seismology more accessible to the general public and thereby help the nation be better prepared for any threats to come. Imamura was both its editor-in-chief and a frequent contributor; *Jishin* appeared regularly until the end of 1944.[41]

Many of the themes that Imamura emphasized in the immediate aftermath of the 1923 disaster remained prominent in his later work. Writing in 1926, Imamura asserted that:

> Those who will overcome the earthquake are not two or three scholars, or a handful of officials, but in fact it must be every single citizen of this nation of earthquakes. If Japan's citizens understand earthquakes, their fear of them will diminish. If their fear diminishes, they can confront them bravely. If they can confront them bravely, they can certainly be overcome. . . . The lack of understanding of earthquakes on the part of our citizens I fear more than the earthquake itself.[42]

This broad goal of educating the public and encouraging their participation in efforts to prevent disastrous outcomes from natural events was one he reiterated in *Stars and Sky, Volcanos and Earthquakes*. Imamura counseled his young readers to imagine the disasters associated with earthquakes as a fierce enemy, one which must be studied carefully in order to defeat it fully.[43] That this was possible Imamura had little doubt. "In fact, I for one believe," Imamura wrote, "that even if seismology didn't make any progress from this day forward, we already possess the means necessary to avert the entirety of the harm caused by earthquakes."[44] Readers familiar with Imamura's earlier works would have found little unusual in his discussion of seismology's development in Japan, in his advice to students about how to react to major tremors, or in his proposals for government policies to mitigate disasters.[45]

Imamura never quite gave up on the idea of using a network of seismometers and other instruments to spot earthquake precursors and warn communities when a potentially destructive one was coming. There was no solid evidence that such precursors existed and thus no way to know whether they could be consistently observed if they did. It certainly seemed reasonable to think, however, that as the powerful geological forces responsible for earthquakes built up over time, changes at the earth's surface of some kind would probably be evident. Reports of changes in elevation along the coastline in the vicinity of the Great Kantō Earthquake prior to September 1, for example, seemed to support such an assumption.[46] One trick would be recognizing those changes at or near the surface as they happened and understanding which ones at which scales might be signaling that a rupture was imminent. Another would be figuring out where to look in the first place.

Imamura came to believe that he had solutions to those challenges. Local geodetic surveys immediately before and after a 1927 earthquake in Niigata prefecture seemed to show that there had been dramatic changes in ground elevation near the epicenter in the weeks leading up to the event.[47] "The possible relationship existing between the topographical changes directly accompanied by earthquakes," Imamura wrote, "and those which were going on previously at the same place, would constitute an important subject in the study of earthquake prediction. In this connection, careful and systematic re-levelling of regions where severe earthquakes may be expected is extremely desirable."[48]

He spoke optimistically about prediction's prospects in remarks at the Third General Assembly of the International Geodetic and Geophysical Union in Prague early that autumn and told reporters in Japan a few months later that once a robust monitoring network was set up, he expected that scientists would be able to predict when an earthquake would occur two or three weeks ahead of time and, in some cases, narrow the window even more, perhaps to just twenty to thirty minutes before it struck.[49]

By 1928, he also believed he knew where it made the most sense to watch for signs of a potentially destructive earthquake. Imamura had decided by that point to focus his efforts on the threat to the Nankaidō district, for two reasons. The first had to do with timing or, more specifically, with how long it had been since the last major earthquake in the area. The historical record of powerful events in the area went back as far as 684, with the most recent ones occurring in 1361, 1498, 1605, 1707, and 1854; all had been followed by destructive tsunami and each had taken many lives. In 1928, the seventy-fifth anniversary of the Ansei-Nankai earthquake was just a year away. Assuming that the next Nankaidō earthquake followed the last one by one hundred or even one hundred fifty years, it wasn't too soon at all to begin preparing for it. This logic was quite similar to the calculations that had led Imamura to make observations about the rising risk to the Kantō district back in 1905.

The other reason to focus on the Nankaidō district was Imamura's belief that he knew what signs to look for in order to tell whether an earthquake was coming and where and how to look for them. His hypothesis was that points along the Nankaidō's Pacific coast would show changes in elevation in the lead-up to a major earthquake like those that had been observed in Tokyo's vicinity prior to the 1923 disaster. For at least as long ago as modern surveying techniques had made it possible to track such things, Imamura argued, the coastline there had been slowly sinking relative to sea level, only to begin a period of uplift shortly before the earthquake. Using geodetic and leveling surveys dating back to 1899, Imamura established by 1929 that parts of the Nankaidō coastline, too, were showing a significant loss of elevation.[50] Assuming that the same processes that had caused the 1923 earthquake were also at work off the Pacific coast, it made sense that there, too, the slow loss of elevation would eventually reverse itself and be replaced by a gain in height relative to sea level at some point before the anticipated earthquake occurred.

In order to be ready for that change, Imamura launched an ambitious and largely personally funded effort to create a network of monitoring stations throughout the region. The first of these sites began operations in Wakayama Prefecture in 1928; over the next several years, Imamura added seven others.[51] At first, he was able to borrow equipment and help from former colleagues and students at the ERI but, after his retirement in 1931, Imamura relied increasingly on an assortment of local educators and officials to keep an eye on the instruments, in exchange for which they received both his thanks and, occasionally, more formal expressions of gratitude in the form of cash. Those expenses and others Imamura paid for out of his own pocket.[52] Once the observation stations were up and running, the plan was to keep them going for as long as it took to pick up signs that a Nankaidō earthquake was imminent. Exactly what those signs might be Imamura never spelled out—his instrumentation wasn't sophisticated enough to measure changes in elevation along the coast—but presumably he expected to be able to spot any anomalies that might show up.[53] In the meantime, Imamura tried to raise awareness of the risk to the region, both in print and during visits to communities he suspected would be hardest hit when the earthquake occurred. He made little headway.[54] Once the war began, prefectural and local administrations would have been hard pressed to attend to the risks that Imamura was describing when other, more proximate ones loomed so large.

Frustrated but stubborn, Imamura kept the project going as best he could, watching and waiting for the warning signs he was sure would appear. It was, after all, only a matter of time.

CHAPTER 2

Embracing Prediction in the 1940s

Imamura Akitsune had been waiting for a Nankaidō earthquake for almost twenty years when one finally struck, in late December 1946. As he had anticipated, the quake was exceptionally powerful; as he had feared, it was also spectacularly and indiscriminately destructive. The M8 event originated two dozen kilometers below the seabed and twice that distance offshore from the southernmost tip of the Kii peninsula.[1] The rupture produced seismic waves powerful enough to be felt hundreds of kilometers away in Tokyo, in towns and villages near the Sea of Japan, and across the interior of Kyushu. Closer to its source, the earthquake completely leveled ten thousand homes and badly damaged another twenty thousand. It sent tsunami racing inland from the Pacific coast in Shikoku, Wakayama, Mie, and elsewhere, where walls of water five and six meters high breached or swept over sea walls, barreled down city streets, and inundated entire villages. Of the 1,300 people who died in the earthquake's aftermath, many were lost to the tsunami. Survivors were still contending with flooded fields, homes (for those still standing), and roadways weeks later.[2] To make matters worse, everywhere the earthquake touched, the war had already been, one way or another. American bombers had targeted communities throughout the Nankai region again and again in the months leading up to the surrender in August 1945, killing thousands of their inhabitants and unhousing many more. This latest catastrophe brought more pain and suffering to the inhabitants of an already damaged landscape.

The Nankaidō earthquake disaster was an important inflection point in the history of the relationships between Japanese earth scientists, the pursuit of earthquake prediction, and speculation about other disasters to come. In the earthquake's aftermath, the number of seismologists and other experts openly in favor of prediction research grew from essentially one—the position Imamura had been in for decades, by that point—to many, as a cohort of prominent scientists from all over Japan collaborated on the creation of the country's first official prediction research project. Wadati Kiyō, Chūji Tsuboi, and the others who participated in that new undertaking knew Imamura and his work, or course, but seldom mentioned either when justifying their own interests in pursuing prediction-related research. The younger scientists made choices after the Nankaidō earthquake that shifted the debate over earthquake prediction's prospects away from arguments long associated with Imamura and the field's past and toward ones that reflected their own perspectives and training and, presumably, seismology's future. The consequences of that shift were far-reaching. In the short term—the focus of this chapter—it yielded the research project mentioned above, a series of very public attempts by scientists to identify Japan's next big earthquake threat, and debates over the values of both. The post-Nankaidō shift also paved the way for the embedding of earthquake prediction research and the assumptions behind it in the earth sciences in Japan. Those processes got under way in the 1960s and are explored in Chapter 3.

The 1946 Nankaidō earthquake would play at least one other role for seismologists in Japan related to the pursuit of prediction and the anticipation of disasters to come. Much as Imamura had described the 1854 Tōnankai and Nankai earthquakes as parts of an ongoing pattern of powerful seismic events in the region to help explain why he feared that others just like them could happen again soon, since the 1970s scientists have turned to studies of what happened in 1944 and 1946 in order to better understand what was likely to take place next in the Nankai and Tōkai districts, seismically speaking. The introduction of plate tectonics theory and other advances in the field gave researchers a much fuller picture than Imamura ever had of the processes that produced powerful earthquakes along the Nankai Trough.[3] Armed with that knowledge, the conclusions those scientists drew from their analyses of the region's history of seismicity were not so different from the ones Imamura had arrived at, decades earlier. One was that the earthquakes in 1944 and 1946 were the most recent iterations of a pattern of seismic activity that went back many hundreds of years, at least. Another was that the locations and timing of those quakes provided important clues as to when and where the district's next major rupture was most likely to occur. The consensus among seismologists in Japan in the 1970s, based on those clues, was that the next earthquake could come at any time, that it would begin offshore of the Tōkai district, and that it would do an incredible amount of damage when it did.

Seismologists in Wartime and Defeat

Imamura retired from the university as the Japanese military was taking its first steps toward a war of unprecedented brutality and suffering, first on the continent and later throughout the region. The invasion of Manchuria in 1931, the start of the war with China in 1937, and the attacks on US and British forces in 1941 unfolded with the patriotic support of most Japanese. And, for the earth sciences community, as it was for most Japanese in the metropole, the war's direct impacts were relatively limited at first.[4] Imamura was able to keep up with the publication of *Jishin*, the journal he had helped found and for which he remained the editor-in-chief, until fairly late in the war, despite shortages of paper and other supplies. His former colleagues at the Earthquake Research Institute continued with their own research and teaching as best they could. After 1941, they had to do so without access to colleagues overseas or to English-language scientific journals.

By 1944, however, even seismologists were caught up in the government's increasingly desperate attempts to mobilize all the nation's resources as the Imperial Army and Navy's losses grew and grew. Researchers and technicians in the Central Meteorological Observatory (the predecessor of the Japan Meteorological Agency) were early targets for conscription and eventually the ERI, too, began to lose staff and was under pressure to develop research projects that somehow supported the war effort. As was the case in other university research centers, some senior professors reportedly did what they could to protect junior colleagues from conscription by involving them in projects that the military seemed likely to value.[5] Unlike the American military, which at one point was reportedly studying whether it would be possible to trigger powerful earthquakes in Japan by selectively bombing known fault zones, neither the Imperial Army nor the Navy saw seismology or the scientists who studied it as all that useful when left to their own devices.[6]

The beginning of the American bombing campaign in late 1944 marked the start of a new phase in the war. In the months remaining before Japan's surrender, in August 1945, most of Japan's major population centers were the targets of repeated air raids. Some scientists at the Earthquake Research Institute moved their labs and data out of Tokyo and into the somewhat safer suburbs, while others remained in the city and kept up the routine of daily travel from home to their offices at the university for as long as they could, choosing the safest routes based on which parts of the city were under threat that day. The American decision to switch to incendiary weapons in early

1945 made the attacks all the more effective and horrific. The fire-bombing of Tokyo on March 9–10, for example, left much of the city in smoldering ruins and as many as 100,000 people dead.[7] Other cities suffered similar fates on a smaller scale. By the time of the attacks on Hiroshima and Nagasaki, only a handful of Japan's urban centers were left unscathed. Many had been targeted repeatedly.

The closing years of the war and its immediate aftermath brought other disasters, as well. Four powerful, destructive earthquakes struck the country in rapid succession between 1943 and 1946, one each year. In September 1943, a powerful, shallow earthquake struck offshore, not far from the city of Tottori in western Honshu, killing more than 1,000. In December of the following year, the 1944 Tōnankai earthquake, a megathrust event along part of the Nankai Trough, devastated communities from the east coast of the Kii peninsula to Shizuoka and points inland. At least 1,200 people died as a result of that event. In January 1945, the Mikawa earthquake took more than 2,300 lives in communities not far from those that had been damaged a little more than a month previously.[8] The fourth event in this series came in December 1946; see more on it, below.

So long as they were in power, Japan's military authorities dealt with the effects of the earthquakes much as they did the Americans' bombing raids, which is to say that they did their best to prevent reports of deaths, injuries, or damage of any kind from reaching the public.[9] A significant share of Japan's aircraft production was situated in areas hardest hit by the earthquakes in 1944 and 1945; some of those factories were badly damaged and a number of workers killed, which only heightened authorities' desire to keep news of what had happened from circulating. The censorship extended to the routine field work of documenting the earthquakes' origins and effects by seismologists. Officials did not take well to strangers asking questions or wandering through the ruins taking careful notes. Even when researchers were grudgingly permitted to conduct field surveys and gather data, they were generally prevented from sharing what they had learned even with other scientists, much less with the public.[10]

Japan's surrender brought the bombings to an end but the hardships that had set in during the war would persist. Tokyo and most other urban centers were in ruins, food was scarce, inflation rampant, the future uncertain on so many registers. This was still true in early 1947, when the Americans first signaled their interest in what Japanese scientists had been doing with earthquake prediction.

The Allied Occupation of Japan lasted from shortly after the surrender until April 1952. Very few of the nation's institutions were left unchanged by the reforms and legislative initiatives launched during that period (although many of these changes were the result of processes already underway well before the Americans arrived).[11] While the emperor's status, the military, and democratization in general were the most visible targets of Occupation-era reforms, the structures and practices of Japan's scientific community also came under scrutiny.[12] It was in the context of that review process that earthquake prediction's status in Japan surfaced as a topic of interest to officials of the Supreme Commander for the Allied Powers (SCAP) and, subsequently, to Japanese earth scientists, as well. One of the results of negotiations between the two groups was a short-lived but fascinating experiment in which Japanese scientists imagined what a long-term project of earthquake prediction research might entail and tested whether it might already be possible to generate reliable, useful predictions, in some circumstances. The next section shows how that experiment came to be, examines how much it owes to Imamura's earlier efforts, and considers some of its implications for later efforts to reduce the uncertainty around earthquakes.

"All for Nothing"

The Nankaidō earthquake was the event Imamura had been waiting for and worried about since at least 1928 and in anticipation of which his makeshift monitoring network had been created. Imamura and his volunteers had kept the project going for quite some time, even as war broke out in China and later against the United States. As was the case even for equipment managed by the ERI, however, wartime rationing and labor shortages eventually made it difficult to keep the network in good working order and, by 1945, most of Imamura's monitoring stations were no longer producing useful data. Imamura lost his ability to keep an eye on the region and watch for signs of the next earthquake—whatever those might be—just as the risk of one happening seemed to be going up. The December 1944 Tōnankai earthquake had been followed just a month later by another near Mikawa; Imamura believed that these events were precursors to the one he feared the most.

As devastated as the country was by the war, and as diminished and exhausted as everyone was left by defeat, after Japan's surrender Imamura still held out the hope that he would be able to resume his watch over the

Nankaidō district. There were many obstacles to overcome. At seventy-five, his physical health was not what it once was and, unable to travel to the monitoring stations himself, he instead wrote to former volunteers to gauge their willingness to continue the work.[13] Money was tight, as it would be for most Japanese in the early years of the Occupation, but Imamura was used to getting by on his pension and, so long as the old equipment was still intact, it might not cost that much to get it up and running again. The need to do so was clear to him, and the sooner they got started, the better.

Imamura's biographer and others who have written about this period in his life highlight its hardships and it isn't hard to see why. In early 1946, the Americans had moved ahead with plans to "purge" former government officials and others associated with the previous regime's militarism. The categories of those subject to removal and other sanctions were broad and often arbitrary, but former seismology professors were generally not among the Americans' targets. Unfortunately for Imamura, however, the Americans were interested in anyone with ties to the former Imperial Army and Navy. Until his promotion in 1923, Imamura's research position at the university had never had a salary attached to it. Between 1896 and 1923, he had made his living teaching math and physics to Imperial Japanese Army officer candidates. As a former military instructor, Imamura was therefore deemed unfit for government employment by the Americans. Given that he had already left his post many years earlier, that dictate alone would have had little effect on him. What did make a difference was that Occupation officials also put an end to the military's pension payments, which in 1946 was Imamura's most important source of income. His pension was worth much more than the paltry sum he was entitled to by virtue of his relatively brief stint as a university professor.[14] The family's financial situation deteriorated significantly once those funds were cut off. Imamura had to put his plans to restart the monitoring project on hold.

In the meantime, though, there was nothing preventing him from once again calling attention to the threat he believed the Nankaidō district faced. In a paper presented to the Imperial Academy (Teikoku Gakushiin) in October 1946, Imamura stated that "it is increasingly felt that the great earthquake just discussed will strike in the near future. This eventuality deserves the utmost attention from an academic standpoint as well as from the standpoint of disaster prevention."[15] A week prior to the earthquake, he wrote directly to local officials in at least one coastal Shikoku town to warn them that a major earthquake was possible and to ask them to fix the local tidal gauge as quickly as they could, to improve the odds of gaining at least some advance warning.[16]

A younger colleague happened to be with him on December 21 and later described Imamura's reaction as reports of what had happened in the Nankaidō region came over the radio that afternoon. "He just stood there at first, listening intently to the news. As soon as it was over, with a deep disappointed sigh, he said, 'Eighteen years of hard work, all for nothing.'"[17] The letter Imamura had sent to Shikoku reached its destination the day after the earthquake.

Newspaper reporters quickly shared this detail and other coverage of Imamura's "prediction" of the Nankaidō earthquake with readers, which is how the Americans would have found out about it.[18] These reports were of particular interest to the staff of the Natural Resources Section of the SCAP bureaucracy, as they were already involved in plans to restructure the sources of support for the earth sciences in Japan, including seismology. If Japanese scientists had found a way to predict earthquakes, the Americans would want access to that knowledge, and, even if they hadn't—yet—the Americans would want to know if a solution was in sight.

The first Occupation-led inquiries into reports of Imamura's success began in February 1947, when SCAP summoned scientists from the Earthquake Research Institute and the Central Meteorological Observatory (CMO) and asked them directly whether Imamura had in fact predicted the December 1946 earthquake.[19] Wadati Kiyō (director of the CMO), Hagiwara Takahiro (ERI), and others among the scientists present at the meeting were all members of a post-Ōmori, post-Imamura generation of seismologists. Wadati was in his mid-forties when the war ended, as were Sassa Kenzō (Kyoto University) and Tsuboi Chūji (University of Tokyo), two other scientists involved in the discussion; Hagiwara was in his thirties.[20] None had been directly involved in the field's earlier encounters with prediction and, although some had studied under Imamura, their primary training was in the hands of other mentors.[21] All were established in their careers and all but Sassa had done their graduate training at Tokyo Imperial University.

The scientists summoned by the Americans responded carefully to questions about what Imamura had done. Their colleague, they suggested, had indeed been investigating the likelihood of an earthquake in the Nankaidō district but, because he had not specified precisely when the event would occur, his warning in 1946 was technically not a prediction.[22] That said, Hagiwara and Wadati took the opportunity to suggest to the Americans that others might yet succeed where Imamura had not. "It was revealed," according to one summary of the meeting, "that earthquake prediction was considered possible by conducting an extensive program for first order leveling, tiltmeter and tide gauge measurements, and study of earth current."[23] This was not a position that Hagiwara or the others were on the record as having supported previously and was, in fact, a close approximation of the methods Imamura had hoped to use to predict the Nankaidō earthquake.[24]

As their discussions with the Americans continued over the next several months, the younger scientists elaborated on these and other possible ways forward for earthquake prediction research in Japan. By the end of the summer of 1947, in a significant shift in the field's priorities up to that point, earthquake prediction had become a high-profile component of Japanese seismology's postwar mission.

Earthquake Prediction and the Occupation

The momentum toward some sort of collective commitment to forecasting and prediction built up among the earth scientists in the months after the Americans first started asking questions about Imamura's warning.[25] Part of the context for those conversations was the ongoing SCAP-led review of Japanese scientific institutions, including the Earthquake Research Institute and the CMO, along with others. Occupation officials may have been after some degree of democratization within academia and the sciences (the baseline for which shifted quite a bit over the course of the Occupation), but they were also looking for opportunities to pare back Japanese government spending. Staffers in the Natural Resources Section, for example, were particularly interested in what they saw as duplication of effort and resources between the CMO and the ERI, and in other redundancies among the various government-funded agencies involved in the earth sciences. For scientists grown used to dealing with the Imperial Japanese military's influence over their livelihoods and research, negotiating with the Americans just a few years later might well have felt familiar.

In the end, the Americans' interventions in the earth sciences were less invasive than Japanese scientists had feared and Natural Resources Section staffers might have have had hoped for. SCAP officials had recruited Beno Gutenberg, an eminent scientist and director of the Caltech Seismological Lab, to come to Japan to advise them on their planned reorganization. If they had expected him to take their side in recommending drastic cuts and sweeping changes, they would have been disappointed, since Gutenberg proved to be quite sympathetic to his Japanese colleagues' concerns during his visit in June 1947. His advice to the Americans was that they leave well enough alone, at least so far as seismology and the institutions supporting it went. The few specific actions he urged the Americans to take were close in content to proposals that the Japanese themselves had come up with and shared with Gutenberg during his visit.[26] Gutenberg's primary interlocutor among the Japanese scientists was Tsuboi; the two had known each other since the latter's stint as a researcher at Caltech in the 1930s. It was Tsuboi who provided Gutenberg with copies of the Japanese proposals and it was Tsuboi who convinced him to propose to the Americans that they also support Japanese research into earthquake forecasting and prediction.

Gutenberg's comments on prediction in his final report were in line with what Tsuboi, Wadati, and the others had been talking about among themselves since spring.[27] Gutenberg categorically ruled out the possibility of a successful prediction at present, given the state of the field. "The claims of certain individuals that they can predict earthquakes are not restricted to Japan," he wrote, "but have been disproved thus far in all instances." "A very few views were expressed," he went on, "that the forecast of earthquakes is practically at hand. However, the seismologists who have longer experience agreed that it will take a long time before forecast with a sufficient accuracy will be possible."[28]

As dismissive as he was of claims that prediction was "practically at hand," Gutenberg did not rule out the possibility that scientists might one day find a way to make it work. To that end, he instructed the Americans that "every effort should be made to furnish the foundations for such forecasts, such as study of changes in level, changes in electric currents in the earth, magnetic elements, in gravity, tilt, and their relationship to the occurrence of earthquakes."[29] That assessment and the suggestions Gutenberg offered in his report were almost identical to those offered by Wadati, Hagiwara, and the others in their meetings with SCAP officials earlier in the year. The Japanese scientists involved in the discussions with the Americans were at their mercy in many respects, yet it was they and not the SCAP officials who were deciding what to do about the earthquake prediction problem going forward.

The earth scientists in Japan who started committing themselves publicly to the pursuit of earthquake prediction in the late summer of 1947 did so in one of two ways (a few did both, as discussed below). The first and largest cohort was made up of participants in the new Earthquake Prediction Research and Liaison Committee (Jishin yochi mondai kenkyū renraku iinkai), which met for the first time in August. It was Japan's first state-sanctioned attempt to coordinate and support forecasting and prediction efforts, and a tangible product of Gutenberg's recommendations to the Americans. Wadati had asked for and quickly received permission to organize a group of scientists to study the prediction problem around the same time as Gutenberg sent in his report. SCAP officials authorized Wadati to "take immediate steps to begin organizing a committee consisting of interested Japanese scientists and organizations for coordinating research on earthquake forecasting problems."[30] The new group's membership included representatives of the nation's leading seismology research facilities and university departments, among them Wadati (CMO), Tsuboi (Tokyo University Geophysics), Sassa (Kyoto University), Nakamura Saemontarō (Tohoku University), and others. Imamura was one of several scientists invited to participate "as authorities in their fields."[31] Among the first tasks the committee set for itself was the production of a detailed proposal to SCAP for an expansive new prediction research initiative.

The second way that scientists expressed support for the idea that earthquake predictions were possible was by making them. Prior to 1947, public warnings from Japanese earth scientists about increased short-term risk from earthquakes were rare events. Even Imamura, early twentieth-century Japan's most prolific raiser-of-alarms about earthquakes to come, was responsible for only a handful of such warnings over the course of his entire career. For many years he was really the only scientist with standing in the field who made such claims in public. Between 1947 and 1949 alone, however, earth scientists in Japan came forward on four separate occasions implicating five different locations to warn the nation that damaging earthquakes would soon occur. All but one of the earthquake warnings of the late 1940s originated with Liaison Committee members.

Kyoto University's Sassa Kenzō drew the public's attention to potential threats to the Kansai District in late 1947, ones that he had detected using precision tiltmeters and other instrumentation installed at locations throughout in the region. Yamaguchi Seiichi of the Home Ministry's Geographical Research Bureau—the only one of the four scientists who was not a Liaison Committee member—shared his concerns about an imminent earthquake in the Kantō region around the same time. He based his claim on observations that the sea level as measured at Aburatsubo on the western tip of the Miura peninsula, not far from Tokyo, had declined rapidly in June, July, and August that year. As Yamaguchi pointed out to reporters, similar changes in sea level—which indicated vertical displacements in the land at water's edge—had been reported around the time of the Kantō earthquake and prior to the 1946 Nankaidō earthquake, too.[32] Unless the levels returned to normal, Yamaguchi warned, a major earthquake could occur in the capital district in the very near future.

The CMO's Inoue Win discussed his theories about potential risks to Fukui and the Chichibu district with his colleagues in 1948, drawing on an elegant analysis that measured "the propagation of earthquake stress or strain waves" after major earthquakes to determine when and where the next ones would occur.[33] Tohoku University's Nakamura Saemontarō, the last to go public with a warning, announced to the press in early 1949 that the city of Niigata should expect a major earthquake within the next one to two months. (Niigata had a population of around 300,000 at the time.) Nakamura had spent part of 1948 on studies of terrestrial magnetism all along the Japan Sea coast and was alarmed by the unusual findings he had recorded. Nakamura interpreted these results—based on historical precedent, he claimed—as the precursors of an imminent earthquake in the area.[34]

Both developments—the existence of the new committee and the uptick in predictions and warnings to the public—were relatively short-lived. The committee's last activities of note were in the spring of 1949, which is about the same time the publicity around Nakamura's warning about a potential Niigata earthquake was winding down. Neither left much in their wake in terms of immediate impacts on seismology or earthquake disaster prevention in Japan. That said, both the committee's actions and the circumstances surrounding the production and reception of the warnings were in their own ways important antecedents to the choices scientists made in the 1960s about earthquake prediction's prospects in Japan.

One straightforward connection between the Occupation-era initiatives and those that followed in the 1960s is that the scientists responsible for the earlier efforts were also the driving forces behind the ones that came later. Wadati, Tsuboi, and Hagiwara are just a few of the experts who played key roles in the discussions about prediction in both eras, for example. The scientific body most closely associated with the pursuit of prediction in Japan from the 1960s on, the Coordinating Committee for Earthquake Prediction, was led by Hagiwara from its establishment in 1969 until he stepped down in 1981. When Hagiwara and his colleagues published the first history of the Coordinating Committee's accomplishments, they themselves traced the origins of contemporary earthquake prediction in Japan back to the Occupation and to the work of the Liaison Committee.[35] Not surprisingly, there are also similarities between the proposal for earthquake prediction research that the Liaison Committee put together in 1947, the ones that formed the basis for the influential "Blueprint" for earthquake prediction research in 1962 and others that followed.

The Occupation-era work on prediction helped set the stage for what came later in some less straightforward ways, as well. In late summer 1947, as mentioned previously, the Liaison Committee solicited proposals for new initiatives in earthquake prediction research from its members and assembled them into a single omnibus proposal. The committee leadership rejected only a few of the ideas they received and, as a result, the document they shared with the Americans took a kitchen-sink approach to exploring prediction's potential. The projects presented to SCAP included plans for the construction of ten new tide gauge stations, the development of nine large survey sectors (rhombi) used to track vertical and horizontal displacement at the surface of the earth, significant improvements to existing geodetic facilities, the purchase of a large number of sophisticated tiltmeters and other measuring devices, and so on; it was a long list. The authors of the various proposals had specific but widely varying ideas about what the pursuit of short-term earthquake prediction would require, and they didn't hold back in their requests for support from the Americans.[36]

Perhaps they should have, since the projects' collective price tag shocked SCAP officials. One estimate put the initial outlay at between 15 and 20 million yen, much more than the Americans were willing to authorize at a moment of general belt-tightening and budgetary restraint.[37] (The entire Ministry of Education research budget in 1947 was just under 72 million yen.)[38] Within weeks of having taken the first steps toward an ambitious research program in earthquake prediction, Japanese scientists were instructed by the Americans that they were free to pursue the initiative as far as they might wish, but they'd have to do so without any new, extraordinary funding. The proposals to construct new monitoring facilities, acquire new equipment, and so on were put on indefinite hold.[39]

This setback is reminiscent of Imamura's experience after the Great Kantō Earthquake, when his efforts to launch a large-scale survey and prediction research effort failed, at least in part because of concerns over its enormous cost. And yet, the circumstances in 1947 were different from those in 1924 in important ways. The committee's very existence testified to earthquake prediction's emerging visibility as a field of research, something that had not been the case when Imamura was lobbying for his expanded observational network. The arguments against funding the committee's proposals in 1947 had not been based solely on concerns about whether they lacked scientific merit. The fiscal situation, not the science, was the primary basis for the Americans' decision. Also different were the nature and number of scientists and institutions involved. Twenty years earlier, Imamura had been the only seismologist making claims in public about prediction's utility. The new committee, in contrast, included many of the country's leading authorities in seismology and related fields; practically all of its premier research institutions and academic departments were represented. With or without the money to make it happen, scientists in Japan had established that they were in favor of moving ahead with prediction-related research, and not on a small scale.

Four Warnings

The new collective enthusiasm for prediction research didn't extend to claims that it was already possible to actually predict earthquakes. As Gutenberg had pointed out to the Americans, a small number of scientists in Japan were confident that successful predictions were either already feasible or soon would be, but most believed that reliable forecasts were still many years away and that it would take longer still to produce short-term predictions, even in a best-case scenario.[40] Nothing happened during the Occupation to significantly alter those ratios or the arguments the two camps used to justify their positions. There were no breakthroughs that might have led scientists to conclude that prediction was going to turn out to be any easier than previously thought.

The decisions by scientists like Sassa, Yamaguchi, Inoue, and Nakamura to come forward with warnings tied to the timing and location of earthquakes that hadn't happened yet highlighted tensions between these two approaches to thinking about prediction in postwar Japan that persisted afterward. Their attempts to reduce the uncertainty around earthquake activity fell outside the boundaries of what most of their colleagues believed was knowable at the time and were scrutinized in that context. Although the responses by other scientists to what Sassa and the others were arguing differed somewhat in each case, they also shared some important qualities. The Liaison Committee's leaders, Wadati Kiyō in particular, took on highly visible roles as interlocutors in the public debates over each of the predictions' implications and, in each instance, actively tried to downplay their significance. Such interventions were not what the committee had been created for but its spokesmen were quick to engage with the media and the public once the predictions and the warnings that accompanied them had begun to attract attention.

This phenomenon was most evident after Sassa's warnings to residents of the Kansai district in December 1947 and then again following news of Inoue's results the following year. Sassa had reached out to warn local police officials after he spotted worrying results from tiltmeters and other instrumentation installed in an unused railway tunnel on the outskirts of Kyoto. He insisted that his goals for issuing the December warning were to alert the public to what he believed was an increased risk and to make officials and average citizens aware of the steps they could take to mitigate against the harm that an earthquake might cause.[41] Sassa had not named a date or even specified a time frame for when the earthquake he feared might strike but, over the next several weeks, the press reported numerous examples of people acting as if it was imminent. Residents of at least one community in the Kansai district were said to have already begun fleeing their homes, fearing a repeat of the devastating December 1946 Nankaidō disaster.[42] A small but noticeable earthquake north of Osaka on December 10 further focused attention on Sassa's statements and heightened the sense of urgency he had sought to convey in his warning to local officials. "We were told by Dr. Sassa that an earthquake would certainly happen," reported the chief of the Kyoto Prefectural Police, "consequently, we are hurriedly organizing counter-measures."[43] In the context of reports of preemptive evacuations, fishermen pulling their nets from the water, and the chief's definitive statement about the need to act quickly, it would appear that many had understood Sassa's warnings as a call to immediate action.

Inoue, on the other hand, never intended for his analyses of "the propagation of earthquake stress or strain waves" to be interpreted as predictions or acted upon as such. He described his findings as merely preliminary when he shared them with the Liaison Committee meeting in June 1948. His colleagues expressed mild interest at the time and encouraged him to come back when he had more evidence.[44] What ultimately made the presentation memorable was Inoue's response to an offhand question; asked by a committee member where his analyses indicated Japan's next major earthquakes would occur, Inoue answered that his model had identified Fukui, a city (and prefecture) on the Japan Sea coast, and Chichibu, a town (and district) west of Tokyo, as the most likely candidates. No one pressed for additional details and neither Inoue nor others present seem to have given the exchange much thought. (SCAP's notes on the meeting make no mention of it.)

A little more than two weeks later, Fukui suffered a massive earthquake, the nation's deadliest since 1923. More than 3,700 were killed and some 20,000 injured; fires in the city burned for five days before they could be brought under control. News reports that Inoue had predicted the Fukui earthquake and identified Chichibu as next in line to be struck appeared soon after. The media frenzy and expressions of dismay by Chichibu residents and officials that followed were on a larger scale than the turmoil that had accompanied Sassa's predictions of only a few months before. Speaking to a packed auditorium at the University of Tokyo in early July, Inoue tried very hard to present his work as speculative and again insisted that his findings were preliminary. When pressed during a question-and-answer session at the end of his talk, however, Inoue admitted that the same calculations that had singled out Fukui before the disaster there suggested that the Chichibu district might be next.[45]

The area's residents did not respond well to this news; some fled outright, while others anxiously sought advice about what to do to protect their communities and homes. Inoue's comments forced the governor of Saitama to cut short a trip to Fukui where he had been inspecting the effects of that district's recent earthquake in order to begin preparing his own prefecture for a similar calamity.

Much as Ōmori had reacted to reports of near-panic after Imamura's warnings in 1905, the Liaison Committee took an active role in trying to reassure a jittery public in 1947 and 1948. Wadati became the go-to scientist for reporters whenever a prediction-related story surfaced; as both chair of the Liaison Committee and director of the CMO, Wadati spoke as a recognized authority on earthquakes and as a representative of the state's own research apparatus.[46] He was also consistent; while never going so far as to question his colleagues' judgment outright, he could be counted on to cast doubt on findings that relied on limited evidence, as he viewed it, and to reassure the public that their worst fears would likely not come true. In early December, Wadati suggested, in reference to Sassa's warning, that "It can't be definitively determined whether such disparate phenomena are actually indications of an earthquake or not."[47] A few days later, in the *Yomiuri* newspaper, he explained the difference between foreknowledge of an earthquake, described as information that a major earthquake was likely, without being able to specify a time or place, and forecasting, which Wadati equated to predicting an earthquake much as one would the weather, within very narrow ranges in terms of time and place.[48] In his opinion, Wadati believed that forecasting might be possible to some degree, at some point, but certainly not at the present time.

"At present," he wrote, "the earthquake research equipments are too poor to foretell the occurrence" of an event. "Plainly speaking, at the present stage of our science, it is impossible to foreknow exactly the place, time, and scale of an earthquake."[49]

After the earthquake in Fukui seemed to have set the stage for another in Chichibu, the Liaison Committee intervened again to help reassure residents that all would be well.[50] By July 1948, Inoue was himself actively trying to convince the press that his own method was flawed, which meant that, although it had singled out Fukui, it had not in fact predicted the earthquake. The results of his calculations and the disaster in June were connected by coincidence, not science, he insisted.[51] At a closed-door meeting of the committee in late July, Inoue insisted that he simply could not get the press to believe him when he tried to explain that his comments about Chichibu had been a mistake and appealed to the committee to speak on his behalf. Chairman Wadati used the blackboard to draft a statement, revising the text with input from other members, and called a press conference for a little after noon. "Committee member Inoue today made it clear," Wadati said, "that he shares the committee's findings that his statistical methodology is at present not very effective in terms of earthquake prediction." Noting that Inoue's work had many merits, and that the committee had urged him to continue with it, Wadati suggested that perhaps it would be best if he shared the elements related to earthquake prediction with the public only after further, thorough review.[52] As had happened in 1947 with Sassa's warnings, these interventions by the committee coincided with diminished anxiety in the affected communities. Once August and September passed without an earthquake in the Chichibu district, the committee's foresight and caution seemed all the more justified.[53]

Setting Prediction Aside

After the brief eruption of warnings that began in 1947, for the next decade and a half Japanese scientists largely abstained from making claims in public about earthquake prediction. The Liaison Committee held its last formal session in March 1949 and the whole undertaking was quietly swallowed up by the newly formed Science Council of Japan (Nihon gakujutsu kaigi) that same year.[54] As Japan's scientists began to reengage with colleagues and developments in geophysics and seismology in the United States and elsewhere in the 1940s and 1950s, they would also have learned that researchers abroad (still) hadn't rushed to embrace prediction in their absence. There had been no paradigmatic change, no discoveries made during the war that the Japanese had somehow missed that had radically changed how scientists elsewhere were thinking about forecasting and the rest. Setting prediction research aside at that point made sense to Japanese scientists; there were plenty of other topics to explore and problems to solve.

That said, the willingness of Sassa, Nakamura, and others like them to go public with warnings that they believed were based on sound science is evidence that some in the field had come to different conclusions about what was known—and knowable—about the uncertainty around earthquakes. It wasn't just Imamura anymore and it wasn't scientists on the fringes of their respective fields. Even as Wadati and those who shared his skepticism pushed back against the specific claims about precursors or predictive methods that surfaced in the late 1940s, they did not rule out the possibility that different efforts might someday produce results they could trust. Beno Gutenberg's reluctance to completely disavow prediction's prospects in Japan was reflected in the care that Wadati and the others took to walk that same line. As the committee ended its work, its members left open the possibility of taking up the prediction question again at some later date; they had ruled nothing out. Tsuboi, for example, was among the researchers whose work in the 1950s gradually left them more optimistic about prediction's prospects than they had been at the start of the decade.[55] Indeed, many of those involved in making or responding to these episodes of prediction in the late 1940s went on to play key roles in the construction of Japan's next attempt at organized, state-sanctioned earthquake prediction research, which began in the 1960s.

One final observation to make about how the contingencies of the Occupation era shaped earthquake prediction's trajectory after the war has to do with Imamura Akitsune's place among them. Reports that Imamura had warned locals that a Nankaidō earthquake was coming had brought first him and then the predictive ambitions of his colleagues to the Americans' attention, but Imamura himself was not a key player in the developments that ensued. Imamura reportedly told his colleagues on the committee how pleased he was to see them finally embrace earthquake prediction as a worthwhile research project; even in his dreams, Imamura said, he had never imagined that he'd live long enough to be part of such an undertaking.[56] For Wadati and the others to have acknowledged that earthquake prediction research was worth pursuing was a vindication of sorts of all the work Imamura had done up to that point in support of that same cause. As he must have soon realized, however, his colleagues were not interested in drawing attention to the similarities between their research agendas for prediction and Imamura's previous efforts along the same lines. The scientists who helped launch Japan's postwar earthquake prediction project knew Imamura and at least something of his history but kept both in the background as they brought their own proposals for how to pursue prediction forward. Alongside the disappointment of having missed his long-awaited opportunity to spot the warning signs he had been so sure would precede the Nankaidō earthquake, it must also have been difficult for someone of Imamura's temperament to watch while others took the lead in promoting prediction's potential, with nary a reference to his contributions. At some point not long after the Liaison Committee's inaugural meeting in the summer of 1947, Imamura resigned from it, having lost whatever enthusiasm he might once have had for what his colleagues were up to.[57]

The earth sciences community was distancing itself from some of its presurrender history in other ways that involved Imamura. In December 1947, although in poor health, Imamura made it a point to attend the first general meeting of the Seismological Society to be held since the end of the war. Imamura had been instrumental in relaunching the society in 1929 and had been its president ever since. He had more or less singlehandedly kept the society's journal (*Jishin*) going until fairly late in the war, despite shortages of paper and other supplies, in part by producing much of its content himself. (Publication ceased only after an American bombing raid in late 1944 leveled parts of Tokyo's Kanda district and with it the facility that printed *Jishin*.)[58] The meeting in December was an effort to relaunch the organization and, as it turned out, to draw a sharp line between the society that Imamura had overseen for eighteen years and the one that members now intended to create. A key step in that process was arranging for Imamura to submit his resignation as president. For a community anxious to disassociate itself from the wartime regime and pursue a more democratic approach to leadership, it made sense that someone with Imamura's background and baggage could not remain at the helm, but the logic behind the decision doesn't seem to have softened the blow all that much. After tendering his resignation and waiting while his colleagues voted to accept it, Imamura's sole comment before walking out of the meeting was to say, "I'll leave you to it then."[59]

That was Imamura's final public appearance. He had been working that winter on revisions to one more research manuscript, even as his health grew steadily worse. Once bedridden and no longer strong enough to write, he dictated what remained to be said to Kawasumi Hiroshi and another of the small group of younger seismologists who had grown close to Imamura in his later years.[60] Kawasumi later wrote of how shocked and angered he was by the professor's circumstances after the war and his situation in 1947, especially. Kawasumi was convinced that Imamura's illness and his death on New Year's Day 1948 both came about because the professor lacked the means to pay for the medical attention that might have saved his life.[61] The fixed income from the small pension Imamura was still entitled to as a retired scientist might once have been enough to make ends meet but postwar inflation had driven prices for basic commodities so high that it no longer came close to doing so. Kawasumi believed that Imamura had deserved better than to die in poverty, his passing noted only in a brief announcement in the newspaper. Later in his own career, as we'll see, Kawasumi made it a point to relate his own efforts at limiting the nation's vulnerability to earthquake hazards to the example Imamura had set.

The earthquake disasters of the 1940s, Japan's defeat, and the occupation that followed put the pursuit of earthquake prediction in Japan on what looked like a new path, with a new set of scientists leading the way. In the short term, that brief "reset" ended up doing more to establish what that younger generation of scientists thought earthquake prediction in Japan should not be about than it did to chart a clear direction forward. References to prewar attempts at prediction were to be avoided, for example, as were most conversations with the public about specific earthquake threats and what to do about them. Wadati, Hagiwara, and the others were as yet uncertain about how to talk to the public about earthquake risk, or to define the role that earthquake prediction might play in Japan in the years ahead. They would return to exactly these same questions in the 1960s, when disasters, earthquake prediction, and science collided once again.

CHAPTER 3

Plans for Disasters and Blueprints for Prediction

On September 1, 1960, the Tokyo Metropolitan Fire Department commemorated Japan's first-ever Disaster Preparedness Day (Bōsai no hi) by putting on expositions of its equipment and techniques at nine different locations across the city, where fire-fighters showed local residents how they would respond should a major earthquake strike. Elsewhere, the public was invited to attend lectures and educational films on such topics as "Architecture and Disaster Prevention," or "Earthquakes and Disaster Prevention."[1] For its part, the fire department would likely have put on its performances that day regardless of the government's recent decision to call attention to disaster prevention in this new format. Tokyo had long set September 1 aside as a day to commemorate those lost in the city's last natural disaster, the Great Kantō Earthquake of 1923. The fire department's annual exhibitions had been part of efforts to mark the anniversary since the early 1950s.[2]

Public notice that a new layer of commemoration might be in the works had first surfaced just a few months before, in June. That's when Nakasone Yasuhiro, director of the Science and Technology Agency and a rising star in the governing Liberal Democratic Party (LDP), told reporters of his plans to call on the prime minister and his cabinet to designate September 1 as Disaster Preparedness Day. To do so would be to create an opportunity to disseminate knowledge about disasters and disaster preparedness to citizens, he suggested, and to help them be ready for calamities to come. His agency had conducted its own study in the aftermath of the recent Ise Bay Typhoon and one of the conclusions it reached was that the country needed to do more to educate the public about the different hazards they faced and to familiarize them with disaster reduction techniques.[3] What Nakasone didn't need to point out to the press or to his colleagues in the cabinet (who approved his proposal expeditiously) was that recent events had raised questions about Japan's vulnerability to those hazards for which there were as yet no good answers.[4] In addition to the Ise Bay Typhoon the previous year, in late May 1960 the M9.5 Valdivia earthquake that devastated parts of Chile was also responsible for a powerful tsunami that struck communities in Miyagi Prefecture and elsewhere along northeastern Japan's Pacific coast, killing 142 and injuring hundreds more. Post-event assessments highlighted the role confused communications, haphazard preparation, and poorly coordinated relief and rescue operations played in amplifying the effects of both disasters. While calamities like these were certainly not new problems for the nation, in 1960 new responses to them were beginning to take shape.

Typhoon Vera

The storm known to English-speaking observers as Typhoon Vera made landfall in Japan around six o'clock in the evening on September 26, 1959, a Saturday. The Japan Meteorological Agency had by that point been monitoring the storm for almost a week, ever since the first signs of a tropical depression were detected in the western Pacific. Reconnaissance aircraft tracked the storm's path and intensity as it turned northward and toward Japan, and provided regular and accurate estimates of its course. They also reported that, at its peak over open water, Vera had sustained winds of more than three hundred kilometers per hour, among the most powerful ever measured in the Pacific.[5] The Japanese did not routinely name typhoons; the JMA instead gave each new event a number, resetting the counter back to "one" on the first of the year. This late September storm was the year's fifteenth.

Typhoon 15 made landfall near the southern tip of the Kii peninsula and moved northwest across Honshu, reaching the Sea of Japan in the very early hours of September 27. There was very little about the behavior of the storm or the nature of the threat it posed that had not been foreseen by meteorologists and shared with officials and the media. The Nagoya Weather Bureau issued its first typhoon report at 6:30 p.m. on the 25th, roughly twenty-four hours before the storm was expected to reach land, warning that the typhoon was very powerful and that its combination of high winds, rain, and an expected storm surge would likely cause a great deal of damage as it approached the coast and then as it made its way inland.[6] Meteorologists were especially worried about conditions in Ise Bay (Ise-wan). That relatively shallow body of water grew more narrow and more shallow at its northern end, where the Kiso Three Rivers flow into the bay from the Nōbi Plain and where the port and city of Nagoya are located. If the surge coincided with high tide in the bay, as forecasts predicted it might, the water there could rise many meters above its normal level and well above flood stage.

Although some of the deaths associated with the typhoon were the result of mudslides and destructive winds well away from the coast, the vast majority of the casualties occurred in and around Nagoya. At its peak, the storm surge moved a wall of water more than 3.5 meters above sea level over the dikes and walls that were all that stood between Ise Bay and the densely inhabited communities on the low-lying plain. Much of the growth in those communities was relatively recent, most residents too newly arrived to have much if any collective memory of past floods or sea wall failures.[7] Caught in the dark, unprepared, and unable to flee as waves of water and debris swept through city streets, thousands drowned in the first few hours of the storm. There had been no order to evacuate from city officials.[8]

The storm—soon referred to not simply by its number but as the Ise Bay Typhoon (Isewan taifu)—did tremendous damage. The number of the dead and missing would eventually reach 5,098; no storm in Japan's modern history had taken a higher toll. The Ise Bay Typhoon remained Japan's deadliest postwar natural disaster until the 1995 Hanshin-Awaji earthquake. Because many sea walls and dikes had been breached by the storm, low lying areas in and around Nagoya (many of them below sea level) remained inundated well after the storm had passed, making rescue and recovery efforts all the more difficult and dangerous. It would be months before the damaged barriers were repaired; the last flooded zone wasn't pumped dry until late December 1959.[9] In the meantime, more than one and a half million people had been affected by the storm, many of them at least temporarily without shelter because their homes were underwater or too badly damaged to safely inhabit. The overall economic impact of the typhoon has been estimated as equivalent to roughly 4 percent of Japan's GNP in 1959.[10]

One facet of this disaster that marked a turning point in the nation's relationship with mass calamity is the role played by radio and television in bringing news and images of what had happened in Mie and Aiichi into the homes and lives of citizens throughout the country. In the aftermath of the Great Kantō Earthquake, magazine and newspaper publishers were quick to disseminate reports and photographs about the disaster in a process that went on for many months after September 1923. The role of the media in shaping the nation's understanding of the Isewan taifu is, arguably, qualitatively different from the practices that had characterized earlier coverage, however. Broadcast media in particular were involved very early on in the repeated dissemination of warnings about the typhoon's path and severity. Both NHK (Japan Broadcasting Corporation, or Nihon Hōsō Kyōkai) and the Nagoya-based Chubu-Nippon Broadcasting (CBC) had installed radio broadcast equipment in local weather bureau offices so that their reporters could continue to transmit the most up-to-date news about the storm as it approached and made landfall. As the power went out across the region, only those residents with access to portable radios would have benefitted directly from whatever information broadcasters could share, at that point, but it was clear that the media could have an important role to play in shaping local responses to disasters for the better. After September 26, print and broadcast media alike were utilized to convey information about relief and recovery efforts to displaced residents of Nagoya and communities nearby.[11]

In the weeks after the storm, both networks devoted considerable airtime to reporting about the catastrophe. Between September 27 and the end of October, NHK radio produced seventy hours of programming on the typhoon and CBC radio another forty-five.[12] Since just about four out of every five households in Japan owned at least one radio in 1959, the potential audiences for these shows were substantial. And while television had not yet come into its own as a truly ubiquitous consumer good—only one out of every five households in Japan had one that year—the nation's networks were very quick to realize that as a medium, broadcast television was exceptionally well suited to conveying scenes of devastation, and of quotidian settings transformed by disaster, and that viewing audiences would find such images compelling. Both NHK and CBC sent television crews into some of Nagoya's most badly damaged wards almost as soon as the storm had passed, and continued to report on the typhoon's impact for weeks afterwards. Between September 27 and the end of October, NHK television broadcast almost thirty hours of programming on the disaster and CBC television more than thirteen.[13]

One theme that was implicit in much of the media coverage of the Ise Bay Typhoon and that soon surfaced in official assessments of the event was that neither central nor local governments had done enough to prepare for it. Some communities—including the city of Nagoya—had never drawn up evacuation plans, much less instructed residents how to respond in the case of an emergency. A national mapping project in the early 1950s had identified the low-lying areas in and around Nagoya that would likely end up under water if the sea walls were breached; local officials apparently never saw the maps and the residents of the communities in those areas were equally unaware of the risks they faced.[14] Some towns and villages in the storm's path had taken steps to prepare for a disaster and had evacuation plans in place. Not surprisingly, those communities suffered fewer casualties than their less well-prepared neighbors, even when the extent of the damage to the built environment was otherwise similar.[15] Such disparate outcomes both highlighted the effectiveness of some forms of disaster planning and preparedness, even in the face of a force as powerful as the Ise Bay Typhoon, and also drew attention to the patchwork nature of the nation's approach to disasters. Newspaper reporters questioned how it was that city planners had allowed housing projects to be built on land that was so clearly in harm's way and why it was that so little had been done to protect those communities ahead of time. If, as one journalist argued, "the city of Nagoya chose the cheaper path to prosperity over the more costly but safer one," was that really the best the country could do for its citizens?[16] What had been largely anecdotal evidence of poor coordination among government agencies leading to bad outcomes for disaster-stricken communities took on new significance with the release of a series of reports that documented, in great detail, many such experiences in the aftermath of the Ise Bay Typhoon.[17]

Dangerous Japan

The Ise Bay Typhoon marked a turning point in postwar Japan's approach to preparing for and responding to major disasters. There had been other calamities in the decade and a half since surrender and this most recent catastrophe highlighted—in a very public way—the growing gap between what many believed could be done to reduce the damage disasters did and the practices that existing policies permitted.[18] Press coverage and survivors' accounts in September 1959 highlighted how poorly prepared many communities had been for the storm and how complicated and incomplete relief and recovery efforts had turned out to be. These were not new problems, in that prefectural governors and others with first-hand experience dealing with disasters had called on the government to address similar shortcomings before. The scale of the devastation caused by the Ise Bay Typhoon is one factor that helps explain the momentum behind the development of a new anti-disaster policy, but others—among them the growth of the state's expertise in risk management—matter too.

Both the new policy and the context in which it was produced shaped how earthquake scientists thought about their role and their work from the late 1950s on. Japan suffered devastating earthquakes and typhoons with bewildering frequency in the mid- to late 1940s. Major earthquakes every year from 1943 to 1946 killed more than a thousand Japanese in each instance. The 1948 Fukui earthquake was, until the Ise Bay Typhoon, the deadliest single catastrophe not related to the war that the country had experienced since the Great Kantō Earthquake. For a country still coping with the lingering effects of wartime scarcity and devastation, even as these other disasters unfolded, the capacity to respond effectively or to take steps to prepare for the next one was limited. Yet the impulse to do something was strong.

The mainstay of the government's approach to dealing with catastrophes in the immediate postwar era was the 1947 Disaster Relief Act (Saigai kyūjohō). The act was a direct response to the problems that local and state officials had encountered trying to provide aid to the survivors of the 1946 Nankaidō earthquake.[19] Prewar disaster relief legislation had left it up to prefectural authorities to decide when to provide help to disaster victims and how much to give.[20] That approach made it difficult to coordinate relief efforts when disasters affected more than one prefecture, as they often did; twenty-five of Japan's forty-seven prefectures reported significant damage from the Nankaidō earthquake.[21] Local governments also lacked the wherewithal or the incentives to fund relief efforts beyond a bare minimum, which was never enough after a significant event.

The 1947 act was much more ambitious. It set out to "protect victims of disaster and maintain social order by causing the Central Government to provide needed relief services on an emergency basis in cooperation with Local Public Entities and the Japan Red Cross, other entities, and the people of Japan."[22] It put prefectural governors in charge of relief and recovery efforts but it also charged them with a much wider range of interventions than had been the case before. In addition to seeing to the basic needs of survivors or evacuees—food, emergency shelter, medical care, and the like—they could also distribute bedding, school supplies, clothing, and other necessities, help with temporary home repairs, and oversee the recovery and disposition of the dead. New arrangements were put in place to coordinate relief not just locally but regionally and nationally, as well.[23] Funding levels for disaster relief went up.[24]

The new act was invoked on 441 occasions between 1948 and 1959, which says something both about the frequency of calamitous events in Japan in that period and about the perceived utility of the law itself.[25] The act was useful, with its detailed lists of what help could be provided, assurances of at least some financial support from the central government, and promises of routine if not necessarily uniform responses to bad events. Like its predecessors, however, the Disaster Relief Act was focused solely on what could be done to help after a catastrophe struck. Little legislative or policy work was done in the 1940s and 1950s to encourage private or public actors to take steps ahead of time that might prevent at least some disasters from happening or limit the harms they would do, if they did occur. Haphazard attempts at disaster planning like those that preceded the Ise Bay Typhoon remained the norm for most communities.[26]

Rethinking a National Disaster Policy

One of the first indications that the Ise Bay Typhoon had convinced policy makers to take another look at the nation's response to disasters came in Prime Minister Kishi Nobusuke's opening remarks at the start of the 33rd Diet, in late October 1959.

> The government is committed to doing everything within its resources to insure recovery from the disaster and the stability of [everyone's] livelihoods. In light of this, I am quite confident that the victims are going to be back on their feet without delay and that the recovery of the damaged areas will move forward. I offer my most sincere prayers with regard to the victims that they not give in to the blows they have suffered but that they rise up in pursuit of a strong recovery.
>
> The government, in light of the recent disaster and with regard to a fundamental disaster policy that would focus quickly on forest and flood control measures, intends to protect and preserve this nation to the fullest possible extent by putting our efforts into developing a lasting approach to disaster prevention, one that is both comprehensive and scientific.[27]

Kishi's language here is indicative of an important shift underway in how policy makers in Japan thought about disasters. Kishi's emphasis on creating something "lasting" (*kōkyūteki*) can be read as a response to long-standing concerns that each new disaster provoked an ad hoc legislative response and that such inefficient and unpredictable practices were likely to continue in the absence of a more regularized set of standards. Similarly, his reference to "disaster prevention" (*saigai yobō*) as the focus of this new endeavor is perhaps even more important, because it signals a willingness to move away from a solely reactive approach to calamities and toward policies that included hazard mitigation and preparedness as worthwhile goals.[28]

From the start, consideration of what to do about disaster policy was entangled in the Kishi Cabinet's increasingly tense relationship not just with opposition parties in the Diet but with the Japanese electorate in general. The merger of the Liberal and Democratic parties in 1955 had ushered in an era of LDP electoral dominance and a decidedly conservative approach to policy making and governance. Kishi, who had been prime minister since February 1957, had attempted to push through a series of unpopular initiatives, among them efforts to reassert central authority over school teachers across the country and to lift constraints on the methods that police could use to maintain order in "emergency situations."[29] Many understood these actions as blatant attempts to roll back Occupation-era reforms, a suspicion that Kishi's own political history only reinforced.

Kishi had served in Hideki Tōjō's Cabinet from 1941 to 1944, first as Minister of Commerce and Industry and, after the ministry's name changed, as Minister of Munitions. After surrender, he was one of many former officials imprisoned in Sugamo Prison on suspicion of war crimes. Never indicted, Kishi was released in late December 1948, one day after the execution of Tōjō (the former prime minister) and six other former defendants at the International Military Tribunal for the Far East. Kishi's successful return to politics in the 1950s, first as a legislator and eventually as a leader of the LDP, was therefore not without controversy. Many continued to associate him with wartime authoritarianism.[30]

In late 1959, as the Diet met to debate relief spending and a new disaster policy in the typhoon's wake, tensions over one of the Kishi Cabinet's other political goals quickly took center stage. As the Allied Occupation of Japan was nearing its end, the United States and Japan defined important aspects of their relationship going forward in a security treaty, signed by both parties in September 1951. The treaty committed the United States to Japan's defense and provided for the ongoing presence of the US military in numerous bases on Japanese soil. It described the purpose of those forces as contributing "to the maintenance of international peace and security in the Far East and to the security of Japan against armed attack from without," but stipulated as well that the troops could be used "at the express request of thc Japanese Government to put down large scale internal riots and disturbances in Japan." The treaty would expire unless both Japan and the United States agreed no later than June 1960 to continue it.[31]

The Kishi Cabinet began discussing changes to the treaty with the United States in 1958. Negotiations continued the following year against a backdrop of mounting public and political opposition to ratification of the new agreement. As had the previous treaty, the terms of the newly negotiated one granted the United States the right to maintain forces on bases throughout Japan. The provisions allowing US troops to be used to put down domestic unrest were removed from the revised text and, on balance, the treaty's new language made Japan's subordination to the United States somewhat less explicit than had the original. Article II spoke of "economic collaboration" in terms more reminiscent of a partnership than of relations between a superpower and client state.

Opposition to the new treaty came from many directions. US military bases and the presence of US troops had been part of Japanese life for almost a decade and a half by the time Kishi's push to ratify a new agreement began. Communities close to the bases had been transformed by their proximity to them, seldom for the better, and ratification of a revised treaty seemed, to many, to be an acknowledgement that the Japanese government was willing to make the bases permanent. The potential use of these bases to launch military operations in support of the United States' Cold War strategies in East Asia was also deeply troubling to a wide cross section of the Japanese electorate. It was as if there was something for everyone to dislike in the new treaty and in Kishi's handling of the negotiations over its revision.

For many months, the work of responding to the Ise Bay Typhoon—at first, in the form of legislators' appeals for additional spending on relief and recovery in stricken districts and, later, as attempts to create a new, comprehensive disaster policy—proceeded in parallel with efforts to prevent treaty revisions from going forward on Kishi's terms. Anti-security-treaty protests began in earnest in November 1959, when Kishi ended discussion on the terms of the revision, and continued until June the following year, when the new treaty officially went into effect. In May 1960, Kishi and the LDP quite literally forced an affirmative vote on the treaty through the Lower House of the Diet; the vote was taken after police called in by the prime minister physically removed protesting opposition legislators from the building and all but carried the Speaker of the House to the podium, where he officiated over a lighting quick ballot. Kishi's tactic was effective but the backlash against his methods was intense. As Yoshikuni Igarashi has argued, Kishi's apparent disregard for the norms of democratic process made a sometimes complex debate over the terms of revision much simpler to parse. In November 1959, anti-security-treaty demonstrations at the Diet attracted 20,000 protestors, not a small number. By June 1960, protests 300,000-strong and nationwide strikes involving millions of Japanese from many walks of life were indicative of the stakes they had assigned to their opposition to Kishi and to the treaty. It was only after Kishi announced his intention to resign and after the new security treaty passed into law in mid-June that the sense of crisis, and the protests, diminished in intensity.[32]

The Anpo protests of 1959–1960 and the political realignment that followed are useful contexts for thinking about the creation of the Disaster Countermeasures Basic Act. Kishi's resignation and his replacement as prime minister by the more conciliatory Ikeda Hayato marked a turning point in Japan's postwar history, one that signified, among other things, a shift away from the long process of recovery from the war and toward a national project of economic growth. Ikeda's announcement late in 1960 of his cabinet's support for an "Income Doubling Plan" was one of many initiatives that distinguished his agenda from those of his predecessors, and one that began to link LDP policies and the expertise of government technocrats to rapid economic growth and all its implications for the everyday lives of average citizens.[33] The Disaster Countermeasures Basic Act took shape in parallel with these developments.

The Disaster Countermeasures Basic Act

It took two years almost to the day from Kishi's statement at the start of the 33rd Diet to passage of the Disaster Countermeasures Basic Act (Saigai taisaku kihon hō). One reason the process dragged on as long as it did was that the struggle over revision of the security treaty and ultimately over the prime minister's tactics in the Diet focused attention elsewhere. Even after those tensions began to wane once the treaty went into effect and Kishi resigned, however, the debates over what a new disaster policy might look like were not quickly or easily resolved.[34] Authorship of different versions of a draft law passed back and forth between subcommittees of the LDP, the Cabinet, and the Ministry of Home Affairs, as negotiations with non-LDP legislators (mostly members of the Socialist Party) and representatives of the many ministries implicated in disaster policy waxed and waned over many months. In the end, it took another devastating typhoon to push policy makers to bring a completed bill to the floor. The second Muroto Typhoon (Typhoon 18, known as Nancy) struck in mid-September 1961, killing more than 200 as it swept across parts of Shikoku and Osaka Bay before making landfall again on Honshu. The Basic Act passed the Diet on October 31, 1961.

The new law embedded disaster in the state and the state in disaster in ways that were unlike anything modern Japan had seen up to that point. The Basic Act represented something very new in terms of how Japan's government and citizens were implicated not just in responding to disasters but in preparing for them as inevitable elements of the country's future.[35] Its signal innovations were the creation of a national network of "Disaster Preparedness Councils" (Bōsai kaigi) and the introduction of mandatory disaster preparedness and recovery planning at every level of postwar Japanese society, from the village to the workplace to the city to the nation as a whole. From the early 1960s onward, Japanese citizens were the objects of and participants in multiple new layers of national and local disaster planning and policy, because of the Basic Act.

Councils could plan, they could consult, and they could be consulted, but the Basic Act did not give them the power to issue directives or wield much authority on their own. The planning process was clearly where the councils were expected to have the greatest impact, because it was there that the long-elusive coordination among the many different agencies and actors already implicated in disaster preparedness could be worked toward. Prefectural councils, for example, would include among their members mayors of towns and villages, the chief of the metropolitan or prefectural police, the superintendent of the school board, "the Commanding Generals of Army Headquarters of the Self-Defense Forces having jurisdiction over the area of the prefecture," "expert members . . . to investigate matters requiring expert knowledge," and others appointed by the governor, who would serve as the council chair.[36] City, town, and village councils would follow the prefectural model and emphasize the participation of representatives of institutions that might be called on to provide help after a disaster, as well as those whose cooperation would be essential for pre-disaster mitigation and preparedness.

The hierarchy of the Disaster Preparedness Plans (Bōsai keikaku) followed that of the councils themselves; at the prefectural and local levels, the emphasis was on providing both an overview of ongoing efforts at prevention and preparation as well as specific descriptions of plans, for example, for the "creation or improvement of disaster prevention establishments, investigation and research, education, drills and other preventive measures, collection and transmission of information, issuance and transmission of information, issuance and transmission of forecasts and alarms, evacuation, firefighting, flood prevention, rescue, sanitation, and other emergency measures and rehabilitation efforts."[37] The list of topics that local and prefectural plans had to address was written in Tokyo; councils could add items to the list if they wished but they couldn't very easily subtract from it. The Central Disaster Prevention Council (Chūō bōsai kaigi), for its part, was also given the unique task of drafting a "long-term comprehensive plan for disaster prevention" for the nation, which, like all the others, would be "reviewed each year in light of research findings, conditions of disasters that have occurred, and the effect of emergency measures taken, and revise it if deemed necessary."[38]

Planning for disasters and thinking about how to prevent them became routine after 1961. The first national disaster preparedness plan was finalized in 1963; prefectures, cities, towns, and villages began to put theirs in place around the same time. As with other aspects of the Basic Act, the planning process and the concerns it reflected remained largely unchanged from the early 1960s until after the Hanshin-Awaji earthquake in 1995. The act did mandate that the plans be reviewed and revised annually, a process that was easiest at the local level but more complicated the larger the jurisdiction involved. The Central Disaster Prevention Council, which was responsible for drafting and updating the national disaster plan, made major revisions to it only once before 1995.[39]

The Blueprint

It isn't clear who first described the new plans to pursue earthquake prediction as a "blueprint," but something about that label resonated within the earthquake science community from early on.[40] The formal title of the 1962 report is *Prediction of Earthquakes: Progress to Date and Plans for Further Development* (*Jishin yochi: Genjō to sono suishin keikaku*), but even before its formal release its authors were referring to it simply as "the Blueprint" (Burū purinto), a shorthand that remained in use in the twenty-first century. The thirty-two-page Japanese version of the report was released in January 1962; an English language translation followed several months later after multiple inquiries from foreign researchers. While the report's arguments were not new, the timing of its publication and its authors' persistence put earthquake prediction in front of policy makers in contexts that were. The early 1960s in Japan saw new attention paid to disasters and disaster-related policies, a concern with the country's vulnerabilities that went hand in hand with rapid economic growth and increasingly tangible markers of a new prosperity. The Blueprint combined cautious optimism about the possibility of forecasting or predicting earthquakes in Japan in the not-so-distant future with descriptions of the investments that would have to be made in order to have a chance of success at either. This framework was durable enough to remain relevant for the several years it took before the government actually committed to the pursuit of earthquake prediction as a national project. That commitment shaped how Japan imagined and prepared for disasters in general and earthquakes in particular for the next fifty years.

It was Wadati Kiyō's idea to take another run at building an earthquake prediction research program in Japan. The last attempt to rally support for a prediction project had elicited enthusiasm from scientists in the late 1940s but not the government funding it would have needed to go forward. Nakamura Saemontaro's 1949 warning of a near-term threat to the Niigata region was the last such prediction-related story for many years to catch the media's attention. It helped that the 1950s was a relatively quiet decade in Japan, seismically speaking. Scientists hadn't forgotten that they had been interested in prediction or their Occupation-era plans to upgrade monitoring equipment in support of a collaborative research program, but they hadn't consulted with each other about what to do next until Wadati organized a meeting to do just that in May 1960.

Why Wadati and the others decided to begin discussing a prediction research program again that spring is not something they later thought needed explanation. The Blueprint itself set the stage for a new prediction research program by pointing out that scientists were already in agreement about many of its elements but did not suggest that the consensus was a recent development.[41] The scientists involved likewise didn't single out a new discovery in the field or major breakthroughs in the technologies seismologists had at their disposal as having prompted them to revisit prediction in 1960. Many were still doubtful about prediction's prospects, although some of the experts who had been most skeptical about prediction research in the past had by 1960 softened their positions a bit.[42] Tsuboi Chūji is a good example of a scientist in that camp; in 1960 he still wasn't sure that prediction could be done but he was also no longer as certain that it could not.[43]

Japan's changing economy in the early 1960s was also part of the context for the scientists' choices, as were other signals that the country was putting the hardships of the war years and Occupation behind it. The International Olympic Committee's May 1959 announcement that Tokyo would host the 1964 Summer Olympics, for example, was one such indicator. Kishi's resignation in 1960 and his successor's insistence that the pursuit of economic growth was postwar Japan's purpose were others. Wadati's group's decision to start work on a proposal to the government to support a large-scale research project in prediction—the Blueprint—came just as Ikeda introduced his Income Doubling Plan.[44] Wadati and his colleagues couldn't have known how quickly or how much Japan's economy would grow over the next decade, of course, but there was no doubting the direction it was heading.[45] Finally, and also relevant to scientists' attitudes toward earthquake prediction research, were Japanese legislators' occasional enthusiastic expressions of interest in it.[46]

Wadati had no trouble attracting scientists to work on the Blueprint. By April 1961, the new Earthquake Prediction Planning Research Group (Jishin yochi keikaku kenkyū gurupu) had about sixty members, twice as many as Wadati had started with.[47] They came from a range of institutions and research backgrounds; there were a good number from the JMA and the ERI (the latter's conference room is where most of the group's meetings were held) but scientists from Kyoto University, Tohoku University, the Geographical Survey Institute, and a number of other labs and facilities joined in, too. There was considerable overlap between the membership of the new Research Group and that of the Earthquake Prediction Research Liaison Committee from the late 1940s. Hagiwara Takahiro (ERI) was chosen to lead the Research Group and, along with Wadati and Tsuboi, was one of its official sponsors. Kawasumi Hiroshi, who had come a little late to the earlier activity around prediction, would take a leading role this time around. Sassa Kenzō, Inoue Win, and Yamaguchi Seiichi were also active participants. Of the group that had offered public predictions in the late 1940s, only Nakamura Saemontarō was not directly involved in this new undertaking.[48]

The process of producing the Blueprint was as important as the final product itself. As the number of participants in the Research Group swelled, so did calls to adopt a wide range of different prediction-related methodologies and the equipment that went with them. Deciding what to put in and what to leave out wasn't easy and it took many meetings spread over many months to arrive at a document that most participants were happy with. Wadati, Hagiwara, and the other primary authors helped make sure that most of the other scientists involved felt they had a stake in the Blueprint's success by accommodating the majority of the serious proposals that came their way. The final version of the Blueprint described eight different research methodologies that the group's members believed might someday lead to a successful earthquake prediction.[49]

The Blueprint's authors offered three overlapping arguments about the pursuit of earthquake prediction in Japan when they finished the report in 1962. The first was simply that it might be possible. "Prediction of earthquakes is an urgent necessity of the nation and is also the final aim of scientific endeavors in the field of seismology in this country," they wrote. "Seismological studies up to the present do suggest possibilities of realizing this aim of earthquake prediction."[50] The evidence offered in support of that conclusion was in the form of reports of phenomena—mostly varieties of crustal deformation—that were said to have frequently preceded seismic activity.[51] The Research Group's language was at the same time carefully calibrated when it came to discussing the nature of the relationships between these phenomena and earthquakes. Commenting on the results of the leveling surveys and other studies of crustal deformation, the group acknowledged that "the idea that some crustal deformation may be detectable before a large earthquake is held by a number of scientists," but went on to admit that there was as yet not enough evidence to support such a conclusion.[52] Similarly, and with regard to reports of anomalous tilt meter and strain gauge readings in the days and weeks leading up to an earthquake, the group again refused to go out on a limb, suggesting only that "more careful investigations will be required before concluding that such variations are really forerunners of earthquakes, since observation of this kind is apt to be affected by weather conditions."[53] None of the methodologies described in the Blueprint, in fact, had yielded evidence that convinced its authors that precursors existed.[54]

The Research Group's second argument was that they knew how to test the hypotheses that posited links between phenomena such as crustal deformation and earthquakes and were prepared to do so. Key to their strategy was the performance of more frequent and more finely grained observations of each one of those phenomena. The Blueprint's authors stressed that the Japanese earthquake science community was all but unanimous in its agreement as to "*what kinds of measurement should be done by what method.*"[55] In other words, the field had arrived at a shared understanding of which phenomena were most likely to be indicators of imminent seismic activity and how they would have to be tracked over time in order to tell whether the sort of patterns that would make prediction possible did, in fact, exist.

The list of potentially predictive phenomena was not long. Scientists in 1962 were interested in learning more about the location and frequency of very small earthquakes, for example, and in pinpointing the location of all the active faults that could be found in Japan. But far and away the most important topic of inquiry for them was the different ways the crust appeared to move—to deform—in proximity to earthquakes. The Research Group's members were anxious to collect as much data as they could about those changes. Most of the methods they agreed were best suited for monitoring crustal deformation relied on practices and technologies already in use in Japan; what the Blueprint did was call for their application on a much larger scale and with much more precision than had ever been attempted before.

One of the problems from the scientists' perspective was that the current leveling and triangulation surveys were much too infrequent to be useful in any practical way for the prediction project. As many as thirty or forty years might pass in some parts of the country between a surveying team's visit, an acceptable interval for map makers but not for earthquake scientists. Geographical coverage was another problem. Where surveys happened was just as important as how often, and here, too, the interests of mapmakers and other agencies responsible for geodetic surveys were not always in alignment with the earth sciences community. The Blueprint's authors argued, for example, that the number of tide gauge stations then in operation was not enough to allow them to adequately measure what was happening along the coasts, or to determine values for sea level with the degree of accuracy they needed.[56] Similarly, tilt meter and strain gauge sites were too few and too dispersed to generate data of the density that researchers sought—information sufficient for them to track deformations at multiple locations at once and thus better understand the relationships (if any) between these changes and seismic activity.[57]

All these problems had solutions. "For the purpose of obtaining information of the earth crust indispensable for earthquake prediction," the Research Group wrote, "*it is required that crustal deformations over the whole of Japan be traced continually through frequent repetition surveys over as large an area as possible.*"[58] Toward that end, they proposed an initiative that would narrow the intervals between national triangulation surveys to ten years and to precise leveling surveys to five. In regions where earthquakes were known to be more frequent, or where the risks to communities seemed highest, surveys would be conducted even more frequently, once every year or two. In addition, the Blueprint's authors called for the construction of more than two dozen new tide gauge stations, a 40 percent increase over 1962 levels and enough for there to be one station "situated approximately every 100km along the coast of our land."[59] Finally, and in parallel with the enhanced geodetic surveys, the Research Group described their plan to build a total of one hundred new stations for the "continuous observation of crustal deformation" using tilt meters, strain gauges, and the like. In most parts of the country, the goal was to have one station for every 100 square kilometers of territory. As with the leveling and triangulation surveys, some districts would require more intense surveillance and in those areas the target was one station for every 50 square kilometers.[60]

The Research Group described these enhanced surveying and observation regimes as "indispensable for earthquake prediction."[61] "This project," they wrote, "aims at founding a gateway to the prediction of earthquakes by obtaining detailed information on crustal deformation and its relationship to earthquake occurrence."[62] "*This work is of uppermost importance*," insisted the Blueprint's authors, "*as it will form the basis of the whole project*."[63]

The clear hierarchy that the report's authors insisted upon—crustal deformation work first, seismographic studies conducted in parallel, and other projects playing supporting roles—only reinforced the argument that they knew what they were doing, that they knew which questions were most important, and that they knew how to answer them. At the same time, the Blueprint was circumspect in the language its authors used to describe what the results of the "prediction project" would be, and left open the possibility that the precursory signals they sought might not exist. "Once this project gets underway," the Research Group suggested, writing here about the geodetic survey piece of what they had proposed, "the crustal deformations as symptoms to earthquakes will be revealed, if they really exist at all, within a fairly short time."[64] Similarly, and in this instance referring to tilt meter and strain gauge studies, the scientists concluded that "the station network will be able to definitely detect forerunners, or the relatively rapid changes in the tilt and strain of the ground immediately prior to earthquakes, if they exist at all."[65] The authors offered similarly qualified assessments of what the other parts of the project might reveal; at no point did they express confidence that the relationships they were looking for were certain to exist and they left open the possibility that they did not. The Blueprint, in fact, never promised that investing in a legitimate research program in earthquake prediction would make predictions possible later. Its authors did provide a timeline for the research program to begin to yield results but the language they used stopped short of suggesting that those results would include successful short-term earthquake predictions.

The Research Group estimated that although some useful information would begin to emerge after five years of work, it was more realistic to think that it would be at least ten before they had enough data in hand to begin producing meaningful analyses of it.[66] The Blueprint's final paragraph features a question sure to have been on readers' minds from the start, the answer to which further clarifies where the scientists expected to be in a decade. "When will earthquake prediction become possible and an efficient forewarning service be available?" the Blueprint's authors asked. "This question cannot be answered now," began the reply. "But if we start the project presented here we should be able to answer the question with sufficient certainty within ten years."[67] This stance is not so different from the goals that the report's authors had acknowledged in the Blueprint's opening pages, where they had described their purpose as getting to the point where they could make the measurements that would allow them to test many of the prediction-related hypotheses already in circulation. They would have to follow the Blueprint to the letter, in other words, just to be able to answer questions about whether or when earthquake prediction might one day be put into practice. What the scientists were proposing in 1962 was not earthquake prediction but earthquake prediction research.

The Blueprint's third and final argument was this: discovering whether earthquake prediction was possible or not would require a greater level of commitment and a different approach to organizing research than anything the nation had attempted up to that point. The days of individual researchers or even single institutions working on prediction in isolation were over. Referring to the different observation strategies under consideration for the project, the report's authors concluded that "*the measurements required would be necessarily of such a large scale that no single institution could adequately afford to operate and, moreover, a thorough consideration is needed regarding our existing scheme of seismic research* for the effective practice of such a nationwide project as required for earthquake prediction."[68] "Deep understanding and ample financial support by the government," they wrote, "is indispensable, in addition to the constant and conscientious endeavors of all the researchers concerned."[69]

Some of the arguments the Blueprint doesn't make are also worth considering. The report does not, for example, link its efforts to any of Japan's previous experiences with forecasting or prediction. Neither Imamura's warnings prior to the Great Kantō Earthquake nor the prediction initiative of the late 1940s are mentioned. The report drew on earlier work related to the study of crustal deformation or other phenomena related to prediction, but it makes no attempt to relate the proposals in the Blueprint to earlier projects that had tried to connect research to actual attempts at prediction. Maintaining a certain distance with the past was made a little easier by the format and presentation of the report in its final, pamphlet form. The Earthquake Prediction Research Group is credited with having written the document; the only members of the group identified by name—indeed, the only scientists whose names appear in the report at all—are Hagiwara, Tsuboi, and Wadati. The three are credited only with having compiled the document, nothing more. Of the three "compilers," only Wadati had been publicly involved with the prediction scares of the late 1940s and his role then had been to rein in his colleagues and to reassure the public that everything would be all right. By stripping out almost all of the scientists' names and by not tying any of the project's components to the institutions or researchers most closely associated with them, the Blueprint also reinforced a message that its arguments were those of the scientific community as a whole and that on this topic they spoke as one. The emphasis on the Blueprint as the product of collaborative work certainly resonated with its argument that the future of prediction research depended on just such group efforts on a much larger scale.

The Americans

Many of the specific interventions the Blueprint proposed reflected conditions that were unique to Japan but its authors recognized early on that the proposal's basic premises were not. Scientists overseas reportedly started asking for a translation of the Blueprint not long after it appeared in Japanese and Ueda Sekiya obliged in short-order with a fine English-language version.[70] The Research Group saw to it that copies were distributed to embassies in Tokyo and more were mailed directly to colleagues overseas. The decision to share the Blueprint's contents with the larger scientific community reflects both Japanese seismologists' understanding of earthquake prediction as a problem that could potentially be solved, eventually, using the methods and tools of science, and their expectation that other scientists would see it the same way.

The American response to the Blueprint in many ways affirmed that the Japanese were right. Jack Oliver, then the director of the Lamont Observatory at Columbia, was impressed enough with what he read that he wrote to Tsuboi to suggest that interested parties from both countries "meet in a conference and exchange knowledge and information they have relating to the problem of earthquake prediction."[71] The Japanese agreed to host the event, and Oliver, along with Hugo Benioff, Leon Knopoff, and half a dozen other American scientists arrived in Tokyo in early March 1964 for the first in what became a series of joint conferences.[72] Wadati, Hagiwara, Tsuboi, Kawasumi, and Asada Toshi were among the dozen or so principal representatives on the Japanese side. Another fifty of their colleagues also participated in the five-day gathering.[73] The conference agenda largely mirrored the Blueprint's areas of focus and featured presentations on crustal deformation, geomagnetism, seismicity and seismic wave velocity, and so on; Hagiwara provided an overview of what the Japanese were hoping to achieve with the Blueprint.

The Americans recognized early on that they had relatively little to offer their Japanese colleagues, so far as exchanges of knowledge and information went. Oliver acknowledged as much in his opening remarks. "Those of us from the United States," he said, "come here, I assure you, with some humility. We know that research related to earthquake prediction is more widespread and in many ways more advanced in Japan than in the US."[74] That gap he attributed in part to the relatively low levels of concern with seismicity in the United States, "where we have many seismologists who have never felt an earth tremor, many areas where earthquakes are hardly a part of reality," compared with what he viewed as a more widely shared understanding of earthquake hazards in Japan.[75] Oliver may have misunderstood the Blueprint's status that spring, since the government had yet to actually fund an earthquake prediction research program, but his explanation for why such efforts had made more progress in Japan than in the United States is noteworthy. His emphasis was on both the conditions that had prompted Japanese scientists to focus on prediction in the first place and on the scientifically sound arguments that supported their research. Writing in *Science* a few months after the conference, Oliver observed,

> Most Japanese think of the earthquake as one of a sequence of events, some of which precede the main shock and have properties which may very well be measurable and indicative of the earthquake to follow. Some Americans, on the other hand, tend to associate the term "earthquake prediction" with various schemes in the realm of astrology, or numerology, or some other form of mysticism or occultism, and thus overlook the possibility that sound scientific principles can be applied to the problem with some hope of success.[76]

Oliver's comments and, indeed, the fact that the conference took place at all are evidence that the ideas behind the Blueprint and the logic of its approach to earthquake prediction were perfectly legible to scientists outside of Japan. The variations between the Japanese and American approaches to earthquake prediction research to date, at least according to Oliver, were products of their different degrees of exposure to major earthquake hazards, not signs of any fundamental disagreements over the potential value of the work. To point out that the Japanese were ahead of the Americans was also to acknowledge that they were both headed in the same direction.

One additional observation about the legibility of the Blueprint and the Japanese prediction research project in the 1960s more generally also involves Americans. Seismologist Frank Press, Beno Gutenberg's successor as the director of Caltech's Seismological Lab and chair of the department of Earth and Planetary Sciences at MIT from 1965 to 1976, for many years led a campaign to convince the US government to fund an American earthquake prediction program modeled on what the Japanese had been doing. Press began to advocate for this idea not long after the massive March 27, 1964, earthquake in Alaska and, the following year, he and other experts urged the White House to launch a ten-year, $137 million American earthquake prediction project.[77] The Johnson administration declined to do so but, as Susan Hough and Robert Stallings have shown, Press was instrumental in keeping the prospect of a large-scale US prediction program alive well into the 1970s.[78]

Press acknowledged from early on that the Blueprint and Japanese efforts to pursue earthquake prediction research were key to what the Americans hoped to accomplish at home. When Hagiwara and a dozen other Japanese scientists arrived in Palisades New York in 1966 for the second United States–Japan joint conference on prediction, Press emphasized how important their work had been to his and his colleague's recent efforts. "We owe a debt to the Japanese and their report on earthquake prediction," Press observed. "The tone of their report was sober and did not oversell the case. It raised all the uncertainties and stressed the need for a long period of research. We adopted this tone for our report as well."[79] They owed a second debt, he said, "to this and the preceding conference in Tokyo, which helped to lend respectability to the subject of earthquake prediction and showed that a broad scientific base was available to implement such a program. The field," Press observed, "is no longer complete guesswork, and now is the time for a major scientific effort."[80]

Backing the Blueprint

Once the initial work of the Research Group was complete and copies of the Blueprint distributed (some twelve hundred copies were printed up), it took time to move the prediction project out of the university conference rooms that had produced it and into venues where decisions could be made about how much of what the scientists wanted to do the nation was willing to support. The first hurdle was getting the Science Council of Japan to agree that the proposal had merit and was worthy of government funding; its approval wasn't finalized until late 1963. Another of the complications Wadati and the others faced was that their plan to add new and better monitoring and research capabilities implicated a long list of different institutes, agencies, and ministries, each with its own priorities.[81] For any one of those units to take ownership of the prediction project ran the risk of denying it the support of the others. The Research Group avoided that potential pitfall by turning to the Geodesy Council (Sokuchigaku shingikai) for help.[82] Although officially an advisory body to the Ministry of Education, the council was made up of representatives from most of the other agencies and ministries potentially affected by the Blueprint, which made it a valuable advocate for the Research Group's agenda.[83] Whatever recommendations the council might produce for the new earthquake prediction research program would have the tacit support of not just the Ministry of Education but of other powerful ministries, as well. Negotiations between Hagiwara and the relevant council members over how to allocate responsibilities and money for the different elements of the Blueprint's research agenda began in the summer of 1963.

The Geodesy Council's recommendations in support of the new project weren't widely shared for another year.[84] The delay was due, in part, to time spent negotiating over spending and priorities but it was also a reflection of the scientists' uncertainty about how the Cabinet and legislators might react to the proposal. Wadati, Hagiwara, and Tsuboi hadn't forgotten what happened the last time they asked the authorities to support a national earthquake prediction research program. Putting the Blueprint into practice was going to be expensive, which was one reason policy makers might veto the whole idea. That there was still no guarantee that anything useful would come of what the scientists were proposing to do was certainly another.

Those potential liabilities were still present when the Geodesy Council released "On the Implementation of the Earthquake Prediction Research Plan" in July 1964 and when the Cabinet agreed shortly thereafter to include funding for the new project in the next year's budget. By then, however, the stakes of any debate over prediction's potential value to Japan had changed dramatically from what they had been a year or even a few months earlier. As of early 1964, the campaign to use the Blueprint to build something tangible and persistent had a fairly low profile, few allies outside the Geodesy Council and the Japanese scientific community, and no particular sense of urgency attached to it. The whole process was unfolding at a leisurely pace.

The June 16, 1964, Niigata earthquake changed all that. Residents of Niigata city and surrounding communities had been fortunate, in the sense that the number of deaths and serious injuries were low, but the disaster's impact was widespread, nonetheless. As images of fallen bridges, toppled apartment complexes, and burning oil refineries appeared on television screens and newspapers' front pages, the country was confronted with painful evidence that recovery from the war and the embrace of economic growth had brought with them new and frightening vulnerabilities to earthquakes. While it is certainly possible that policy makers would have gone forward with a prediction research program regardless, the earthquake in Niigata likely hastened their approval of the Geodesy Council's call to action.[85] Chapter 4 examines how what happened in Niigata moved the Blueprint's agenda forward and how it helped focus attention on Tokyo's vulnerabilities to a future earthquake, in particular.

CHAPTER 4

The Niigata Earthquake and Kawasumi Hiroshi's Theory

The June 16, 1964, Niigata earthquake (M7.5) caused damage to communities all along the northwest Japan Sea coast, including many in Yamagata and Akita prefectures, but it was Niigata's capital that was the hardest hit. Just under half of all the buildings in the city sustained significant damage in the quake, either as a result of the initial tremors themselves or from the subsequent flooding, which inundated large swaths of the city in water and mud.[1] Of the twenty-six people who died as a result of the earthquake, fourteen were residents of Niigata, and eleven of that number fell victim in Niigata City.[2]

Some observers noted, quite rightly, that the Niigata earthquake was much less destructive than the one that had devastated Fukui in 1948 (in which more than 3,700 had perished) and that, while some of the city's relatively good fortune had to be attributed to the timing and circumstances of the quake, lessons learned from past events also played a role in protecting citizens from some of its worst effects. There were only a small number of serious fires in Niigata, for example. The earthquake struck just after one in the afternoon when most residents would have been done cooking lunch, which greatly reduced the number of open flames that had to be dealt with once the tremors began. Because it was summer, there were no kerosene heaters to worry about, which further limited the number of likely sources of ignition. Widespread flooding throughout the city, though damaging in its own right, also did its part to extinguish some fires and contain others. The relatively good news that the number of fires was small was offset by the huge size of just one of them, the conflagration in the Shōwa Oil refinery complex, discussed below.

It is also the case that relief and recovery efforts in 1964 were vast improvements over what had happened after the Ise Bay Typhoon. One reason for this was that the government's response was shaped for the better by the provisions of the newly enacted Disaster Countermeasures Basic Act. By 3:00 p.m. on the day of the quake, Prime Minister Ikeda had convened an emergency meeting of the Central Disaster Prevention Council and, in quick succession, he invoked the Disaster Relief Act, dispatched Home Affairs Minister Akamatsu to the scene, and established a Niigata Earthquake Emergency Disaster Response headquarters.[3] First aid stations and shelters were in place in Niigata City before nightfall.

While these outcomes compare favorably to the country's other recent encounters with disaster, the Niigata earthquake was disruptive and troubling in ways that the events in Fukui or after the Ise Bay Typhoon were not. The nation itself was different in ways that mattered for how the disaster in Niigata would be understood and responded to. Japan in 1964 was well-embarked on Prime Minister Ikeda's "Income Doubling Plan" and its pursuit of high-speed economic growth. The very symbol of that growth (and its pace), the "Kodama" bullet train (*shinkansen*), was on schedule to make its inaugural run between Tokyo and Shin-Osaka station later that year, an achievement overshadowed only by preparations for the event it was built for, the 1964 Tokyo Olympics.[4]

As had been the case in 1959, photographs, television, film, and, of course, radio and print journalism carried narratives about what was happening in Niigata to the nation. Reports from Niigata City began almost simultaneously with the earthquake itself. NHK's broadcasting facilities there were without power only briefly before emergency generators kicked in. The Broadcasting System of Niigata (BSN, Niigata Hōsō) was not so lucky, but it, too, managed to go back on the air within a few hours.[5] The contemporaneity and fullness of the news coverage was itself a reminder of how much Japan had changed since it had last confronted a natural disaster on such a large scale. Fewer than one in four Japanese households owned a television in 1959; by 1964, there was a unit in just about nine out of every ten.[6]

Three Sites

The most widely shared images of the earthquake's aftermath reproduced scenes from three emblematic sites of Niigata's catastrophe—a burning oil refinery, precariously tipped apartment buildings, and a collapsed bridge—and all speak to the ways in which the disaster highlighted the nation's vulnerability. Each site was rendered in many different formats and from many different perspectives—there is no single, iconic portrayal attached to any of them—and each was ubiquitous in the print and film coverage of the disaster. The *Asahi* newspaper chose an aerial photograph of the Shōwa refinery fires for the front page of its evening edition on June 16; the *Yomiuri* used a similar image in its initial coverage of the earthquake.[7] Photographs of the fires reappeared repeatedly over the next several weeks, as did those of the Shōwa Bridge and the Kawagishi-chō apartments.[8]

Images and stories associated with these three scenes from the aftermath of the Niigata earthquake quickly became the most widely evoked and recognizable symbols of the larger disaster. All three are from Niigata City and all three have continued to serve as a sort of visual shorthand for that calamity's impact on the nation. Depictions of the first site typically feature massive clouds of black smoke towering above the brightly burning Shōwa Oil refinery complex on the city's northeast side, between the city center and the airport. The industrial park housed a densely packed assortment of petrochemical plants,

Figure 3. Shōwa Bridge in the foreground; smoke rises from fires at the Shōwa Oil (top left) and Narisawa Oil (top right) complexes on June 16, 1964. Photo by Mainichi Newspaper/AFLO.

Figure 4. Kawagishi-cho Apartment Complex, Niigata City, June 17, 1964. Photo by Mainichi Newspaper/AFLO.

warehouses, and residential neighborhoods, some of the latter where company housing for plant workers was situated. Shōwa Oil's two refinery units dominated the complex and it was one of that company's massive storage tanks that caught fire first, shortly after the first tremors began. Explosions soon rocked other parts of the plant, as burning oil spread throughout the site.

There was very little that local firefighters could do to extinguish the blaze. It took them almost five hours to navigate their way through flooded and damaged streets to reach the plant and to begin an organized evacuation of the adjoining neighborhoods. There wasn't a single foam-dispensing fire truck in the entire prefecture, and nothing else was effective against the conflagration then underway in the refinery. Reinforcements and specialized firefighting equipment arrived from Tokyo the next day but it wasn't until late on June 18 that they were able to mount a coordinated attempt to at least keep the fire from spreading into other parts of the plant. Over the next several days, firefighters, SDF soldiers, and plant workers battled to bring the blaze under control, even as the heat from burning oil weakened fittings and seals on tanks and pipelines, adding more fuel to the fire. In the end, the decision was made to just let the fires in and around Tank 1103 burn, there being no other viable options. It took until July 1—360 hours after the fires began—for the last of the flames to go out.[9]

The second of the earthquake's touchstones was a cluster of three- and four-story apartment buildings in Kawagishi ward, not far from the banks of the Shinano River. There were eight units and, although none had collapsed in the usual sense, the earthquake left each tilted and askew, so that together they resembled nothing so much as rows of dominoes caught in the act of tipping over. Most of the buildings were just a few degrees off the vertical plane but one in the middle row leaned sharply to the north and another adjacent to it—Unit 4—came to rest at a 70-degree angle. A prefectural employee who had returned to his top-floor apartment that day to have lunch with his wife described the building's toppling over as starting soon after the first tremors hit and then proceeding slowly and almost silently. Luckily for them the heaviest of their possessions were on the same side of the room as the direction of the unit's descent, so they were not at risk of injury from within their own apartment. "It felt," he recollected, "just as if we were being swallowed up by the earth. We had our backs pressed up against one of the walls of the apartment, which just kept moving closer and closer toward the ground. Our pots and dishes came flying at us from the other side of the room. Eventually we just lay there on our backs, at wits end." Once the building stopped moving, he and his wife discovered that their fourth floor window pointed almost straight down at the ground, by then almost close enough to touch. They opened the window, lowered themselves through it, and walked away.[10]

Unit 4 was the most visible casualty of the liquefaction that occurred in sites throughout the city as the earthquake's first tremors arrived. Liquefaction occurs when saturated, loose, and sandy soil is subject to rapid, sustained stresses, as can happen during earthquakes. Under the right circumstances, seismic activity can cause soil and water to interact with each other in ways that radically change how they behave in combination. Liquefaction transforms previously stable soil into a slurry that flows as a thick fluid would and will no longer fully bear the weight of objects on its surface. Owners of cars and trucks parked on ground that undergoes liquefaction, for example, often return to their vehicles to find them submerged up their axles, if not deeper, in soil that has since returned to its normal, solid state. In Unit 4's case, enough of the ground beneath the building's footings underwent liquefaction that there was no longer enough force being exerted from below to keep it fully upright. Once the apartment house began to lean in the direction of least resistance, the soil underneath it continued to give way slowly, flowing around the footing walls and filling in the spaces where the building had once stood. One of the reasons that other structures in the same apartment complex didn't suffer the same fate is that most had full basements; Unit 4 did not. That additional mass and surface area below ground appears to have been enough to keep those other properties more or less upright.

Liquefaction was responsible for a great deal of the harm to the built environment that the earthquake left in its wake. Niigata City is built atop a thick layer of sandy alluvial deposits where the Shinano and Agano rivers flow into the Sea of Japan. Centuries of canal building and flood control projects in and around the city had repeatedly changed the course of the two rivers and made it possible to both claim new land from where the Shinano and Agano had once run and to build in locations protected from the river by dikes and other earthworks. What had not been clearly understood before the Niigata earthquake but was abundantly clear afterward was that the sandy, saturated soil in and around the city (and the Echigo Plain more generally) was a near-perfect medium for liquefaction to occur and that land reclaimed from the rivers was especially prone to the phenomenon. The Kawagishi-chō apartment complex, for example, was constructed close to the banks of the Shinano River; other buildings atop similar terrain throughout the city tilted as the apartments had done, with many also suffering major structural damage as a result. Roughly one in four of all the concrete buildings in the city sustained significant damage, due mostly to the failures of their foundations.[11]

The spectacular failure of the Shōwa Bridge (Shōwa Ōhashi) became the third powerful image of the Niigata disaster. One of four in Niigata City crossing the Shinano River, the bridge was essentially brand new in 1964 and had opened to traffic only a month or so before the earthquake struck. It was one of several major construction projects the city had undertaken in preparation for the nineteenth annual National Sports Festival, which Niigata had been chosen to host in winter, spring, and summer sessions over the course of 1964.[12] Winning the opportunity to host the Sports Festival was something of a coup—it had never been held in Niigata before—and because it was widely seen as a warm up of sorts for the Olympics, it attracted more attention and participants than usual. The spring competition ran from June 6 to 11 and brought more than 18,000 participants and spectators (including the emperor and empress) to the city, making it the best attended Sports Festival in more than a decade. The summer session that was set to open on August 1 was expected to attract another three or four thousand visitors.

Niigata's preparation for the Sports Festival was a smaller-scale version of what Tokyo was doing to get ready for the Olympics. The city completed construction of a new four-thousand-seat sports stadium in 1960 and of an entirely new span over the Shinano, the Yachiyo Bridge, in 1962. The new Shōwa Bridge was the third built on the same site, replacing a wooden span completed in 1950, which was, in turn, a bigger and better version of the original 1931 structure. The iteration finished in 1964 was quite different from earlier designs in its use of a system of steel and concrete piers to support a dozen concrete slabs, together spanning more than 300 meters from one end of the bridge to the other. While not nearly as picturesque as the elegant arches of the much older Bandai Bridge further downriver, the new structure, like the Kawagishi-chō apartment buildings just a short walk away, reflected ideas about urban life and construction practices alike that were taking hold all over Japan. That bridge, and that apartment building, could have been part of any cityscape in Japan in 1964.

Which was not a reassuring thought in the aftermath of the earthquake. The ends of four of the Shōwa Bridge's twelve spans slipped entirely off their supports and dropped into the river below; a fifth span separated at both of its ends from the piers supporting it and promptly sank to the bottom of the river. Eyewitnesses described watching in disbelief as a car raced across the bridge just as it was coming apart, luckily reaching each new segment just as the one behind it collapsed.[13] The bridge failed relatively slowly—everyone who was on the bridge when the earthquake struck seems to have made it off in one piece—but definitively. None of the other bridges over the Shinano River were as badly damaged. The Yachiyo Bridge to the west was impassable but managed not to fall into the river, while the venerable Bandai Bridge remained more or less intact. It was the only bridge in the city center left safe to use.

Although some of the structural damage to the bridges was the direct result of the tremors, the systemic misadventures that rendered them unusable were eventually blamed on liquefaction. The piers holding up the Shōwa and Yachiyo bridges didn't fail so much as go along for the ride when the ground in which they were embedded shifted. Both banks of the Shinano were destabilized by the sustained vibrations of the earthquake and in some locations the soil began to flow toward the center of the river, taking the piers with them. At either end of the Bandai Bridge, the ground simply sank by more than a meter, all in a matter of minutes. The soil under the new sports stadium—built atop reclaimed land a stone's throw from the river—also shifted and sank, buckling the stadium floor, rupturing expansion joints, and leaving large cracks in the structure's distinctive sweeping reinforced concrete beams.[14]

What happened in Niigata introduced a new set of contexts for thinking about what the country should do next. The disaster drew attention to the fact that long periods of relative seismic quiet could come to a dramatic end at any moment, for example. More than a century had passed between major seismic events in that part of Niigata and there had been no slow build-up, no warning signs that another event was imminent before the earthquake struck. If Niigata could experience a massive earthquake, so could just about any other city in Japan.

To that point, it was hard to ignore the ways in which Niigata's circumstances looked a lot like what one would find in many of Japan's other cities in the mid-1960s. The similarities of building styles and materials, and the shared priorities in development and infrastructure projects that characterized Japan's cities at that moment were, in part, a byproduct of the wartime bombing that had leveled so much of the nation's urban landscape.[15] Those communities that had set out to rebuild themselves at more or less the same time reached the era of high-speed growth having a great deal in common with each other. This is not to argue that Japan's cities all looked alike but rather that, by 1964, many factors had come together to make it easier than ever to see that they were vulnerable to disasters and to earthquake threats in common and worrisome ways.

The particulars of Niigata's vulnerability were also important. The relatively thick layer of alluvial soil that was the source of much of the structural damage in Niigata could also be found in and around the main island's other major cities: Tokyo, Yokohama, Nagoya, Kobe, and Osaka among them. The earthquake demonstrated that, in such settings, even structures built using the most up-to-date methods and materials were far more vulnerable than had previously been thought. The ubiquitous images of the collapsed Shōwa Bridge and of the tipped-over Kawagishi-chō apartment buildings, as well as those of the many damaged roadways, residences, and schools carried that argument into the homes of television viewers and newspaper readers throughout Japan. The practices spelled out in the Building Standards Act and in other regulations had been designed to address all the important ways that experts thought the built environment and seismicity would interact with each other. The Niigata earthquake cast doubt on the assumptions that had shaped those practices and suggested that there were threats that had not as yet been taken fully into account and, thus, had not been and could not have been prepared for. The Shōwa Oil refinery fire called attention to another aspect of the nation's vulnerability to harm in the mid-1960s, namely the emergence of new risks associated with high-speed growth and affluence.

Finally, the earthquake in Niigata in June left people with plenty of time to think about what could happen before or during the Olympics in Tokyo in October. The National Sports Festival, the new stadium and other construction projects completed for the festival, and the visit to Niigata by the emperor and empress all in some way echoed the preparations and plans for the much grander events that would happen once the Olympics were underway. The disaster forced the cancellation of the next session of the festival, which would have convened in Niigata in August and, while there was never a possibility that the Olympics would be postponed or cancelled as a precaution, there was also very little that could be done in the intervening months to address new concerns about Tokyo's vulnerability.[16]

Kawasumi Hiroshi and the Countdown to Tokyo's Demise

It was in this post-Niigata context of heightened concerns about the nation's vulnerability that earthquake scientists were able to intervene in the public discourse about disasters in ways that had significant, long-term effects on Japan's approach to dealing with seismic risk and hazards. These interventions centered on two claims. The first argument called for significant new investments in earthquake prediction research. The Niigata earthquake gave new life to the Blueprint, the 1962 research proposal that had until then failed to generate much enthusiasm apart from the scientists responsible for it. During Diet testimony about the disaster, Wadati and Hagiwara consistently described their plans for research into earthquake prediction as the most promising of all possible scientific responses to the devastation in Niigata. At that point, no money had been allocated toward the project, despite all the work that had gone into generating a proposal and getting it in front of policy makers. Before the Niigata earthquake, the prediction project had stalled and was going nowhere. After the earthquake, the scientists received the first of what would be many years of funding for earthquake prediction. This is a topic we will return to.

The second set of arguments to gain credence post-Niigata also had to do with thinking about future earthquakes. Where the cohort represented

Figure 5. Kawasumi Hiroshi in Fukui City after the earthquake, July 3, 1948. Photo by Mainichi Newspaper/AFLO.

by Wadati and Hagiwara wanted to search for clues that would help them understand when earthquakes were likely to occur, this alternative set of ideas operated on a different register. The primary spokesman for this other agenda was Kawasumi Hiroshi, who, as director of the ERI in 1964 and an early supporter of earthquake prediction research, was both an unlikely insurgent and someone whose opinions could not easily be discounted.[17] During testimony before a Diet committee investigating the Niigata disaster, Kawasumi announced that his own recent research had led to a discovery about the seismicity of the southern Kantō district. According to Kawasumi, his analysis of the historical data on earthquakes in the region had revealed

Figure 6. Kawasumi Hiroshi (seated, with pencil and map) and others from the ERI in Fukui City after the earthquake, July 3, 1948. Photo by Mainichi Newspaper/AFLO.

that the likelihood of major, destructive events in the Kantō district increased on a regular, sixty-nine-year cycle and had been doing so since at least the ninth century. Based on his calculations, Kawasumi said, Tokyo would begin entering the next phase of greatly increased earthquake risk in the late 1970s.

Kawasumi's analysis and the conclusions he drew from it were not necessarily in conflict with what the other prediction researchers wanted. His emphasis on periodicity and on risks specific to the Kantō district did not call into question the basic tenets of the prediction project, namely that careful observation might reveal earthquake precursors. Nor did his findings make successful prediction any less necessary; despite the precision its label implied, the sixty-nine-year cycle described the likely timing of the next major event only in terms of increasing risk over the long term. The prediction researchers' goals focused on achieving warnings that would work in the short and medium term, and were in no way made irrelevant by Kawasumi's discovery.

Kawasumi's work did, however, make demands on the scientific community and on policy makers that the prediction researchers' agenda did not. Kawasumi's warnings about the probability of a major earthquake occurring in the southern Kantō district emerged alongside a growing body of research and of other narratives describing what would happen to Tokyo if it did. Institutions like the Tokyo Metropolitan Fire Department were among those that put out detailed studies of the potential effects of another Great Kantō Earthquake; NHK was one of several media outlets that published their own speculative accounts of Tokyo's destruction in the late 1960s and early 1970s. Finally, experts in Japan were quick to use images and analyses of the 1971 San Fernando earthquake to once again draw attention to Tokyo's vulnerabilities. The sections below show how Kawasumi's theory helped create a new understanding of earthquake risk for the capital district alongside those developments.

Kawasumi Hiroshi came to Tokyo and the First Higher School from Suwa, in Nagano Prefecture, with his heart set on studying geology. Kawasumi had been at the Higher School for only a year when the Great Kantō Earthquake struck. The experience of being in Tokyo for that event as an eyewitness to destruction had a profound effect on him. Later in life he often brought those experiences up to emphasize why he was so worried about what the next major earthquake could do to the city.[18] At the time, it led Kawasumi to decide to abandon his earlier plans to focus on geology at university and to turn instead to seismology.

One might think that the earthquake in 1923 would have had a similar effect on others of Kawasumi's age group and that Tokyo Imperial University's specialization in seismology would have attracted many dedicated students. This was not the case; in 1924, the first year that seismology had been authorized as a distinct course of study at the university, not a single admitted student listed it as his first choice. Most students with an interest in the earth sciences gravitated toward fields with a more practical aspect to them, like mining and geology. In its second year, seismology did somewhat better and was able to admit five students, one of whom was Kawasumi. Of that entering class, three ended up shifting to other departments entirely, and one other went to work for the Geological Survey of Japan after graduation, doing work unrelated to earthquakes.[19]

Which left Kawasumi as the only member of his graduating class to choose seismology as a career. He may well have had second thoughts at first, given his limited prospects for gainful employment right out of college and the long road ahead that a career in research entailed. Kawasumi spent his first year after graduation working at the ERI's Mitaka observatory as an assistant, watching over the equipment and keeping track of whatever the instruments had to report. It was lonely and tedious work. It was also very much in keeping with institutional norms that expected young researchers to be familiar with all aspects of their chosen field and that valued a slow, orderly rise through the academic hierarchy. By the time he received his doctorate, in 1937, Kawasumi was fully a member of that community. Hired first as an assistant professor, he was promoted to full professor at the ERI in 1944. He served as the ERI's director from 1963 to 1965, when he retired from the University of Tokyo.

Kawasumi's career and personal history matters to our understanding of how he influenced popular and policy makers' responses to disaster after Niigata in several ways. Although I wouldn't go so far as to call Kawasumi a disciple of Imamura Akitsune's in the usual sense—Imamura's retirement from the university in 1931 meant that the two officially overlapped in that setting for a relatively short period of time—it is clear that Imamura played an important role in shaping not just the younger scientist's research interests but also his sense of why they did the work they did.[20] Like Imamura, Kawasumi's research was organized around close analysis of the records of past earthquakes, their impact on the built environment, and the application of that knowledge toward disaster prevention.

Kawasumi's early area of expertise focused on calculations of earthquake magnitude and, more specifically, on estimates of the magnitude of events for which there were no modern seismographic records. In articles published in *Jishin* in 1943 (just before his promotion to professor and also just before receiving the less welcome news that he was being sent to Sumatra to help search for oil deposits), Kawasumi described how it was possible, first, to assign an intensity to past earthquakes using the current CMO intensity scale, based on careful study of the written and physical records of events dating from as early as the sixth century and, second, to use that number to derive a standard measure of magnitude at a given distance from the epicenter.[21] Why this was important became clearer in the early 1950s.

By then, Kawasumi was very close to completing the work he had started more than decade earlier—in part at Imamura's urging—which was to finish the search for documentation of past earthquakes and then to update that long list with data on the epicenters and magnitudes for each of the 342 events included in it. The earliest event on the list took place in 599; the last in 1949.[22] What that list made possible was the creation of a series of detailed maps that described the likelihood that any given location in Japan would experience an earthquake of worrisome intensity (in the three categories of 5 and above, 6 and above, and 7 and above) over time frames reaching seventy-five, one hundred, and two hundred years into the future. What would in present day usage likely be referred to as seismic hazard maps were simply called "Kawasumi Maps" throughout the 1950s, when engineers and architects put them to use to help comply with the new Building Standards Act's requirements for earthquake resistant structures.[23] His focus on the local effects of earthquakes also meant that Kawasumi was a frequent participant in site surveys following major events. Soon after the 1946 Nankaidō earthquake, he and his team of researchers spent more than a month conducting interviews and performing site surveys in the communities most affected by it. Kawasumi led a similar undertaking after the 1948 Fukui earthquake. (See Figures 5 and 6 for images of Kawasumi on location in Fukui.) At the time of his Diet testimony in July 1964, he had only recently returned from an extended survey of the effects of the March 16 Great Alaskan (Good Friday) earthquake. Images from that disaster were fresh in his mind both during his brief helicopter tour of post-earthquake Niigata in June and during his appearance before legislators the following month.[24]

Kawasumi saw himself as continuing work that Imamura had started. He was well versed in his former teacher's wartime efforts to monitor the Nankaidō district in the hope of "catching" a precursor for the quake that Imamura was certain was coming and knew, as well, how deeply disappointed Imamura was to have nothing to show for so many years of work.[25] Kawasumi described himself as acting in Imamura's stead as the leader of the Nankaidō site surveys and post-earthquake analysis, which is in part a recognition of the event's significance to him and in part an indication that at least some of Imamura's agenda was now Kawasumi's, as well. By that I mean that the younger scientist continued to think, as Imamura had, in terms both of the cyclic, periodic nature of seismic activity in at least some parts of Japan and of the implications of that periodicity for disaster prevention and preparedness.

Kawasumi didn't set out in search of evidence of seismic periodicity in the capital district; the discovery of the sixty-nine-year cycle was the byproduct of a project that was focused on other questions entirely. In the 1950s, the Japanese government and cultural institutions of all kinds had begun identifying works of art, buildings, sculpture, and the like for designation under the new Law for the Protection of Cultural Properties. One of the objects selected for designation was the massive bronze "Great Buddha" statue on the grounds of the Kōtoku-in in Kamakura and, as part of that process, engineers and other specialists were brought in to assess how vulnerable the object was to various hazards. Earthquakes were one of those hazards and Kawasumi was asked to calculate how much force the most powerful earthquake that was likely to occur in the district would exert on the statue. Once that figure was known, engineers and architects could do what they did best, and strengthen the structure in ways that would allow it to survive such a worst-case scenario.[26]

In order to answer their question, Kawasumi turned to the data about past earthquakes in the region, some of which he had helped compile, and to the detailed reports on the damage to the statue caused by the Great Kantō Earthquake, most of which Imamura and his survey team had written in that disaster's aftermath. The historical records yielded a list of thirty-three major seismic events in the Kamakura district, the first dating back to 818. Using the magnitude and intensity estimates for those events, in conjunction with Imamura's description of the 1923 earthquake's effects, Kawasumi calculated that it was unlikely that the statue would ever be subject to a force greater than the one it had experienced on September 1, 1923. His advice to the specialists preparing to reinforce the structure therefore was that they could go forward using the recorded intensities and effects of the Great Kantō Earthquake as their baseline; their repairs should be strong enough for the statue to survive a repeat of that earlier event.

The conservation efforts went forward according to Kawasumi's recommendations and were complete by 1961. In his report to the committee, Kawasumi, almost as an aside, pointed out the unusual regularity with which the thirty-three damaging earthquakes in his data set had occurred. He included his analysis of this pattern—in which all the major earthquakes in the district between 818 and 1960 repeated in cycles of almost exactly sixty-nine years—in an eight-page appendix to the official report describing the repairs to the statue.[27] No one noticed, and Kawasumi himself seems to have made only a limited effort to draw attention to his findings. Apart from the appendix, he published only one article on the subject (in a rather obscure insurance industry journal in 1963) and there is no evidence that he spoke publicly about it prior to his testimony before the Lower House's Special Committee on Disaster Countermeasures (Saigai taisaku tokubetsu iinkai) in early July 1964.[28]

Years later, Kawasumi insisted that one of the reasons he waited several years before going public with his sixty-nine-year thesis in 1964 was because he feared that it might trigger a panic, for which there could be no adequate response. His anxiety over public panic was one that several of his predecessors in the field had shared, starting with Ōmori early in the century and continuing into the postwar attempts by Wadati and others to rein in un-vetted earthquake predictions in the late 1940s. Kawasumi himself had watched carefully for signs of public distress after the Tokyo Metropolitan Fire Department's release in 1961 of its report describing the potential effects of a major earthquake in the capital district.[29] "Investigation of Damage to the City of Tokyo from a Major Earthquake Disaster"—about which more later—painted a grim picture of what was in store should such an event occur. Kawasumi had been encouraged when the study's publication hadn't provoked a panicky response from the public, he recalled, but he still held off drawing attention to the results of his own research until his appearance before legislators in July 1964.

It is entirely possible that Kawasumi arrived at the Diet that day intending to explain his sixty-nine-year theory and confront legislators and the media with its implications but it is also possible that he had no plans to do either.[30] The exchange that elicited Kawasumi's description of the sixty-nine-year cycle came near the end of the lengthy committee session, by which time he had responded to almost a dozen questions from legislators about the Niigata earthquake and its implications for the rest of the country. It wasn't until the Japan Communist Party's Hayashi Hyakurō (1912–1992) asked a series of pointed questions about probabilities and prediction that the sixty-nine-year cycle eventually came up.[31]

Kawasumi and Wadati alike had, over the course of their testimony that day, been cautiously optimistic about earthquake prediction's long-term prospects while insisting that the field had not yet reached a point where useful predictions were possible. If anything, Kawasumi argued, earthquake predictions were likely to do more harm than good in the near term, given how limited the field's knowledge currently was. Hayashi pressed him on this point. Was it within the capabilities of seismology at present, he asked, to offer up an estimate of the probability that an earthquake of a certain strength would occur in a given region? Could they offer a formula along the lines of "once every several decades," for example?[32]

Kawasumi's response was measured. He pointed out that earthquakes are phenomena that occur over a very long time span, much longer than the relatively brief period for which humans have been keeping track of them. Given how limited the records were, one couldn't rely on them for a truly accurate picture of what to expect in the future. That said, Kawasumi suggested that the state of the field did allow for the calculation of probabilities under some circumstances. His own work and, to some extent, that of Imamura before him had involved the analysis of historical records to map earthquake frequency and magnitude onto locations throughout Japan, he noted. Hayashi pressed him to go further. How worried should they be in the near term about an earthquake doing harm to Tokyo, Yokohama, Chiba, and the communities in their vicinity? How likely was it that a major event would strike soon? "Can you, in light of the evidence and the probabilities," Hayashi asked, "testify in good conscience that there will not be an earthquake in this area in the near future?"[33]

Kawasumi's response to Hayashi's question had two parts. In the first, he pointed out the recent timing of the most powerful earthquakes in the vicinity of the capital district—the M8.4 Hōei event in 1707, the back-to-back Ansei earthquakes in 1854 and 1855, and of course the Great Kantō Earthquake in 1923—suggested that it would be reasonable to expect a major earthquake to affect Tokyo and other major cities along the Pacific once every hundred years or so. He also noted that since it had been forty-one years since the Great Kantō Earthquake, "I don't think I would say that you should not be concerned about the Kantō district."[34]

It was only at this point that Kawasumi started talking about periodicity and elaborating on the risks he feared the capital district now faced.

> When I performed a statistical analysis on the periodicity of damaging earthquakes in the Kantō area, and more specifically near Kamakura, sixty-nine years came through very clearly—so much so that the evidence showed that the probability drops significantly if you move even one year in either direction. Of course, these are just statistics and we should also consider cases in which the term is one hundred years or so but, in fact, that sixty-nine-year cycle has repeated itself again and again. We are, in my opinion, getting very close to the point at which we should be thinking seriously about earthquake policies for the Kantō district.[35]

Hayashi did not ask Kawasumi to elaborate on how he had determined that sixty-nine years was the right time span—neither did anyone else that day—but he did press him to say more about what he thought ought to be done to prepare for the threat that Tokyo apparently faced. Kawasumi had a lot to say on that topic and, given the way that Japan's large-circulation dailies reported his testimony, much of it came as an unpleasant surprise to his listeners.

One of the points Kawasumi managed to convey especially effectively was just how worried he and his colleagues were about the threat to the capital district. "I myself experienced the great earthquake of 1923 as a First Higher School student," he told the audience, "and my whole purpose in setting out to do this research was to protect the capital city from harm. I worry about this day and night."[36] Invoking the Great Kantō Earthquake to explain why he chose to become a scientist is an interesting choice, in that it not only made a clear connection between his research and the practical pursuit of the public good, but it also encouraged listeners to see the threat to the city as Kawasumi did, as just as real and just as terrifying in the present day as what happened in 1923. His description of why he was so worried borrowed both from his personal experience as a witness to the city's near-destruction forty years ago and his scientific expertise. "If the Kantō earthquake were to happen again, there will be fires and, unless the residents of the city really cooperate and put them out—unless they are able to extinguish at least half of all the original fires," he said, "then the Kōtō district, as well as Mikawashima, Okū, and towards Senjū, that entire area will once again be turned into a sea of fire. Within five hours [of that happening], it is unlikely that anyone at all would be left alive."[37]

The Theory

Kawasumi's investigatory methods combined sophisticated mathematics and fairly straightforward pattern recognition. As he described it to legislators in 1964 and then frequently to other audiences thereafter, the analytical work that went into making a list of the major earthquakes in the Kamakura era eventually focused his attention on patterns in the data and on a surprising degree of consistency in the time between events. One of the qualities that made his arguments so compelling was that the evidence was accessible and could almost speak for itself. No one needed an advanced degree to make sense of Kawasumi's list of earthquakes and dates, especially once the press began presenting his theory to the public via easy-to-understand tables and figures.[38] Kawasumi was quite good at explaining himself in simple language, when he needed to, but there is no denying how persuasive just a few columns of numbers could be. The newspapers covering Kawasumi's warning

Table 1. Kawasumi's Timeline vs Intervals Between Actual Earthquakes, 818–1970

Kawasumi's Timeline (69-year intervals)	*Difference from predicted date*	*Dates of actual earthquakes*
818	0	818
887	-9	878-11-01
956		
1025		
1094	2	1096-12-17
1163		
1232	9	1241-05-22
1301	-8	1293-05-28
1370		
1439	-6	1433-11-07
1508	-10	1498-09-20
1577		
1646	1	1647-06-16
1715	-12	1703-12-31
1784	-2	1782-08-23
1853	0	1853-03-11
	1	1854-12-23
	2	1855-11-11
1922	0	1922-04-26
	1	1923-09-01
1991	?	?

A table that includes the data on which this one is based appears in "Magunichūdo 7.5 (2)," *Yomiuri shinbun*, September 2, 1970.

presented his data in simple tables. One column listed the dates for some of the major earthquakes known to have affected the Kamakura district, from 818 until the most recent, which struck in early 1924. In another column, they used Kawasumi's transcription of the series of dates you'd get if you began with the year 818, added 69 (years) to it to get 887, and then kept leaping forward by sixty-nine years until you reached the early twentieth century. The sixteenth step in that series was 1922. To make the point even easier to grasp, some published versions of the two lists included a third intervening column, which showed readers how many years separated the dates of the actual major earthquakes from the dates in the 818+69 series.

The match is not perfect. After 818, in fact, the dates of significant earthquakes coincide exactly with one of the years in the +69 series only twice, in 1853 and 1922. But part of the argument's appeal is that the match up doesn't have to be perfect for it to remain persuasive, since the central premise of Kawasumi's claim is not that earthquakes occur like clockwork every sixty-nine years but rather that the likelihood of a major event increases significantly the closer one is to a "sixty-ninth anniversary" year. As evidence in support of this claim, Kawasumi pointed out that the sixty-nine-year cycle had so far repeated sixteen times since 818, the year of the first recorded earthquake on the list, and that major earthquakes had occurred in close proximity to the southern Kantō and to those "anniversary" years on no less than eleven occasions. Close in terms of timing in this context meant that the earthquake happened within what Kawasumi defined as "the danger period," which statistically speaking encompassed the thirteen years leading up to or immediately following one of the anniversary years.[39] For example: although there were no major recorded earthquakes in the area in 1715, an "anniversary" year (the thirteenth since 818), the devastating Genroku quake of 1703 still fit the pattern that Kawasumi was describing. Yes, it happened a dozen years before what would have been an exact match with the sixty-nine-year series, but the model easily accommodated variations like this. According to Kawasumi's calculations, it was roughly four times more likely that a major earthquake would occur in the district during a "danger period" than at other times.[40]

For many observers, it was the years near the end of the long list of dates that made up Kawasumi's evidence that would have made the strongest impression. 1853 marked the start of the fifteenth sixty-nine-year cycle on Kawasumi's list and 1922 was the beginning of the sixteenth. It doesn't take much imagination to see the similarities between the timing predicted by Kawasumi's model and what residents of the district had actually experienced, first in 1855 as a result of the Ansei-Edō earthquake and then again in 1923 (sixty-eight years later) when the Great Kantō Earthquake struck. It didn't much matter that the locations of the earthquakes—their epicenters and thus presumably the mechanisms that produced them—were not the same. Kawasumi was not arguing that there was a single geological process at work in the region, one that somehow stored up energy for a cycle that lasted sixty-nine years before releasing it and starting over again. He was instead drawing attention to patterns in the frequency with which large, damaging events occurred, without necessarily attempting to explain why the patterns existed.

Why these patterns mattered to present-day inhabitants of the area was obvious enough but Kawasumi helped by spelling it out as clearly as he could. The odds were very high that Tokyo, Yokohama, and the rest of the southern

Kantō region would experience another major seismic event relatively soon. The actual start of the next sixty-nine-year cycle wouldn't be until 1991, but the next dangerous phase—which is when the likelihood of a disastrous earthquake occurring would begin trending sharply upward—would begin much earlier, in 1978. The district was fourteen years away from that marker in 1964, a length of time that made the problem of the coming earthquake both proximate enough to take seriously and far enough off that it was possible to imagine being able to do something about it. Had Kawasumi's calculations suggested that the next disastrous earthquake in the district would likely strike in thirty years, or in three, no doubt the public's interest in the model and in its implications for their own lives would have been different. As it was, the "sixty-nine-year theory" and the countdown to 1978 shifted how experts and the public thought about the district's future and the trajectory of anti-disaster planning in important ways.

Another reason that the "sixty-nine-year theory" likely stayed relevant for as long as it did was that it accommodated evidence that at first glance looked as if it might contradict Kawasumi's basic claims. He had found no record of significant seismic events in the district around five of the sixty-nine-year anniversaries, even allowing for the thirteen-year buffer on either side of those years.[41] It was also the case that powerful earthquakes occurred with some frequency well outside the "danger periods." Although most newspaper and magazine accounts of the theory emphasized only the earthquakes that matched up well with the patterns predicted by Kawasumi's argument, his own published accounts included a much longer list of almost three dozen powerful events between 818 and 1930, many of which did not.[42]

Neither the particulars of the sixty-nine-year theory nor Kawasumi's broader purpose in sharing it with the public were undermined by acknowledging that earthquakes did not abide by rigid schedules. The risk of an earthquake was never zero, he argued, and the probability of one occurring every sixty-nine years like clockwork was far from certain. What the evidence did seem to suggest was that the likelihood that a damaging earthquake would occur in the southern Kantō district did seem to go up, more often than not, in accord with those sixty-nine-year cycles. As imperfect as the model was, it would have been hard to disagree with Kawasumi's point that the patterns he had identified were good enough reasons for people to start paying attention to what needed to be done to protect themselves and the city, just in case. To the extent that Kawasumi's broader purpose was all about preparing the district and its residents for a disastrous earthquake, there was no downside to the idea that risk might vary over time but would almost never diminish to the point that it was safe to ignore it.

Once he went public with it in 1964, Kawasumi's "sixty-nine-year theory" was a surprisingly effective catalyst for keeping the earthquake problem in front of the press and policy makers. It was the most commonly deployed framework for thinking about the risk earthquakes posed to Tokyo and the southern Kantō district well into the early 1970s, by which I mean that it both explained why journalists and the public ought to be concerned and provided a starting point for responding to the risks it described. The series of studies of Tokyo's vulnerability to a major earthquake that followed the Niigata quake and many of the anti-disaster initiatives developed in response to those studies' findings were shaped from the start by Kawasumi's arguments, as we will see.

CHAPTER 5

The Matsushiro Earthquake Cluster

The former castle town of Matsushiro is about a half hour by car from Nagano City and is nestled in the flatlands between two promontories of modest mountains extending from a more imposing range to the southeast. In the mid-1960s the town was home to around 20,000 people. It was by then also a regular destination for scientists and technicians from the JMA and other research institutions throughout Japan drawn to the town by the chance to work in the massive tunnel complexes that extended deep into and beneath the neighboring mountains. The tunnel systems had been dug near the end of the Second World War as part of the Japanese military's plan to relocate the emperor, his family, and government and military officials from Tokyo to these underground facilities to establish an Imperial Headquarters less vulnerable to American bombs and harder for the Americans to reach were they to invade. The facility was never used for its intended purpose but by the time Japan's surrender was announced the huge work force assembled for the project—many of them conscripted Korean laborers—had finished excavating some ten kilometers of tunnels and rooms, distributed across three separate sites.[1]

Some of the qualities that had made the location attractive to the military also appealed to seismologists. The tunnels were far away from big cities, factories, or most other sources of seismic noise, and some of them were quite deep, extending one hundred meters or so underground. At those depths, not only were instruments reasonably well isolated from extraneous signals caused by human activity at the surface but it was also relatively easy for technicians to maintain a constant temperature in the rooms where the instruments were located. For several decades after the end of the war, there were few better locations in Japan to install seismometers, tilt meters, or other instrumentation, if what you were hoping to detect were minute changes in the lithosphere.

The JMA started putting instruments in the tunnels as early as 1947 and maintained a facility there from then on, investing in new and better equipment over the years. The Matsushiro Observatory reached two milestones at the beginning of August 1965, when the JMA announced that it had started collecting data from its newly installed one-hundred-meter-long strain seismometer and that it had become the first observatory in Japan to participate in the Worldwide Standardized Seismographic Network. The WWSSN was a US-led Cold War initiative to detect underground nuclear testing anywhere on the planet and, if nothing else, being asked to join was a good indication that Matsushiro's equipment was considered to be among the best of its kind in Japan.[2]

The new seismometer went online on August 1; on August 3, instruments at the observatory started detecting multiple small earthquakes, too weak to be felt by anyone on the surface. This was not a normal occurrence and even more puzzling was the discovery that whatever was going on seemed to be originating quite close by, somewhere beneath Matsushiro itself. At first, the technicians weren't sure that what they were seeing wasn't just a glitch in the instrumentation but, over the next few days, the intensity and frequency of the earthquakes both ratcheted upwards. By August 7, some of the earthquakes were powerful enough for Matsushiro's residents to remark upon and, over the next week, the total number of recordable events per day climbed past one hundred, then past two hundred, then three. By November, there were days when the observatory tracked more than two thousand earthquakes in a single twenty-four-hour period. Late that month, the town experienced the first quakes with force enough to seriously rattle buildings in town, damaging some. Even on days without that level of drama, residents were subject to an unnerving series of tremors, as unpredictable in their timing and severity as they were relentlessly focused on Matsushiro.[3]

This went on for almost five years. The cluster was most active in the period between August 1965 and the end of 1967, during which time instruments recorded some sixty thousand earthquakes with an intensity of Shindo 2 or above.[4] A pronounced drop in the frequency and severity of the quakes began in February 1966 and led to speculation that the cluster was coming to an end, but in March that trend reversed itself and April 1966 turned out to be the town's worst month yet. A similar cycle of peaks and valleys was repeated several times over those first two years, so that when the number of tremors and their intensity began again to decline in 1967, residents would have been right to question whether the relative quiet that had returned to the town would last. Up to that point each drop-off in activity had been followed by another flurry of earthquakes and, although the scientists who had descended on Matsushiro could report that each new cycle was coming from a different location under the town and that lingering volcanic activity was responsible for the tremors, it was 1970 before they declared the event officially over. Fortunately for the town, the decrease in activity that had begun in 1967 did continue into 1968 and for most residents the last few years of the cluster's hold over Matsushiro were less frightening than the first few had been.

The Matsushiro Earthquake Cluster began while memories of the catastrophe in Niigata were still fresh, but what was happening in Nagano was a different kind of earthquake disaster than the county was used to. At first, the cluster was more a curiosity than a cause for concern. It was not obvious that Matsushiro's residents would require assistance from outside, or what forms help might take if they did. The harm being done by the clustered quakes was hard to see or quantify and physical injuries to residents were few. Matsushiro didn't much look like a disaster zone, at least not in the early phases of the cluster, and at times its residents seemed to be going about their lives much as they had done before the earthquakes had begun.

The longer the earthquakes went on, however, the more obvious the damage became. Cracks large and small began to appear in local buildings, as did other worrying indications that the rest of the built environment was slowly losing its battle with seismicity. There was also no denying the toll the earthquakes were taking on the townspeople themselves. The stress of living day in and day out with the frequent tremors began to disrupt residents' mental and physical health. People complained that actual tremors and anticipation of those yet to come left them unable to sleep, night after night. Life during the day was only slightly less surreal. At the annual sports festival in 1966, towns folk participated in the tug-of-war and wheelbarrow races under the professional scrutiny of an Iwanami Studios cameraman, there to get footage for a documentary on the earthquakes. He captured the joy on the faces of the children (and on those of some of the adults too) as they competed or watched as others did, but these images would in the end be juxtaposed against those of spectators staring off into the distance, exhausted, as an earthquake's deep rumble echoed in the background. More on this unusual film below.[5]

The role that JMA experts and other scientists would play in Matsushiro and the role that Matsushiro would play in shaping Japan's nascent earthquake prediction program also took some time to come into focus. The ERI sent a team and a bevy of portable seismometers to Matsushiro, where they joined JMA staffers and other researchers drawn by the opportunity to collect good data on not just one earthquake but hundreds of them. One problem they ran into early was deciding what to say to residents about the possibility that these swarms of small earthquakes would eventually produce one big one, or when they would end. JMA personnel issued a series of contradictory announcements in October 1965 about the likelihood that the cluster would produce one or more large earthquakes. The first announcement, from JMA staff in Nagano, reassured residents that they had nothing to worry about.[6] The second, this one from Kimura Kōzō, recently installed as head of the JMA's Earthquake Section in Tokyo, came a day later and warned residents that "Earthquake activity has recently increased significantly," and that "in light of past examples, there is at this time the possibility that a mid-level earthquake capable of doing localized damage will occur."[7]

In Kimura's defense, he was attempting to draw a distinction between the "big earthquake" the local observatory staff had been focused on, which he agreed was unlikely, and a potentially destructive "mid-level earthquake," which Kimura feared might happen, but those nuances were not obvious to everyone else.[8] News of Kimura's announcement spread quickly in Matsushiro and gave rise to enough unease that the mayor used the town's household intercom network to get as many people listening as he could before offering a short, sometimes humorous speech in which he made his opinion clear, which was that they should continue to put their trust in the local experts over what they were hearing from Tokyo. He would personally be getting in touch with the Meteorological Agency to complain about the harm its confusing pronouncements had caused. Don't worry, he told them, and be sure to show up for Sports Day tomorrow.[9]

Incidents like that one highlighted a growing tension between what residents (and the public more broadly) expected scientists to be able to tell them about the risks ahead and what experts felt they could safely say. One of Mayor Nakamura's remarks in early 1966 offers a glimpse into what he and other residents were experiencing. Legislators met with the mayor during his visit to the capital in February 1966 and asked him what it was that he wanted most—what would be the most useful thing they could provide for him? What they expected him to ask for was firefighting equipment, or materials to reinforce local schools—something along those lines—but here is how one of Matsushiro's representatives in the Lower House of the Diet described the mayor's response in a meeting of the Special Subcommittee for the Promotion of Science and Technology:

> You know what the first thing out of his mouth was? It was "I want science" [*gakumon ga hoshii*]. I think this is a very telling response. How is it in this day and age when the Soviets can manage a soft landing on the moon for the purposes of space science, and we know what the dark side of the moon is like, that we don't understand what is happening, I don't know, even two or three kilometers beneath the surface of the earth?[10]

The implications of what was going on in Matsushiro for Japan's nascent prediction initiative were not lost on members of the Diet, or on functionaries of the many ministries with a potential role to play in its development. Hagiwara, Kimura, Wadati, and others were frequently called to testify before Diet subcommittees as the crisis in Matsushiro unfolded. The focus of discussion in those sessions moved easily from the situation in Nagano to the status of earthquake prediction. It made sense to the participants to see these problems as closely related and to wonder whether moving forward with earthquake prediction research was also a way to help address the very real needs of Matsushiro's rattled population.[11]

Many of the scientists and institutions engaged in Matsushiro were among those who had taken the lead in drafting the Blueprint and the first five-year research plan. What their experiences studying the cluster and interacting with people in the district seems to have done is reinforced the belief that with proper instrumentation and resources, prediction was a problem they could solve. And no wonder—for a while at least there was probably no place on the planet as closely surveilled for seismic and a whole host of other activities as the Matsushiro district, which meant that scientists there and at research centers throughout Japan were able for the first time ever to begin to get a taste of how much information a comprehensive, sophisticated sensory network could yield.

One other good indicator of how the Matsushiro cluster highlighted earthquake prediction's potential comes in the form of the short educational film *Jishin yochi e no michi, Matsushiro jishin o megutte*, or, to use the title it went by in English, *Approach to Prediction of Earthquakes, Matsushiro Earthquakes*.[12] Shot by documentary and newsreel veteran Kuwano Shigeru in late 1966, the twenty-seven-minute film was done in close cooperation with the ERI and the JMA's staff at the Matsushiro Observatory. It included overviews of some of the explanations then being offered for seismicity and earthquakes in general and for the cluster in Matsushiro in particular, and acknowledged that there were still a great many questions on these topics that had yet to be answered. The film organized itself around three narrative threads that were distinct from the descriptions of what might be responsible for the earthquakes themselves.

The first of these reflects Kuwano's longstanding interest in grass-roots experience and the agency of average people, intertwining scenes of the damage being done to the built environment of the town with glimpses into different facets of everyday life during the cluster. The town's sports festival introduces Matsushiro's residents as resilient in the face of the constant tremors. Viewers see images of broken thresholds, homes braced to protect them from further damage from tremors, and long fissures in the earth near Mt. Minakami, among many others. Kuwano and his crew took up a post on the second floor of a local inn and set up a pair of cameras facing the street. The team stayed with the cameras in shifts for two weeks, ready to trigger them as soon as tremors began. The footage they took of a series of Shindo 3 and Shindo 4 earthquakes—on some days there were several such events—shows about what one would expect, namely images of the buildings next to the inn twisting and swaying, electric wires swinging back and forth, and raucous noises as furnishings, shutters and anything else not securely tied down began to rattle and shake. What were perhaps the most striking images of that sequence were filmed on the morning of September 14, 1966, as a Shindo 4 earthquake struck the town. Kuwano's camera captures the shaking buildings and utility poles as the earthquake is underway but it also shows the behavior of people who were walking or bicycling down the street as all this was going on, presumably unaware that they were being filmed. The striking aspect of that particular scene is that almost everyone in it continues on as if nothing at all unusual was occurring around them. One woman who had been standing beneath the eaves of a building facing the street steps away from it and into the middle of the road, but a pedestrian passing near her continues on unperturbed, as does the mailman on his bicycle and traffic at the intersection down the block. Kuwano later remarked that there were several episodes like that one, during which the building he and his companions were in would be shaking violently while people on the street below appeared to be all but oblivious to the tremors and the noise.[13]

Not all of the town's encounters with earthquakes were as benign but Kuwano waits until near the very end of the film to show viewers what residents stood to lose. The film crew was on hand when a large section of a mountainside across from Mt. Minakami's southern exposure began to give way, triggering a sizable landslide that swept away orchards, buildings, and everything else in its way. The film shows viewers houses being torn apart by the immense wall of earth moving down the mountain and, for a few seconds only, they also see and hear the reactions of two women as they watch in disbelief, on the verge of tears, as the destruction unfolds in front of them. It is such a small part of the film but, by including it, Kuwano guarantees that viewers will see for themselves the forces at work in Matsushiro and perhaps take a moment to imagine what it would be like to live in the shadow of that threat day after day.

The second narrative thread focuses on the different ways that scientists might be able to use what they were discovering at Matsushiro to one day successfully predict earthquakes. The film documents laboratory tests in which cylinders of rock are subjected to greater and greater force in a hydraulic press, demonstrating both that even solid rock has elastic properties and that once the limits of that elasticity are reached, it will shatter without warning. Building on those results, the film turns to a series of experiments in which large trays are filled with a mix of liquid pine resin, gravel, and steel wool. In each experiment, a technician waits for the resin-mixture to fully cool and harden, then attaches a microphone to the resin-bed before activating the heat source embedded in each. What the experiments reveal is that the adulterated pools of hardened resin shatter in stages, with small, barely audible micro-fractures preceding the bigger, more violent breaks. The point being that, since the earth's crust is also heterogeneous, just like the trays of pine resin (whether pine resin is a valid proxy for the earth's crust is perhaps beside the point), it might be possible to detect the same sort of micro-fractures and other signals given off before the primary, and most energetic, rupture occurs.

To be clear, nowhere in the film is it suggested that prediction was possible at that moment and no promises are made that it would be soon. The emphasis throughout *Approach to Prediction of Earthquakes* is instead on the new knowledge being created in Matsushiro and how Japanese science had already begun to make use of it. Immediately following the scenes in the film depicting homes being destroyed by the landslide, for example, viewers are told that, despite the large scale of the event, no one had been seriously injured. Why? Because ERI personnel had determined well ahead of time that a landslide was likely and, after warning the authorities, had guided everyone in the threatened areas to safety with plenty of time to spare.

The third narrative thread puts the scientists themselves on display. They appear not just in scenes depicting experiments in the laboratory but are shown performing their duties in Matsushiro, which are many and complex. Viewers see a lone technician driving over bumpy country roads—a placard above the car's license plate reads "University of Tokyo Earthquake Research Institute"—on his way to check on seismographs installed all over town. We accompany him as he inspects each instrument, records its data, and restarts its observation cycle. Another team of researchers, also from the ERI, is shown making their way up a steep mountain road in a truck, en route to sites where they will use a geodimeter to measure minute changes in the distances between observation markers on the mountains surrounding Matsushiro.

Approach to Prediction of Earthquakes ends by connecting Matsushiro's plight, scientists, and the promise of prediction one final time. Viewers are presented with a close-up of Hagiwara Takahiro and, as he stares thoughtfully into the distance, they are urged to think about some of the same problems that the professor himself must be pondering. "What would happen," the narrator asks, "if densely populated cities were assailed by a large earthquake?" Brief clips from stock footage of contemporary cityscapes, many of them presumably in Tokyo, appear: a roadway packed with traffic, the bullet train racing along its tracks, an aerial view of the city center, tall buildings as far as the eye can see. Each one of those short vignettes is followed in the film by a single still image of similar landscapes, but as they have looked immediately after a major recent earthquake. Footage showing a large refinery or chemical plant surrounded by an urban sprawl of homes and apartment buildings, for example, is juxtaposed against photos of Niigata's burning oil tanks. The cheerful soundtrack that accompanies the "before" images of fast-paced urban life drops out completely when the stark pictures of the earthquake aftermaths appear on screen, so that viewers encounter them in sudden silence. In case audiences missed the point, the narrator reminds them that, "As a step toward preventing disasters, accurate methods to predict earthquakes must be established as soon as possible."

Nakasone's Advice

The Matsushiro earthquake, the Ministry of Education's film, and frequent press coverage of the earthquake prediction project speak to the important conversations with the public that had begun around prediction and the nation's plans to put it into practice. That said, those plans were off to a slower start than its supporters had hoped for. The Geodesy Council had presented its version of the Earthquake Prediction Research Plan to the Cabinet in 1964 as a project that would unfold in five-year increments. The Cabinet accepted the Council's overall recommendations and funding for the new project of measurement, data collection, and searching for precursory phenomena, finally got underway in 1965.[14] When the scientists compared the goals they had spelled out in the first five year plan with funding levels and hiring at the end of 1967, however, the results were disappointing. The plan had called for spending roughly a billion yen each year to acquire, install, and maintain the new instrumentation and analysis infrastructure scientists had said they needed to pursue earthquake prediction research. Those new facilities would need technicians to support them and the data being generated by all that new instrumentation would only be useful if enough researchers were in place to analyze it. The plan's authors had imagined being able to hire in increments of around one hundred new positions a year. Three years into the plan, the various ministries and universities involved in it had together allocated 748 million yen in new spending toward the prediction project, or, about a fifth of the funding the scientists had anticipated having access to. And instead of the 321 new technical and research positions planners had hoped for, they had been permitted to fill only forty-eight.[15]

The project's prospects took an unexpected turn in 1968, just as the Matsushiro event was winding down. Three powerful earthquakes rattled Japan early that year. They struck far apart from each other and not in quick succession, but each was deadly. The Ebino earthquake on February 21 in Kyushu (three dead) was followed on April 1 by another off Kyushu's east coast (one dead), and the third happened on May 16 off the coast of Hokkaido at Tokachi-oki. It left fifty-two people dead or missing. At M7.8, the Tokachi-oki quake was quite close in size to the Great Kantō Earthquake, and much more powerful than the one in Niigata just a few years prior.[16] The prolonged and violent shaking generated by the Tokachi-oki event did considerable damage to many three- and four-story buildings in and around Aomori, Hokkaido, and Iwate prefectures. Like the scenes of Niigata's collapsed bridges and fallen over apartment complexes, the images of modern, steel-reinforced concrete structures rendered dangerous and uninhabitable in 1968 made a powerful impression. It was not difficult to imagine similar scenes and similar tragedies unfolding in Japan's other cities and towns.[17]

It wasn't that unusual for Japan to record three or four powerful earthquakes in a year but, as had been the case with the Matsushiro cluster, the existence of the earthquake prediction initiative raised questions about the nation's tolerance for uncertainty that had not been asked before. Some of these questions were about precursors to the recent earthquakes and whether researchers had identified any, while others were in service of improving prediction's capabilities ahead of the next damaging event. An editorial in the *Asahi* newspaper shortly after the Ebino earthquake reminded readers that, some years back, a group of Japan's seismologists had said quite clearly that if they could just install the one hundred new observation stations they needed then, in ten years' time, they would be able to say how much longer it would take before it would be possible for them to predict earthquakes. "In other words," the editorial went on, "earthquake prediction itself is no longer the issue; we're at a point where the issue is just when its day will come."[18] Hagiwara and his colleagues might well have qualified that statement but they would likely have agreed with the point the editorial was trying to press, which was that prediction was not being given the support it needed. Referring again to the observation stations that the scientists had said were a prerequisite for prediction, the paper quoted an unnamed but "eminent" seismologist's lament that, "Not even 10 percent of them have been built. Compared to so-called 'big science' and the like, our budget amounts to almost nothing." Meanwhile, concluded the editorial, "while Ebino's townspeople are putting up their tents in the bamboo forest and making do with half-frozen miso soup against the cold, they wait for the professors to lend a hand."[19]

Readers expressed similar sentiments about earthquake prediction and their expectations of it. In her letter to the *Yomiuri* a week or so after the Tokachi-oki earthquake, a woman from Shizuoka asked, "We're at the point where science has advanced so far that maybe we'll even be able to travel to other planets, but earthquake prediction can't be done with any reliability?"[20] A forty-five-year-old civil servant writing around the same time had this to say: "As I read about the destruction caused by the Tokachi-oki earthquake and watched scenes of it on television, it was clear to me how fearsome earthquakes can be. But at the same time I couldn't stop myself from getting angry. One might say that since it was a natural phenomenon that there was nothing to be done and I guess that would be that, but really, we still can't predict an earthquake the size of the one that caused the destruction of Tokyo in 1923?"[21]

Public interest in earthquake prediction was a promising development, but it took a politician to show the scientists how to get the funding their project needed. Transport Minister Nakasone Yasuhiro (formerly head of the Science and Technology Agency) began laying the groundwork for a significant increase in government support for prediction after the first two earthquakes that year. The *Asahi* quoted him in April declaring that, "It has been indicated to me quite strongly by people in the field that a national network for earthquake forecasting (*jishin no yosoku*) is needed. My intent, therefore, is to develop a five-year plan to be implemented with this year's budget that will deal in full with these facilities."[22] The *Asahi*'s coverage makes it sound as if Nakasone was putting forward an idea for an entirely new initiative. He was not—he knew about the existing earthquake prediction plan and the efforts to get an observation system built. But as he pointed out to his colleagues in the Diet, those initiatives had stalled. Nakasone was proposing that they start fresh with a new five-year plan with new funding arrangements and with the JMA—which happened to be part of the Ministry of Transportation—taking the lead.[23]

May 16 was a Thursday. On Friday morning, the day after the Tokachi-oki earthquake, Nakasone pitched his plan for a reinvigorated earthquake prediction initiative to colleagues in the Upper House of the Diet. He repeated his performance in the Lower House later the same day and convinced the rest of the Cabinet to support his proposal in time for news of it to make it into the evening papers. Nakasone's explanations of prediction's basic premises in those performances were in line with the arguments in the Blueprint, the first five-year plan, and any number of public statements by Hagiwara and his colleagues since 1965.[24] His synopses of why the search for precursory signs was promising and why scientists needed the instruments they did to do a good job of it weren't bad. He was also more or less in line with the position most scientists had taken about when or if earthquake prediction would be possible.[25] In his post-Tokachi-oki descriptions of prediction's prospects, Nakasone also stressed that it was going to take a long time and a lot of money before the country would be able to start putting prediction into practice. He mentioned a budget of 10 billion yen, or a billion yen a year for each of the ten years of the project, which no doubt sounded generous but was no more than Hagiwara and the others had estimated they would need right from the start. The point of the two five-year plans, Nakasone continued, was first to build the infrastructure for all the observations the scientists had said they would need—he described creating a mesh-like network of sensing devices throughout Japan—and second to use all that newly gathered evidence to decide which approach to detection worked best. "If we thoroughly implement this ten-year plan," he assured his colleagues, "then I'm told that at the end of it we will have a very good idea of which method we can trust to be the most reliable at catching earthquakes [*jishin o haaku*]."[26]

Nakasone's intervention was a stroke of luck for Hagiwara and the others. Under normal circumstances they would have had to wait until 1969 to introduce a second five-year plan and accelerating that timeline by a year allowed the project to make a new appeal for support that was backed by the transport minister himself. On that front, Nakasone gave the scientists very specific guidance about how to obtain the funding the project needed. Morimoto Ryōhei, who was involved in the prediction initiative at the time, had this to say many years later about what that was:

> The advice we were given at the time by Transport Minister Nakasone was to the effect that, with a research plan [*kenkyū keikaku*], the best we could hope for was funding measured in the millions of yen. If we did it as an implementation plan [*jisshi keikaku*], however, very large budget allocations measured in the tens of millions of yen or more would be possible. In reality, it was a research plan, but in order to get the funding we needed, we would have to drop the two characters for "*kenkyū*." We persuaded the [Geodesy] Council committee members of this and in 1969 the Earthquake Prediction Plan was released as a national project.[27]

When the first five-year proposal was put forward in 1964 it was for an "Earthquake Prediction Research Plan," and Hagiwara and others never shied away from telling the public or policy makers that their focus then and for the foreseeable future was on research.[28] When the second proposal was released in July 1968, however, it had become simply "A Plan for the Promotion of Earthquake Prediction" (Jishin yochi no suishin ni kansuru keikaku). The director of the Geodesy Council made no mention of the new plan's new title when he introduced it but he does acknowledge that the council had modified the trajectory it had originally endorsed for the project back in 1965. "While realizing that the prediction of earthquakes is a complicated and difficult problem and that many basic research issues remain," the director wrote, "at this point, in order to be able to put it into practice and in light of both the progress in recent research and appeals from society, we [the advisory council] have determined that pushing the plan one step forward is a necessary and indeed valid thing to do."[29]

The change in labelling was significant enough that it merited explanation when the first installment of the earthquake prediction project's official history was written in 1979. Referring to the decision to go ahead with the new framework in 1968, seismologist Rikitake Tsuneji observed, "It is important to note that the characters for 'research' in the 'Earthquake Prediction Research Plan' were dropped, making it an 'Earthquake Prediction Plan.' That is to say, if the role of the first plan was one of collecting foundational data in support of earthquake prediction, the role of the second changed in favor of trying our luck at putting prediction into practice."[30]

Rikitake wrote his description at a moment when enthusiasm for earthquake prediction was at its peak, in Japan and in other settings as well. As of 1979, the project still hadn't produced a successful prediction, but many scientists were more convinced than ever that they would soon do so. From that perspective, the decision to emphasize the project's pursuit of prediction as a tool that scientists could actually use certainly seemed justified. Funding for the project went up after 1968 and, although it never approached the billion-yen-a-year target Hagiwara and the others had hoped for, the pace of hiring and equipment purchases was much better than it had been.

The Coordinating Committee on Earthquake Prediction

The JMA's Kimura Kōzō remembers it being his idea to create the Coordinating Committee on Earthquake Prediction (Jishin Yochi Renrakukai, or Yochiren). According to Kimura, he approached Hagiwara sometime in the spring of 1968 with the suggestion that they take a model that had worked well for them in Matsushiro and adapt it to the earthquake prediction program as a whole.[31] Kimura was referring to the Hokushin Area Crustal Deformation Information Liaison Committee (Hokushin Chiiki Chikaku Katsudō Jōhō Renrakukai), which had been set up in Matsushiro in April 1966. That committee brought together representatives of all the different agencies and universities active in the area at the time and established ground rules for the sharing of data with each other and for announcements to the public.[32] It had worked reasonably well at the time and Kimura saw no reason why they couldn't replicate something like it on the national level, for more or less the same purposes.

The Yochiren's creation in 1969 was one of the final pieces in the construction of Japan's earthquake prediction regime. After considerable bureaucratic infighting, the new committee was set up as a private advisory body to the director of the Geographical Survey Institute, which was part of the Construction Ministry—that is, not the JMA, which is where Kimura had assumed it would be housed. (The Geographical Survey Institute's claim on the Yochiren derived in large part from its existing focus on geodetic studies and its claims that it was best equipped to oversee the expansion of the national monitoring network.) Who was in the Yochiren would likely have been more or less the same no matter which organizational chart it was spliced into. The committee started out with thirty permanent members, a number that changed little after its first meeting in April 1969. Of those, about half were scientists with university appointments and the rest were representatives of the JMA, the Geographical Survey Institute, the National Disaster Science and Technology Center (Kokuritsu bōsai kagaku gijutsu sentā), and other government-run entities. Through the first decade of its operation the committee always had six or seven scientists from the University of Tokyo and the ERI on it. The other national universities—Hokkaido, Tohoku, Nagoya and Kyoto—were also represented.[33] Members were volunteers and part-timers, meaning that they did the work of the committee in addition to their regular jobs.

Once on the committee, its members generally served for long stretches. Appointments were in two-year terms but over the first decade and on into the second there was relatively little turnover among their number. Academics were slightly more likely to stay put than their ministerial counterparts but several of the JMA, the Geographical Survey Institute, and Disaster Science and Technology Center representatives who were there at the first meeting were still on the roster in the late 1970s. Even when a member left a position at one university to start a new job somewhere else, they tended to keep their spots in the Yochiren.[34]

The selection of the committee's leadership also reflected a strong preference for stability in the ranks. Its members elected Hagiwara Takahiro as chair at their first meeting in April 1969 and then reelected him for five more terms after that.[35] When Hagiwara stepped down in 1981, he was succeeded by Asada Toshi (1919–2003), who, like Hagiwara, had been on the committee from the very start. Asada served a ten-year term as chair, as did his successor Mogi Kiyō (1929–2021), who at that point had himself been a member of the Yochiren for eighteen years. Between 1969 and 2012, every one of the committee's five chairmen had been trained at the University of Tokyo. There are many factors that went into the University of Tokyo's grip on that position and I mention it here not as a critique of that institution but as another facet of the committee's identity. Its structure replicated the educational hierarchy that shaped many other aspects of academic and government appointments at the time, one in which the University of Tokyo stood head and shoulders above the other national universities, and the national universities in general were widely regarded as more prestigious than any of their public or private counterparts. Scientists outside of that cohort—and there were many—were not wrong when they described the committee as a closed shop.

The formation of the Yochiren was an important moment in the history of earthquake prediction in Japan; together with the new five-year plan Nakasone had helped promote, researchers were finally in a position to begin building the instrumentation network they had long sought and to start sharing data in ways that made the search for precursors more likely to yield results. It is worth taking a moment to consider how the Yochiren actually functioned and the work it did.

Committee meetings settled into a fairly standard routine within a few months of its inaugural session. Before the year was out, Hagiwara decided to start holding press conferences after each meeting to give reporters a chance to ask questions and to give himself a chance to shape how the news coverage would turn out. Early on, the committee was meeting monthly but in 1970 one formal session every three months became the norm; those meetings tended to go on for three or four hours each. That frequency seemed to be sufficient for normal business, which consisted primarily of a series of detailed reports on what had been observed before and after any recent earthquakes and summaries of new survey data from all over the country. Meetings in the aftermath of larger earthquakes were different, naturally, and could go on much longer to accommodate all the presentations on the agenda. In June 1974, the first meeting after the previous month's powerful Izu Peninsula earthquake, members heard fifteen separate reports related to that event alone on topics ranging from fault models for the event, records of the aftershocks, post-quake ground tilt surveys, and so on. That was a seven-hour meeting.

The committee concerned itself solely with what could be learned about earthquakes as geophysical events. It neither studied nor reported on earthquakes as disasters, by which I mean events with sometimes catastrophic effects on people's lives and the built environment. What an earthquake did or did not do above ground was irrelevant to the question of whether the physical processes that produced it could be identified beforehand. To be clear, I'm not suggesting that the scientists on the committee or anywhere else didn't know or care about what had happened (or might happen) to their fellow human beings; obviously that was not the case. Earthquakes like the one in Izu in 1974 received the attention they did in part because of the damage they inflicted and not merely as an object of scientific curiosity. But it is still important to be clear about how the committee defined its role and its mandate. The committee had concluded that it wasn't its job to think about better disaster preparedness policies or earthquake-resistant building techniques or even whether a successful prediction could help avoid a disaster. In this it hewed quite closely to the parameters spelled out in the Blueprint and many times since that had described the tools and methods that Japan would need to take up prediction as a scientific problem.

A second aspect of the committee's work is that it did not create a forum for open debate or even discussion of the problem they were trying to solve. Participants shared data and alerted each other when what were thought to be anomalous changes were spotted, but there were no mechanisms in place to allow for outside evaluation of their findings or for the use of metrics of any kind to judge whether they were getting any closer to their goal.[36] The atmosphere at the committee meetings was formal, orderly, polite. The Yochiren hadn't been set up to serve as a forum for the resolution of debates over whether prediction worked or not; it was there to facilitate the exchange of information and to weigh in on the implications of anomalous findings when other scientists brought them to members' attention.

The third aspect of the committee's performance in the 1970s and beyond that we should discuss is its role in publicly identifying those locations in Japan that it had determined were most at risk for a major earthquake and thus most in need of close study and observation. Under the new five-year plan, if instruments or other modes of observation reported unusual changes in the indicators that seismologists associated with potential earthquake activity, then the committee was empowered to designate that area for additional monitoring and oversight and to ask the universities and agencies participating in the earthquake prediction effort to provide the necessary support. "Empowered" might be too strong a word, since the committee could ask for help from any of the research centers or ministries represented in its membership but it had no authority over any of them. It could make requests but couldn't issue orders. Even so, here was something that hadn't existed before—a single, centralized, government-sanctioned body tasked with monitoring the nation for signs of earthquakes yet to occur, with access to the data that might be relevant to that work.

There were two pathways that would lead to intervention by the committee and three classifications it could then use to describe the different degrees of observation that its members believed a location warranted. In the first scenario, the committee would designate a site or a district for observation if it received reports of an anomalous phenomenon—perhaps unexpected results from a geodetic survey or, to take a more recent example, news of frequent, localized, small earthquakes like those that had marked the beginning of the Matsushiro cluster. The second, more common approach to designation, was preemptory; parts of the country with "a history of major earthquakes, active faults, frequent earthquakes, and the Tokyo area" were all candidates for labeling as "special observation districts" (*tokutei kansoku chiiki*). ("Urban centers" later replaced "the Tokyo area" as a qualification for designation.) These areas would, from the beginning, be the target of more aggressive monitoring than would be normal in baseline districts. Although not exactly test fields, the special observation districts were expected to be at the center of prediction-related research and the places where the most effort would be invested in techniques like micro-seismic sensing, continuous crustal deformation observations, and the use of deep borehole drilling to situate even more instrumentation.[37] If, once all this additional surveillance was in place, research or observation revealed unusual phenomena along the same lines as described earlier, then the committee could move the district's status from "special" to "enhanced observation" (*kansoku kyōka*), triggering the addition of still more attention and analysis.

The third and final designation that the committee could apply was that of "intensive observation district" (*kansoku shūchū chiiki*). Only parts of the country already undergoing "enhanced observation" would normally be candidates for escalation to this level of scrutiny and concern. The implication of each of these steps along a progression of ever-more-finely grained study was that the further along you went, the closer you were to the event that had been the point of the attention all along, the actual earthquake. Intensive observation would begin once Yochiren members decided that the unusual phenomena being detected were in fact "associated with earthquakes."[38] At that point, as many mobile research teams as necessary would converge on the location in question and add their capabilities to those already in place. As for what would happen next, the plan offers only this rather cryptic description: "research and observation in the forms listed above will be performed at high density and an attempt will be made to put earthquake prediction into practice." Readers are left to imagine for themselves what trying to put earthquake prediction into practice under those circumstances would entail. At one end of a spectrum of possibilities, it could take the form of a warning that met all three of the standard requirements for a successful prediction and would specify when and where the earthquake would occur and how powerful it would be when it did. It is reasonable to think that this is the outcome that most people expected to see happen sooner or later.

Media coverage when the 1968 plan was announced made quite a fuss over these different stages of designation. News reports noted that the capital district had already been chosen for special observation status, before the committee had even met.[39] An observer might well have concluded that the plan's authors had given a lot of thought to how they would use these categories but when the first opportunity to do so came in 1969, committee members realized quickly that they had not. The problem wasn't so much with the categories themselves as it was with deciding which criteria mattered most for each and then figuring out how to map these categories onto physical space in ways that made sense. That mapping process mattered because it would define where any additional monitoring efforts would be concentrated and where the scientists would look for the clues they sought, and because anyone living in the vicinity would naturally be interested in finding out whether they were inside or outside the boundaries of this newly defined zone. The committee's decisions about where to designate—and why—were among the most public interventions it was possible for it to make.

The scientists did have criteria in mind for selecting parts of the country for special observation but there was a diverse set of ideas within the Yochiren about how to apply them. Should the number of areas be large or small? Would parts of the country that were already under close observation by university researchers and observatories automatically become special observation zones? One scientist took the straightforward position that what mattered most was whether or not a given district had a record of powerful earthquakes in the past. Another suggested narrowing that field down to locations where an earthquake was most likely, based on an understanding of their periodicity.[40] The JMA's Sekiya Hiroshi made a similar proposal, which was that they select places where they knew damaging earthquakes had occurred in the past but that hadn't had a powerful quake since 1869.

When the Yochiren met in early 1970 to finalize its choices, its members ended up taking a fairly broad view of the qualifications for "special observation" status. Hagiwara drew his colleagues' attention to recent analyses that had estimated the intensity of the shaking associated with the 1854 Tōkai earthquake and the committee readily agreed to designate that district for special observation.[41] They did the same for seven other areas: parts of Akita and western Yamagata, northern Nagano/southwestern Niigata, the Biwa Lake district, eastern Shimane Prefecture, Iyonan, Hokkaido's eastern region, and the Osaka-Kobe area. Based on a series of reports of elevation changes on the Miura Peninsula and other unusual readings, the committee also decided to "upgrade" the capital district's status from "special" to "enhanced observation," making it the only part of the country to be so designated.[42]

The system that the Yochiren devised to single out some parts of the country for extra monitoring helped define the prediction initiative for the next forty years. The committee revised the list of nine districts of interest it settled on in 1970 only twice before the whole designation structure was abolished entirely in 2008. Neither of those alterations changed its character or purpose all that much. In 1974, the Tōkai district's status was upgraded from "special" to "enhanced" and, in 1978, the committee launched a full review (*minaoshi*) of its selection and designation process. Conducted over many months, in the end the review did little more than confirm that the committee thought that it had been on the right track all along. The new, post-review maps of designated districts in 1978 looked a lot like the old ones. The committee did add coverage to two parts of the country that hadn't been among those discussed earlier: one ranged across parts of western Nagano and eastern Gifu and the other extended along the Miyagi and Fukushima coasts. The explanation given for adding the former had to do with the distribution of known faults throughout that area, increases in seismic activity, and the recent history of large earthquakes in neighboring areas. As for Miyagi and Fukushima, it is puzzling that the area hadn't been added earlier, given its long history of destructive offshore and coastal earthquakes.[43] A little more than two months before Miyagi and Fukushima were given special observation status, in fact, an M7.4 earthquake originating off Miyagi's coast had killed twenty-eight people, injured many more, and damaged or destroyed some 7,400 buildings and homes. This was the first time since the designation system had been put in place that a major, destructive earthquake had occurred in an area of Japan that the committee hadn't already said it was watching closely. The committee's decision to focus more closely on Miyagi and Fukushima may have happened anyway but the earthquake was also a factor in its designation, at that point.

The Yochiren's announcement in 1978 that it had decided to put more districts under special observation was front page, above-the-fold news the next day and was accompanied by detailed descriptions of how the coverage had changed and why. The addition of the two new zones, according to one headline, meant that altogether the Yochiren had designated roughly one quarter of Japan's entire surface area for some form of prediction-related observation.[44]

The Yochiren's spokesmen argued that it was the progress they had made in data collection and developments in earthquake prediction research over the past eight or so years that had encouraged them to revisit their earlier choices. They knew more now than they had in 1970 about where and how to observe for the purposes of the earthquake prediction program. These latest changes, in other words, were improvements to a system that continued to serve an important purpose and would presumably continue to evolve.

No changes of note were made to the observation zones between 1978 and their abolition thirty years later.

CHAPTER 6

Tokyo's Disastrous Futures

Average Japanese began to worry much more about disasters in general and earthquakes in particular over the course of the 1970s than they had before, according to the government's own public opinion surveys. That various arms of the state took the interest they did in trying to gauge how worried citizens were about the uncertainties they faced, and how they might react in the event of a catastrophe, is itself an indicator that disasters had joined education, housing, pollution and the like as topics about which people were expected to have opinions, ones that needed to be heard.

The government's first foray into assessing attitudes toward disaster came in 1951, when it posed this question in one of its national surveys: "Our country often experiences major earthquakes and major fires, typhoons, and damage from floods, but which worries you the most?" Asked to pick just one, 42.5 percent of the survey's respondents chose "fires"; the rest of the responses were split almost evenly between typhoons and floods (26.7 percent) and earthquakes (26.8 percent). Asked to indicate which potential misfortune concerned their households most on a daily basis, 40 percent said fire. At just 2.2 percent, earthquakes were less worrisome than theft (17.3), illness (15.8), or traffic accidents (9.5).[1]

It was twenty years before the Prime Minister's office again asked citizens for their thoughts on disasters. In the early summer of 1971, a few months after the San Fernando earthquake, the government conducted its first ever survey devoted entirely to "disaster preparedness awareness" (*bōsai ishiki*) among residents of Tokyo, Nagoya, Kyoto, and Osaka.[2] While many of the questions focused on what city dwellers planned to do after a catastrophic event, others gave respondents a chance to describe what they saw as the main threats to themselves and their families. "When you hear the term 'disaster' [*saigai*]," the poll asked, "what comes to mind?" "Earthquakes" was the answer 45.5

percent of the time, ahead of fires (43.5 percent) and storms (28.7 percent). "Of those different disasters," the survey continued, "looking ahead, which one worries you most?" Earthquakes again led the list (at 42 percent); fires were well behind (at 32.5 percent).[3] And, in a question that highlighted the role recent reports and scenarios in the media involving future earthquakes might have been playing in shaping public discourse about disasters, pollsters asked, "Have you recently come across references to the possibility of a major earthquake occurring, or of the danger to big cities if one were to happen, in newspapers or on the television?" More than 88 percent replied that they had; 56 percent of respondents also acknowledged that those dangers had also come up at least once in recent conversations with family members.[4]

Questions about earthquakes and other disasters were included in the government's regular surveys on fire fighting for the rest of the decade; in the 1980s and 1990s, national polls focused primarily on disasters and disaster reduction were conducted every few years. What the results of these surveys suggest is that the public's worry over earthquakes remained at more or less constant levels for some time. In 1976, for example, 57 percent of respondents reported that yes, the possibility of a major earthquake or the damage it would do had been on their mind lately.[5] Two years later, in response to a similarly phrased question, 66 percent of those polled replied that they had indeed thought recently about the dangers a powerful earthquake would pose.[6]

This chapter shows how scientists and other experts began in the 1960s to draw attention to Japan's vulnerabilities to powerful, destructive earthquakes. They did so in part by setting aside questions about whether earthquakes could or could not be predicted in the short term in favor of a narrative about historical patterns and statistics. Kawasumi Hiroshi helped make this framework for thinking about uncertainty accessible to the general public and, in doing so, convinced many Japanese that Tokyo and its neighboring communities were entering a period of greatly increased danger. At the same time, Kawasumi, his colleagues in the Tokyo Metropolitan Disaster Prevention Council, and others in the media also offered persuasive, sophisticated analyses of the hazards that people could expect to face in the event of a major earthquake. Together, these developments raised difficult questions for policy makers, for scientists, and for families, all of whom were hard pressed to imagine how best to respond to this state of affairs. In this chapter, we will examine a series of efforts in the 1960s and 1970s to protect Tokyo's residents, or at least some of them, from the earthquake that Kawasumi had warned them was coming.

Investigations of Damage

In his Diet testimony in 1964 after the Niigata earthquake, Kawasumi had claimed that parts of Tokyo would "once again be turned into a sea of fire" in the event of a major earthquake and that "within five hours [of that happening], it is unlikely that anyone at all would be left alive."[7] He wasn't just guessing about the effects of another Great Kantō-like earthquake. The Tokyo Metropolitan Fire Department's report *Investigations of Damage to the City of Tokyo from a Major Earthquake Disaster* drew on findings from studies that it had commissioned starting in the mid-1950s but didn't share with the public until 1961. *Investigations of Damage* began with a careful analysis of what had happened to the city in 1923, drawing on the reports that Imamura and others did in the earthquake's aftermath. The more speculative work that went into the study involved translating the outcomes of that earlier disaster to the city Tokyo had become by the 1960s. Researchers turned to tax records, architectural plans, geological surveys of the depth of the alluvial layer in different parts of the city, numerous site visits, and an array of other data to construct as detailed a model as they could of Tokyo's built environment. They then used that model to painstakingly calculate what would happen, building-by-building and neighborhood-by-neighborhood, if the city were to be struck by an earthquake just as powerful as the one that had leveled much of it on September 1, 1923. The report's findings more than justified Kawasumi's fears.

Investigations of Damage was hardly sensationalistic—its tone is matter of fact, its language dry and technical—but even with all the caveats its authors attached to their findings, its conclusions were alarming. Even if local citizens put out 60 percent of all the fires that the earthquake started (a slightly higher percentage than had been achieved in 1923), the Fire Department expected that it would have to face as many as 264 other fires in its immediate aftermath.[8] After taking into account the projected damage to fire equipment and hydrants, disrupted radio communications, blocked or gridlocked roads, and the fires' rate of spread, researchers calculated that firefighters would reach only about two-thirds of those fires. The report's final estimate was that just over an hour after the earthquake struck, more than one hundred major fires would still be burning in the city.[9]

One of the things that is most striking about these findings—and which Kawasumi picked up on and amplified—was that a handful of neighborhoods accounted for almost all of the unchecked conflagrations. The investigators' models projected that about half of the fires in the city would be in just two wards—Kōtō and Sumida. If nearby Arakawa, Adachi, and Edogawa were included, the ratio jumped to three-quarters.[10] This scenario is what Kawasumi had in mind when he referred in his testimony to parts of the city being consumed in "a sea of fire" in the event of an earthquake. That imagery is apt, given the Fire Department's description of where and how fast parts of Tokyo would burn. An hour into the event, the unchecked fires were expected to have burned through in excess of 900,000 square meters of the city; after two hours, 2.7 million square meters would be gone; after five, more than 13 million. Four-fifths of that burned-over landscape was in the handful of lower city neighborhoods that Kawasumi drew attention to in the Diet.[11] While there was some good news in the report—better building practices in the city's central business districts meant that they would likely sustain less damage than they had in 1923 in the event of an equally powerful earthquake—the takeaway message described a city that was more vulnerable in 1960 than it had been some forty years before.

Kawasumi's testimony in 1964 caught the attention of reporters, who made sure that his warnings made it into the major papers. As coverage of the disaster in Niigata faded from view that summer, reminders of the threat to Tokyo and nearby communities lingered. An industrious reporter for the *Yomiuri* newspaper sought out Hamada Minoru (1902–1974), a professor in the engineering faculty at the University of Tokyo and one of the lead authors of *Investigations of Damage*, for his thoughts on Kawasumi's statement. Hamada was one of the country's leading experts on fire-resistant construction and had been since he first started studying it as part of Japan's preparations for potential air raids prior to the war.[12] After noting that he himself had only recently returned from his own trip to Niigata to study the effects of the earthquake, Hamada told the reporter that he had come away from that visit more convinced than ever that the models he and his team had used to describe what would happen to Tokyo had been accurate.[13] The threat to the city, in other words, was as real as Kawasumi had warned it would be.

Soon after his appearance in the Diet, Kawasumi was asked to head up a newly formed committee in the Tokyo Metropolitan Disaster Prevention Council, one whose sole purpose was to come up with countermeasures against earthquake hazards. After Kawasumi retired from the University of Tokyo and from his post as director of the ERI in 1965, he devoted himself to the work of the committee. His efforts in that new capacity included a series of initiatives, working with the Tokyo Metropolitan Fire Department and other institutions, to use urban planning tools and redevelopment projects to create fire- and flood-resistant neighborhoods, especially in at-risk wards like Kōtō. The idea that Tokyo was vulnerable to a major earthquake and that steps needed to be taken to address that threat set the stage for the debut of multiple government initiatives over the next several years, among them a series of further studies of potential effects of earthquake disasters in the capital. Detailed analyses of the city's vulnerabilities appeared one after the other between 1967 and 1970. The Tokyo Metropolitan Fire Department released the results of its second comprehensive study of the damage the city could endure in the event of a major earthquake in 1967; Tokyo's Disaster Prevention Council published a similar survey the following year and, one month after its release, the director of the national Fire and Disaster Management Agency (Shōbōchō) commissioned his own analysis of "the steps to be taken with regard to the fires and other damage that would happen if an earthquake on the same scale as the Great Kantō Earthquake were to occur in the Tokyo area."[14]

The three studies made similar arguments. They all, for example, linked their concerns about the risks to Tokyo back to Kawasumi's timeline.[15] The periodicity theory (*shūkisetsu*) is, in this context, treated as a given, as a way of thinking about the risk to the district that is neither all that complicated nor up for debate. The studies also shared a common argument that the city was becoming more vulnerable, not less, over time. There were a few bright spots—the Fire Department noted that the numbers of fire stations, fire trucks, fire hydrants, and fire fighters had all risen since 1961, for example—but on balance the news was consistently grim.[16] The Fire Department estimated that although fire fighters would put out more fires in absolute terms in 1967 than they would have been able to in 1961, the number of conflagrations they would not be able to extinguish or contain would also increase. More of the city would be left in ruins as a result.[17]

There were also new hazards that had to be taken into account. In 1968 more than four hundred buildings in Tokyo were ten or more stories tall, for example, and the number of new underground shopping centers and pedestrian passages in the city had also grown rapidly in recent years. No one was sure how those as-yet-untested structures and the people in them would fare in the event of a powerful earthquake but each was associated with novel hazards—pedestrians who managed to escape the pitch-black, smoke-filled passageways below ground would be lucky to avoid the cascades of broken glass falling from shattered skyscraper windows onto the sidewalks and streets above.[18] Along similar lines, researchers flagged the rapid increases in the volume of gasoline and other fuels in the city on any given day as another worrisome problem. The capital's residents consumed twenty-one million kiloliters of petroleum products in 1968 (almost twice as much as they had in 1965), which meant that roughly 15 percent of Japan's total oil imports were used to satisfy demand in Tokyo alone that year.[19] In 1961, the work of transporting all that fuel from storage depots to gas stations, factories, and the like was performed by Tokyo's 706 registered tanker trucks. Seven years later, the city relied on more than three times as many—2,195 tankers—to manage the distribution of gasoline and other fuels.[20] These were just some of the new hazards that observers flagged in the late 1960s.

The work that engineers and scientists were doing to estimate the devastation an earthquake would inflict on Tokyo reached the public in a variety of formats. The media covered the official reports themselves and summarized their findings, of course, but a number of mainstream news outlets also produced their own detailed "what-if" accounts of a potential earthquake's catastrophic effects on Japan. These speculative documents refracted data and arguments from the official, technical studies into familiar scenarios from everyday life. The *Yomiuri*'s "Magnichūdo 7.5" series of articles in 1970 is one example of the mainstream media's contributions to this genre; the *Asahi*'s 1973 "Earthquake Archipelago" (Jisshin rettō) series is another.[21] Some of these pieces conveyed fairly straightforward warnings and were meant to encourage residents to learn about the hazards they faced and prepare accordingly. Others—though not necessarily less accurate in their depictions of what could happen in the event of a major earthquake—rendered the next earthquake real for readers in well-crafted, dramatic accounts of what it would be like to experience it in person.

NHK embraced both approaches when it published its assessment of the risks and hazards the nation faced. *Prepare for an Earthquake: Your Disaster-Preparedness Countermeasures* came out in the summer of 1971.[22] The half-dozen or so journalists who contributed to it produced what was arguably the most comprehensive, non-academic survey of the earthquake problem then in print. Part "how-to"—one whole section was devoted to questions like, "What should I do if I'm at home (or at work, at school, or out and about) when an earthquake strikes?" or "What do I do if I'm driving?"—*Prepare* offered a survey of current thinking on how the country and its citizens could best defend against the effects of a major earthquake.[23] *Prepare* also devoted a chapter to what its authors called a "hypothetical" account of "The Great Tokyo Earthquake," but insisted that this "sketch of the events on the day our nation's capital Tokyo falls into ruin" was not at all a work of fiction.[24] It drew from an array of recent studies and surveys by the Tokyo Metropolitan Disaster Prevention Council, the Tokyo Metropolitan Fire Department, and seismologists, all sources that others could check if they so desired. "While one might suggest that it imagines the worst possible outcomes," the authors acknowledged, "these are in fact events with a very high probability of actually occurring *in reality*."[25] In a gesture that both linked Tokyo's next disastrous earthquake to the last one and reminded readers of *Prepare*'s scientific underpinnings, a reproduction of the famous seismogram from September 1, 1923, appeared on the preface's opening page.[26]

Prepare follows the experiences of the members of several families before, during, and after an earthquake that devastates Tokyo, in four separate scenarios. Readers were to imagine the events unfolding over the course of a mild winter's day—a Saturday in December—sometime in the 1970s. City sidewalks are crowded with shoppers and with people taking care of year-end errands; the roads are likewise full of cars and trucks. The weather is mild for the season—which also helps explain why so many people are out and about—and a light wind stirs the pall of smog that hangs over the city. There are no warnings that anything unusual is about to happen.

The scenarios are set in four different urban environments, each reflecting what members of typical middle-class and blue-collar families might be doing that day. In one, "Shoichi" was hard at work in his office on the seventh floor of an office building in the Ginza district; his wife and son, meanwhile, were making their way through a crowded underground shopping mall at lunchtime. "Tadashi," the head of the other family (NHK was not known for challenging gender norms) was on the subway heading home after working a long overnight shift; his wife "Hanako" and their newborn child were waiting for him in their modest apartment in one of the Kōtō district's low-lying neighborhoods. The two-story, wooden building in which they lived was just one of the many that lined their narrow street.

Prepare's four narratives take place simultaneously. In each, the respective family members survive the initial effects of the earthquake's powerful tremors more or less unscathed, but are then threatened by all the hazards associated with their particular settings. Each story ends badly for all involved and each implies that none of the family members survived, while not actually depicting their deaths. The stories' authors provide vivid images of them trapped and helpless, and evoke the mounting terror each felt as the end neared. Shoichi's story ends in his office-building's stairwell, packed to immobility as everyone at work that day tried to flee at once. Someone higher up in the queue tripped and, "as if a dam had broken open, people spilled down the stairs like an avalanche. The human wave came from above, one step at a time."[27] Shoichi's wife and son faced a similar fate, swept along in a panicked crowd trying—and failing—to escape from the smoke-filled underground shopping arcade.[28] Tadashi, meanwhile, was trapped even further below the surface. He and fellow subway passengers try to make their way from their stranded train to the nearest station, when a flash flood swept the others away, leaving Tadashi clinging to the top of one of the subway cars. As he watches, the water level in the tunnel rises higher and higher; the narrative ends as the water reaches Tadashi's perch near the tunnel's ceiling.[29] Tadashi's spouse Hanako, and their infant, meanwhile, flee from their home in the Kōtō district just ahead of the fires that engulf their apartment building. Evacuation routes are impassable and the failure of the barriers protecting the delta neighborhoods from the rivers surrounding them sends a huge wave of water sweeping over the delta. Hanako's story ends when the building she has taken refuge in catches fire—the parts of the district that were still above water had continued to burn, flames spreading from one structure to another. "Clutching her now silent baby to her chest, Hanako whispered, 'Mama's here, mama's here,' words that were now without any meaning at all."[30]

The San Fernando Valley Quake

Later it would be said how lucky it had been that the earthquake struck so early in the morning and not during rush hour or at some other time of day that would have seen more people out and about. The fault that slipped beneath the foothills of the San Gabriel Mountains just after 6:00 a.m. on February 9, 1971, produced an earthquake powerful enough (M6.6) to damage older buildings in Los Angeles proper, some 30 miles away. Closer to the epicenter, communities at the northern end of the San Fernando Valley were hit hard. The dramatic collapse of both several major freeway interchanges nearby and key buildings in more than one of the region's medical centers—of all places—led residents to imagine how much worse the harm might have been. As it was, the earthquake was directly responsible for more than sixty deaths, injuries to some two thousand, and roughly half a billion dollars in damage to local buildings and infrastructure.[31]

Much as the Niigata earthquake had focused attention on newly visible hazards to Japan's urban spaces, the San Fernando earthquake again highlighted for Japanese observers the continued vulnerability of even the most modern cities. Media coverage of the Los Angeles quake featured photos of long stretches of elevated highway that had plummeted onto roadways below, some segments still intact despite the long fall. The public also saw images of the rubble-strewn grounds of what was left of the Veterans Memorial Hospital and of the brand-new Olive View Medical Center building, from which entire multistory sections had peeled away. A few of the component parts of the building stayed precariously upright even as the rest had toppled over but there was no avoiding the fact that the new and supposedly earthquake-resistant structure had not weathered the earthquake well.[32]

Survey teams from Japan began arriving in Los Angeles within days. Tokyo's governor, Minobe Ryōkichi, assigned Kawasumi to head up a team of half a dozen or so experts from Tokyo; the central government and a few other cities dispatched teams of their own.[33] It wasn't that Los Angeles resembled cities in Japan all that much—it did not. But Los Angeles and cities like Tokyo had features that were new in some of the same ways and had built skyscrapers, freeways, and other infrastructure projects with at least some attention to the risks that earthquakes posed. The teams from Japan were also interested in California's disaster response efforts—in how the city and its residents had fared in the earthquake's aftermath. Had there been panic? Had people known what to do, and had help arrived in a timely manner?

According to Kawasumi and his colleagues, the news from Los Angeles was mixed. On the positive side were findings that pointed to the overall success of contemporary earthquake-resistant architecture and engineering techniques. Schools generally fared well, since even older buildings had been reinforced under the requirements of the 1933 Field Act. Veterans Memorial Hospital, which is where most of the deaths occurred, had been in use since 1926, well before more stringent building codes for non-school structures had gone into effect in California. The damage to the just completed Olive View Medical Center, on the other hand, was worrisome, since its design had reflected up-to-date earthquake-resistant engineering and should have fared better than it did. Kawasumi's take on what happened at Olive View, however, was generally positive in terms of its implications for earthquake engineering in general and for buildings in Japan in particular. His argument was that Olive View's design was fundamentally fine, given the severity of the earthquake the structure had been expected to withstand. The problem wasn't the engineering, it was the builders' understanding of the nature of the earthquake hazard in the area. Put another way, Kawasumi was suggesting that had engineers known what they were up against, they could have adopted their designs accordingly and ended up with a structure that might well have withstood even February 9's powerful shocks.[34] One of the implications of this argument was that the better the understanding of the underlying seismic mechanisms and geology in any given location, the better prepared architects and engineers would be to build safe structures.

For Japanese audiences, such observations were good news in the sense that they validated arguments that Kawasumi and others had been making for some time, namely that there was much that could be done to mitigate earthquake hazards. Also reassuring were claims that building standards in Japan were already more rigorous than the ones in California and that Japanese highways and bridges subject to the same forces that had been so destructive in Los Angeles would likely have been left intact.[35]

If there was some good news to take away from the San Fernando earthquake for Japanese policy makers and the public, the overall effect of that disaster was to draw even more attention to Tokyo's vulnerabilities. Headlines in the Japanese press immediately after the earthquake described Los Angeles as "A Modern City Weak Against Earthquakes," which readers no doubt read to suggest that their own modern cities—and Tokyo especially—were better equipped to deal with seismicity than LA had been.[36] The press did not maintain that attitude for long and reporting quickly began to use descriptions of the disaster in LA as preludes to discussions of the risks and hazards that Tokyo faced. As survey teams returned from California, they were quick to emphasize that however good Japanese engineering might be, an earthquake as powerful as the one that had rocked LA would still have done serious damage in Japan. Southern California's relatively firm surface geology, the good coordination and communication among the different medical and rescue efforts that followed the event, and LA's sprawl had all helped keep casualties to a minimum, but none of those factors were necessarily relevant to city dwellers in Japan.[37] Major cities like Tokyo and Osaka would likely have fared quite poorly against an equally strong quake in their vicinity, the experts said. Not only would the transportation gridlock be much worse than anything LA had experienced but there would also be more fires and fewer opportunities to bring them under control. The more experts and average citizens alike learned about what had happened in LA, the more reasons they had to worry about their own circumstances.

They didn't have to start from scratch, of course. The studies of Tokyo's vulnerabilities that the city's Disaster Prevention Council and other agencies had produced in the late 1960s—studies that Kawasumi had already paired with his warnings about the coming Kantō earthquake—were part of public discussions of what the San Fernando earthquake might mean for Japan right from the start. LA's disaster brought the Tokyo Metropolitan Fire Department's grim estimates of how many fires an earthquake would trigger, how many would rage out of control, and how much of the city would eventually be burned to the ground, back onto the national stage. Kawasumi described those scenarios in detail during Diet testimony a few days after February 9, 1971, and brought them up again in appearances before legislators in March and May that year. The press reported on those sessions, repeating the expert's estimates of how much damage Tokyo and other local communities could expect, and pursued their own stories on related topics. Reminders of the hazards Tokyo faced were seldom out of the public eye for long that year.

One thing that was different between how those hazards had been framed before February and how they were discussed after the quake in LA was that disaster experts began to talk openly for the first time about the numbers of people they thought might die in Tokyo's next major earthquake. Prior to this, estimates of how much damage the city would suffer had focused on just about everything but the human cost. The Tokyo Metropolitan Fire Department and the city's Disaster Prevention Council had both produced detailed accounts of how many fires would break out, where, their rate of spread, and so on, in a grim litany of blocked roads, risky evacuation routes, and burned-out neighborhoods. For all their detail, however, those studies had not said how many people were likely to be injured or killed as a result of the earthquake's effects.

Those estimates were first made public two days after the San Fernando earthquake. The *Asahi* told readers on the morning of February 11 that, although the figures had never been officially released, experts in the field believed the number of casualties from the next major Kantō earthquake would be very high. According to calculations in one of the scenarios modeled by the Disaster Prevention Council in 1967, between 550,000 and a million people would likely perish in Tokyo alone.[38] In his Diet testimony that same day, Kawasumi offered up the same figures as part of an appeal to legislators to do more to prepare the city and the nation against disasters.

"The way things are now," Kawasumi said, speaking in reference to Tokyo's defenses against a major earthquake, "the number of residents who will not be able to get out of the city may be as high as 560,000. They will probably not escape being burned to death. Such was the scientifically derived conclusion of the Earthquake Section of the Disaster Prevention Council."[39]

That number—560,000 dead—became a touchstone of sorts in many of the analyses and discussions of Tokyo's vulnerability that followed Kawasumi's testimony. Kawasumi himself brought it up again during another Diet appearance in March and it appeared regularly in press accounts of the earthquake threat for several years thereafter.[40] The number matters as an absolute metric to help members of the public gauge their odds of survival, as a way to track the ups and downs of the city's efforts to prepare for the next earthquake, and as an indicator of another degree of acceptance of the experts' mastery of the science of disaster. Up to that point, Kawasumi and his colleagues in Tokyo's Disaster Prevention Council and elsewhere had offered precise, "scientific" findings on just about every aspect of the disasters to come, except answers to the questions that arguably loomed largest, which, of course, had to do with the human scale of the anticipated harm. To describe the next massive disaster as one that would take more than half-a-million lives was to intervene in the public's understanding of what was at stake in ways that a narrative about property damage could not.[41]

Tokyo Prepares

The central government and Tokyo's launched two ambitious anti-earthquake disaster initiatives in the late 1960s and early 1970s, both targeting the capital. One was a project to build a half-dozen disaster shelters (*bōsai kyoten*) at different sites in the Kōtō delta, each large enough to harbor tens of thousands of refugees in flight from fires and other hazards in an earthquake's aftermath. The other, Tokyo's 1971 Earthquake Disaster Prevention Ordinance, put the city on a new footing so far as expectations about the catastrophes in its future were concerned. Both measures' legacies were still visible in the city and in the contours of its preparations for disasters many decades later. This section shows how Kawasumi's theory about the timing of Tokyo's next major earthquake and the growing number of descriptions of what such an event could do to the capital help explain why officials acted and why their interventions took the forms they did.

The Disaster Shelters

Although officials decided early on to make shelters their answer to the problems that the Kōtō district possessed in such abundance, exactly how the shelters would work and even what it was they were supposed to accomplish were questions they turned to outside experts, Takayama Eika and his student Murakami Suminao, to help answer. Takayama had been a middle school student in Tokyo when the Great Kantō Earthquake struck. He credited that experience with sparking his lifelong interest in anti-disaster planning and with influencing his decision to study urban planning at Tokyo Imperial University.[42] He trained there under Hamada Minoru and others and, by the early 1960s, was a prominent figure in his field; the Komazaka Olympic venue was one among many major projects he had been responsible for by that point in his career.[43] Murakami Suminao began studying with Takayama in 1962 in the University of Tokyo's newly formed Urban Engineering Program, where, at his professor's urging, he focused from the start on disaster prevention. It was an unusual direction to take—Takayama hadn't asked any of his other students to specialize in planning for disasters—but it meant that Murakami was the natural choice to work on the Kōtō delta projects a few years later.[44]

Both Takayama and Murakami wrote about their involvement in the various stages of planning for the Kōtō delta shelters as if what they were doing was quite different from anything else they'd attempted in their careers up to that point. Murakami in particular returned again and again to his conviction that the lives of the delta's residents had been placed in their hands and that their work as planners and urban engineers would be judged on whether or not everyone survived the next earthquake.[45] It struck Murakami that, up until this moment in postwar Japan's history, urban planning and the work that he and others like him did were expressions of and responses to economic growth, to production in one form or another. The problem that needed to be solved in Kōtō, however, was on a different register entirely. With tens if not hundreds of thousands of lives at stake, the question Murakami and the others now faced was how to apply the tools of urban planning and design to these new goals of protecting residents from hazards and saving lives.

Kawasumi's timeline comes up often in accounts of their work on the shelter project and, along with the Fire Department's studies and others that followed, had convinced them of two things. One was that they could not hope to protect the district in the usual sense, if what that meant was using the tools at their disposal to keep the fires at bay and homes and factories intact. The other was that the next earthquake's effects on the delta could be predicted accurately enough to justify investing in the complicated, costly, and disruptive countermeasures they believed were the only way to protect the delta's inhabitants.[46] From the start of the process, these urban planners and other experts therefore focused less on fireproofing or other strategies designed to render existing structures less vulnerable and more on a different approach, one that involved creating new spaces—and new structures—that would provide safe haven as the rest of the community burned.

The first designs for the delta's disaster preparedness shelters (*bōsai kyoten*) were released in 1965, a second version followed in 1967, and a third and final iteration two years after that. The first plan in 1965 began with an attempt to calculate how many shelter sites the district would need to accommodate the district's 800,000 residents and where they would have to be located so that a typical family—fleeing on foot—would be able to reach at least one of the sites within a half hour of leaving home. Planners determined that they would need at least sixteen separate shelters carefully distributed throughout the district so that residents would never be more than 1.5 kilometers away from at least one of the sites.[47] There was also the design of the shelters themselves to think about. Assuming that the district's population would be divided more or less equally among the sixteen sites (or roughly 50,000 people each) and that every person would require roughly two square meters of space once inside, then each location would need at least ten hectares of land just for the evacuees. After consulting with fire expert Hamada Minoru, who insisted that no location was safe unless it included a sufficiently wide buffer zone between everyone on the inside and the heat, smoke, and flames they were fleeing from, planners added another fifteen hectares to each site, more than doubling each hub's footprint.

The shelters' designers took the buffer premise one step further by assigning it a vertical component, not just a horizontal one. The planners conceived of each of the sixteen shelters as ringed all along their perimeters by tall, fire-resistant apartment complexes or office buildings. One of the functions of these structures would be to serve as fire breaks, as barriers to the encroaching flames, sparks, and heated gases that had brought such deadly outcomes to so many of those who had sought shelter in open spaces in 1923. Although this idea had not yet been subject to much real-world testing, the planners theorized that sufficiently tall buildings on the shelter's perimeter would direct heat and smoke up and over the site's interior space, protecting those inside from the worst of their effects.[48] The apartment buildings were also essential to the planners' efforts to embed disaster preparedness into the structures of everyday life and to make the shelters present and accessible, day in and day out, whether as a location where many people lived or as open space configured as parks or playgrounds.

The prospect of building the shelters into the community—or recreating the community as a shelter—was one city planners continued to embrace as their designs evolved. By 1969, their original scheme to construct sixteen small shelters had been dropped in favor of a proposal to build six much larger ones.[49] The six sites together would have enough room to accommodate hundreds of thousands of people, if need be. Even what was ultimately the smallest single shelter in the plan—in Shirahige-East (Shirahige-higashi),

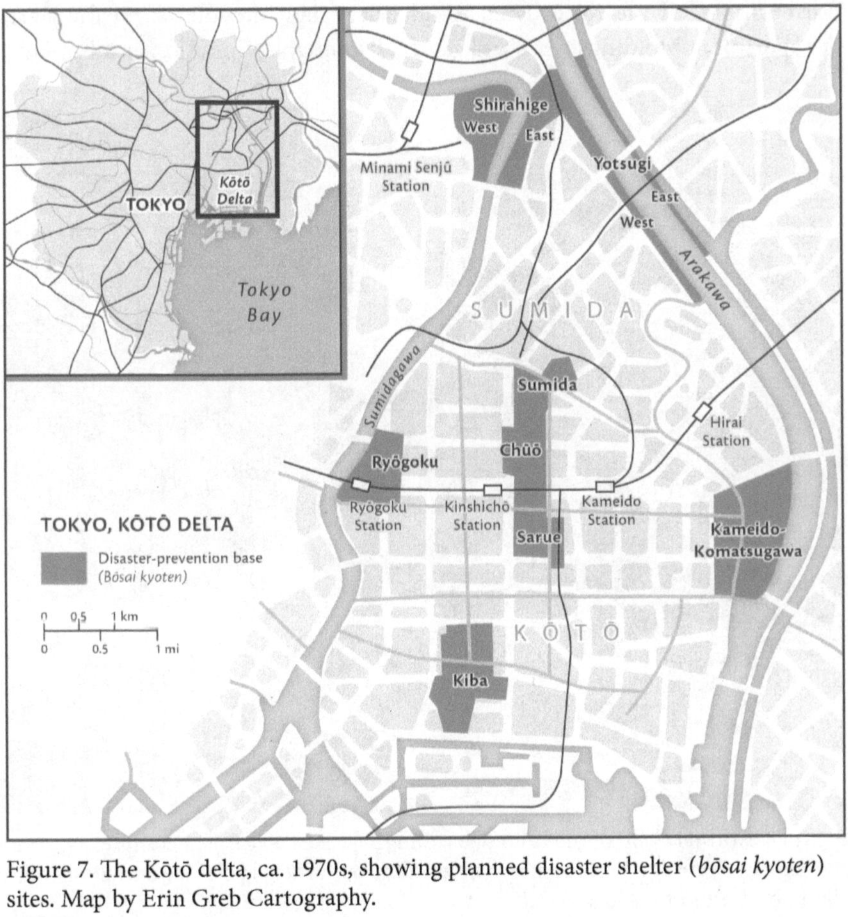

Figure 7. The Kōtō delta, ca. 1970s, showing planned disaster shelter (*bōsai kyoten*) sites. Map by Erin Greb Cartography.

along the banks of the Sumida River at the northwestern edge of the delta—was designed to harbor 80,000 evacuees. (The largest complex among the six was to be built near Ōjima-Komatsugawa, on the east bank of the district. It would take up more than 114 hectares and would safely hold 200,000.) The six locations were distributed such that someone setting out from just about anywhere in the district and walking at a slow pace could reach at least one of the sites within thirty minutes.[50]

The first phase of construction began at the Shirahige-East site in December 1975 and, in the summer of 1982, the new facility was finally ready for the 1,900 families who had been waiting to move in. Their apartments were housed inside eighteen high-rise units lined up in a row that zig-zagged for more than a kilometer alongside one of the main thoroughfares at the northern end of the Kōtō delta. As planners had intended from the start, the apartment buildings were wide and tall enough to function as a fire- and

Figure 8. Northern end of the Kōtō delta. Shirahige-East apartment complex is top-center; Shirahige-West park is visible on the opposite bank of the Sumida River. Photo by the author, 2016.

Figure 9. Entrance portal, Shirahige-East apartment complex. The doors may be sealed to protect those inside. Photo by the author, 2016.

smoke-proof barrier between the interior and the worrisomely flammable neighborhoods nearby. Giant metal doors flanked portals at the barrier's base; once closed (they could be operated remotely) the entire complex could be sealed off from whatever threats the exterior posed. Until then, water nozzles (also remotely operable) mounted over the entryways would help protect anyone still trying to make their way inside from encroaching heat and smoke. Nestled between the high-rises and the Sumida River to the west was a well-kept, spacious park, complete with baseball fields, tennis courts, ponds, and trees. The site in its entirety took up almost twenty-eight hectares (seventy acres); it even had space set aside for stockpiles of food, water, and other supplies sufficient to meet the needs of everyone inside, for days if need be. Its scale was striking in the context of the rest of the built environment in the Kōtō delta and Tokyo's lower city in the 1970s and early 1980s, where one- and two-story shops and houses were the norm. Nothing else in the vicinity of the shelter looked anything like it.

Tokyo's Earthquake Disaster Prevention Ordinance

Kawasumi believed that it was the San Fernando earthquake that convinced the central government and Tokyo's to start thinking about the seismic threat to the capital as a whole and not just the Kōtō delta.[51] The Central Disaster Prevention Council released its *Guidelines for the Promotion of Major City Earthquake Disaster Countermeasures*, which acknowledged that Japan's urban centers were at risk from earthquakes and called for new approaches to land use and city planning in light of those threats, a few months after the disaster in LA.[52] Tokyo issued its own Earthquake Disaster Prevention Ordinance (Tokyo-to shinsai yobō jōrei) in October 1971, based in large part on recommendations that Kawasumi had drawn up after leading the city's survey team to LA earlier that year.

The ordinance offered a blunt assessment of why the city was so poorly prepared for the next earthquake disaster. "Tokyo's development as a city has taken place without regard for its safety," it read.

> As a result, a brittleness with regard to earthquake (and other) disasters has become embedded in the city's structure. In order to protect Tokyo from the disasters caused by earthquakes, we must act quickly to put all necessary measures into place.
>
> That earthquakes are a natural phenomenon goes without saying, and yet most of the calamities they cause are man-made ones. In light of which, human wisdom, skill, and effort should be able to either prevent earthquake disasters before they happen or keep the damage they do to an absolute minimum.
>
> This ordinance expresses the commitment of the people of the city and the city itself to draw on that wisdom and bravery, and it expresses the accord between the people of the city and the city itself to act together in order to protect Tokyo from an earthquake disaster.[53]

If one of the ordinance's arguments was that the capital was poorly prepared to defend itself against a powerful earthquake, another was that the problem could be addressed. Tokyo was at risk not because of choices made by its current residents (or the current leadership, clearly) but instead because not enough attention had been paid in the past to "safety" (*anzensei*) as something a city owed to its inhabitants. Because the worst of those hazards resulted from the choices institutions and people had made about how they interacted with the environment and with each other, it must also be within their power to rectify those mistakes and create a safer city.

The Earthquake Disaster Prevention Ordinance described an ambitious approach to responding to the threats the city faced. This it did in part by adopting almost all of the recommendations that Kawasumi's team had drawn up when they returned from LA, in one form or another. The new directive called for the inspection of and, if needed, the strengthening of police and fire stations, as well as hospital and school buildings (Article 22), prohibited the use of private automobiles during evacuations (Article 42), and even called for the installation of strong motion seismometers in buildings designated by the governor (Article 26). The ordinance also fully embraced Kawasumi's team's insistence on the importance of creating and preserving open space.

Minobe's administration didn't limit itself to the lessons from LA, however, and introduced a number of provisions that hadn't been on the team's list, including several that dealt directly with evacuation sites and the routes that would deliver fleeing residents safely to them. The ordinance required the governor to identify evacuation zones and routes throughout the city and then to make sure that they remained safe as the city continued to develop. Where existing sites or escape routes were insufficient, the city was obligated to do what needed to be done to improve them. Many of the other worries that studies in the 1960s had highlighted—glass falling from tall buildings, the willy-nilly distribution of hazardous substances throughout the city, and so on—were likewise addressed in the new ordinance.

Before the new guidelines went into effect, there hadn't been a coordinated approach to planning for such an event, other than whatever provisions the Disaster Countermeasures Basic Act had made for post-disaster response. After the Earthquake Disaster Prevention Ordinance was in place, Tokyo operated under a new and different set of assumptions about what could be done to protect the city and its residents from an earthquake's effects and about the distribution of responsibility for those protections, from the highest levels of the metropolitan government on down to neighborhood associations and individual families. Those assumptions included something close to a consensus about how much damage a major earthquake would do (potentially a tremendous amount), how much could be done to mitigate the harm, and the necessity to act quickly, before it was too late. The ordinance bundled these elements together and made them visible to the public to an extent that was unprecedented in Tokyo's history.

Kawasumi's Theory under the Spotlight

Kawasumi's "sixty-nine-year theory" about the earthquake risk to Tokyo was resilient.[54] There was no denying its hold on the public imagination, even though his colleagues in the field were at best ambivalent about his claims. At a press conference they called in late June 1973 to announce a third five-year earthquake prediction plan, Hagiwara Takahiro, Asada Toshi (University of Tokyo, Geophysics), and the JMA's Suehiro Shigeji were brought up short when a reporter posed the following question: "According to the sixty-nine-year period theory, we are getting close to the time when a major earthquake is likely to strike Tokyo. So why is it that this earthquake prediction plan doesn't touch on this problem at all?"[55]

Hagiwara's response, after a long pause, was to say that "It's because we've never taken [that argument] as a given."[56] A journalist who witnessed this exchange wrote later that Hagiwara and the others seemed torn between the impulse to criticize the sixty-nine-year theory and their considerable respect, if not affection, for Kawasumi himself, who had passed away only a few months before. Hagiwara's careful choice of words managed to avoid saying anything directly critical of either the idea of the sixty-nine-year cycle or Kawasumi. This was a common strategy within the scientific community, for several reasons. One reason is that, by the time Kawasumi's colleagues began publishing critiques of his thesis, the concept and its implications were thoroughly embedded in both public discourse about earthquake risk and in disaster preparedness planning for Tokyo and nearby communities. Whatever else those in the field might want to say about earthquake prediction or about the risk to Tokyo, they inevitably found themselves having to respond in some way to the set of expectations and preparations that Kawasumi had helped create.

There is another quality to Kawasumi's theory that helped shape how other scientists engaged with it, which is that, although his evidence and the conclusions he drew from it were both covered at length in popular media from 1964 on, many years passed after his initial announcement before an account of his analysis appeared in scientific journals. Some of his data and reasoning had appeared earlier in the appendix of the (rather obscure) 1961 report on measures to conserve the figure of the Great Buddha in Kamakura, but it wasn't until almost a decade later that he published a more thorough description of his evidence and methods.[57] Two articles in 1970, one in *Soil Mechanics and Foundation Engineering* and the other in *Chigaku Zasshi* (*Journal of the Earth Sciences*), offered the first full accounting of his work written for other scientists.[58] The first published critiques of his analysis likewise didn't appear in print until 1970.

Seismologist Usami Tatsuo, a professor at the ERI just as Kawasumi had been, was the most vocal critic of the sixty-nine-year thesis within the scientific community.[59] Usami's own research had made him familiar enough with the data that Kawasumi had used to calculate the periodicity of major earthquakes in the region and with his statistical methodology to take issue with both. Usami argued, in part, that the list of historical earthquakes used to determine the risk to Tokyo was incomplete; since Kamakura was a political and cultural hub long before Edo was, it was entirely possible that there had been many more significant seismic events in what was now the capital district that never made it into the historical record. He also pointed out that several of the earthquakes that were on Kawasumi's list would have been felt only in Kamakura and were irrelevant when it came to calculating the risk to Tokyo. When Usami did his own calculations of earthquake periodicity for Tokyo, his results looked nothing like Kawasumi's. He instead found evidence for a nineteen-year segment and more for another that was thirty-five years long. Usami's more complete and precise data yielded results that were neither consistent nor elegant, qualities that Kawasumi's thesis had possessed in abundance.[60]

Critiquing Kawasumi on his own terms did little to diminish the theory's visibility or, arguably, belief in its utility. The *Yomiuri* ran a series of articles in late 1970 devoted to a close-up look at Tokyo's vulnerability to a major earthquake. One of the first topics the series dealt with was Kawasumi's thesis, which was presented to readers along with a brief history of the professor's career and lifelong commitment to earthquake research. A table captioned "The Kantō Earthquake sixty-nine-year thesis and the facts" provided in one column a list of the major recorded earthquakes in the Kantō district, beginning in 818. A different column showed readers when earthquakes "should have happened" according to Kawasumi's theory and a third displayed the difference between what the theory had predicted and when earthquakes actually occurred. (See Chapter Four, Table 1, for a reproduction.) Depending on how one interpreted the results, they either showed that Kawasumi's claims were without merit—since there was seldom a perfect fit between the date his theory assigned to a seismic event and when one struck—or they showed that the theory was close enough often enough to be useful.[61] The last line in the "The Kantō Earthquake sixty-nine-year thesis and the facts" table read simply "1991" (which fell sixty-nine years after 1922). Where the dates of the actual event and the number of years that it differed from Kawasumi's claim would have appeared there were only question marks.

The *Yomiuri* did let readers know that other scientists had reached different conclusions than had Kawasumi. Usami's analysis of major earthquakes near Tokyo, for example, was reported to have produced evidence of multiple possible periods. But, rather than suggest that Usami's findings raised some significant questions about whether Kawasumi's approach was a reliable method for estimating when the next major earthquake would strike, the *Yomiuri* concluded instead that because Usami had found periods between events that were shorter than sixty-nine years, the district might be at risk well before Kawasumi's "danger time" arrived in 1978.[62] Usami's work, in other words, was described by the *Yomiuri* as building on, rather than challenging, Kawasumi's arguments about periodicity.

The theory benefitted by having Kawasumi as its spokesman. He was its public face and, more often than not, references to it in the press and other venues described the ideas behind the "sixty-nine-year period" as "Professor Kawasumi's theory." That association almost certainly helped make the argument seem more plausible. Kawasumi himself didn't seem to benefit personally in any way from promoting the theory or by advocating for more and better anti-disaster policies. He never seems to have used it, for example, as a justification to ask for more funding for projects of which he was part, or to promote his own career in any way. Kawasumi's sole official appointment once he retired from the University of Tokyo was as director of the earthquake section of Tokyo's Disaster Prevention Council. In any case, the arguments he was most interested in making were not at all dependent on whatever status an appointment in the Tokyo Metropolitan Government might have conferred. He made frequent appearances in the print media after 1964, far more regularly than in any other period in his lengthy career up that point. He was still giving interviews in the months before his death, at age 68, in late 1972.[63] The media's coverage of Kawasumi portrays him in an avuncular light, as a personable, slightly rumpled professor compelled by both his discoveries and his sense of responsibility to the public to take on a role he would not otherwise have sought for himself.[64]

If there were downsides to Kawasumi's association with the theory, one was that direct advocacy for it more or less ended with the professor's death. The sixty-nine-year period argument had never been publicly identified in public or in the field as the product of anyone's work but Kawasumi's. There were no collaborators, institutional allies, or even devoted graduate students to pick up where Kawasumi had left off, or to elaborate on the research that had yielded the patterns in the first place.

That said, echoes of Kawasumi's efforts to make the threat to Tokyo impossible to ignore continued well after his own participation in that struggle came to an end. In an article published a few years after Kawasumi's death, in which Usami was otherwise critical of the sixty-nine-year theory, he made an interesting observation about its impact. "This theory," he wrote, "motivated Tokyo to take counter-measures for prevention of disasters due to earthquakes," some of which, including stockpiling food and water for the millions of refugees an earthquake would create and the identification of safe evacuation routes, were already under way. Others, such as the construction of a half-dozen massive shelters in parts of the city most at risk from fire and flooding, were at that point still in the planning stages. The conclusion that Usami drew on the basis of what had been accomplished so far was that "the intention of Dr. Kawasumi has begun to be realized."[65]

There is an intriguing tension at work in Usami's assessment, one that is also evident in other reflections on the theory and the uses to which it had been put. Kawasumi's obituary in the *Asahi* in late 1972 was devoted almost entirely to his advocacy of the sixty-nine-year theory and to an acknowledgement that it had been subject to growing criticism in recent years. It concluded with these sentiments: "Setting aside for the moment whether the theory is valid or not," the obituary's author opined, "by promoting the sixty-nine-year theory as he did, Kawasumi raised the alarm about otherwise all-too-easily forgotten natural disasters. No one can deny that he deserves a tremendous amount of the credit for the progress that has been made in anti-disaster policies for earthquakes and in research into earthquake prediction."[66] In this reading, too, it doesn't matter much whether the theory offered an accurate description of the risk to Tokyo or not. What mattered was that it pushed policy makers to protect the city and to support science that presumably rested on more solid ground than the sixty-nine-year theory ever had. After Kawasumi's death, some argued that he himself hadn't been all that convinced by his own arguments so far as periodicity was concerned. The *Asahi* suggested that Kawasumi had admitted to those closest to him that he knew for some time that the foundations of his theory were weak and that he didn't think it would hold up to close scrutiny for very long. "I'll probably end up under attack by generations to come, my reputation as a scholar erased," one report has him observing.[67] "None of that means a thing to me," he went on, "so long as the periodicity theory led to progress being made in anti-disaster polices."[68]

Kawasumi is on the record expressing similar sentiments, if not quite so candidly, during testimony before an Upper House subcommittee on anti-disaster policies in 1971. Responding to a request that he elaborate on the status of the sixty-nine-year theory, Kawasumi eventually began talking about what it might mean if it turned out that he had been wrong about periodicity, whether a little or a lot. He shared several scenarios with the legislators but the question that was key to each was this: under what circumstances would it be safe to do nothing (or very little) to try to protect Tokyo and the rest of the Kantō district from the damage a massive earthquake might do? If all his assumptions and calculations were correct, for example, and it was more than 99 percent likely that destructive earthquakes near Tokyo occurred with a sixty-nine-year interval, then there seemed little doubt that the right thing to do—indeed, the necessary thing to do—was to take steps to mitigate the harm that the city's residents might suffer. If his calculations were slightly off or if he had missed some data and the probability that he was right fell to, say, 70 percent, Kawasumi argued that the same imperative to act applied.[69] "Even if it turns out that I was mistaken about the proof of my theory or some other issue," he argued, "that is a responsibility that I alone will have to bear." But when it came to the question of whether or not to institute policies to deal with the earthquake hazards, the costs of getting it wrong would ultimately have to be borne by the entire nation. "I may have made a lot of mistakes," Kawasumi went on,

> but the real problem here is that the less we do to put policies in place, the greater the destruction is going to be. It could well be that predictions based on my theory will be off by a bit and will happen a little earlier or a little later [than expected]. Those possibilities cannot be ruled out. But even if that is the case, the nation can't very well pass up the opportunity to put policies in place that will make a difference, can it? . . .
>
> If I'm wrong, and we keep going along without any major earthquakes, my own sense right now is that we should be happy to have had this opportunity to do more to prepare. For me, I'd be very grateful indeed if it turns out that this sixty-nine-year interval, the interval I've been referring to, never materializes at all. It would be fine with me if my theory were to fall apart, to just fall totally apart with no harm coming to anyone, or to fail on a smaller scale and have the timing pushed back. But, if we were to do nothing in the

> way of policy during that delay and just keep putting it off, I believe it would be absolutely terrible if a destructive earthquake were to happen in the future under those circumstances. It is to prevent such an outcome that I am compelled to ask you to do everything in your power to pursue these policies in as organized and timely a manner as possible.[70]

The fears about Tokyo's future that Kawasumi helped articulate became the scaffolding around which the next several decades of discussions about preparations for the next disaster were built. Tokyo's era of waiting for its next disastrous earthquake lasted longer than the Cold War, longer than the Meiji period, and, as of 2023 almost as long as the Shōwa era. Before Kawasumi shared his "sixty-nine-year theory" with legislators, what studies there were of the harms an earthquake might do to the capital district had not yet been linked to a convincing argument that one was likely in the short term. Kawasumi's warning in the summer of 1964 provided one and, from that point on, the city's future has never not had the threat of an earthquake disaster in it.

CHAPTER 7

The Making of the Tōkai Earthquake

Most earthquake disasters come to our attention after the fact, their histories a record of the damage done and of the recovery that follows. The Tōkai earthquake was different, in that it was the focus of intense study and had a powerful impact on how Japan planned to manage future earthquake risks, all before the actual seismic event that gave the disaster its name ever occurred. When Ishibashi Katsuhiko first raised the alarm about a potential Tōkai earthquake in 1976, he and several of his fellow scientists suggested that it could happen as soon as the next day, or it might be a decade or two or three before it struck. Residents of the Tōkai district and, arguably, the rest of the country, too, spent the next forty years waiting for the real one to arrive. As of 2023, it had not. This chapter shows how the Tōkai earthquake came to be, both as a geophysical event that scientists felt they could describe with reasonable precision and as a public construct—an as yet imaginary disaster on the cusp of becoming all too real.

The possibility that a powerful earthquake would strike offshore of the Tōkai district had been on the Yochiren's radar since its formation. Ishibashi's 1976 warning followed several years of discussion within the Coordinating Committee on Earthquake Prediction and in public forums about how much of a threat such an event might pose and when to start worrying about it. The Yochiren added parts of the Tōkai district to its list of areas needing "special observation" in 1969, based in part on evidence brought to its attention by one of its members, the ERI's Mogi Kiyō. Mogi had used Geographical Survey Institute measurements going back to the 1890s to show that parts of the western edge of the Izu Peninsula and other locations along Shizuoka's coastline were shifting inexorably toward the continent. Some locations appeared to have moved a meter or more in that direction over the course of seventy years.[1] The conclusion that Mori drew from that data was that sections of peninsula and coastline nearby were being compressed, squeezed between a length of Honshu's unmoving mass on one side and the inexorable force generated by the slow spreading of the sea floor on the other. Eventually, Mogi argued, somewhere within that zone of mounting pressure, the crust would shatter. Unlike most of the other parts of Japan that the Yochiren was thinking about designating for close observation, the zone Mogi was talking about had a history of producing M8-class earthquakes.

When the popular magazine *Shukan Gendai* described the hazards facing the district in December 1969, it used more or less the same framework that other journalists had been using to talk about what would happen in Tokyo when the next big quake struck there. Many residents of Shizuoka, Aiichi, and the Tōkai district more generally had suffered through powerful earthquakes relatively recently, in 1944 and 1946, and, although neither of those events were widely known outside the area (censors prohibited reporting the extent of the damage or even the exact location of the 1944 earthquake), many in the affected towns and villages remembered the events all too well. One of the points that journalists and other observers emphasized in 1969 was that, as in Tokyo, communities in the Tōkai district were vulnerable now in ways that they never had been before. Crowded cities, new industries, *kombinato* (industrial complexes, often petro-chemical), elevated roadways, fast trains, and any number of other products of high speed growth would all be at risk from an M8 event. As in the detailed scenarios imagining Tokyo's destruction, these same technologies and conveniences of modern life would turn into formidable hazards in the event of an earthquake. And just as Kawasumi and others had argued that a major quake in Tokyo in the 1960s would do more harm than the one in 1923 had, the *Shukan Gendai* put it to readers that Mogi's earthquake would make the disasters of 1944 and 1946 look mild in comparison.[2]

After Mogi made his case and the Yochiren designated the Tōkai district for monitoring, the next time the public was directly confronted with the prospect of a major earthquake in that area was in 1973. In late April of that year, Rikitake Tsuneji and Hagiwara Takahiro both testified before the Lower House's Special Committee on the Advancement of Science and Technology (Kagaku gijutsu shinkō tokubetsu iinkai) about the Yochiren's activities and the state of earthquake prediction in Japan, in general. The Yochiren was getting ready to ask the government to fund a third five-year earthquake prediction plan, so Hagiwara and Rikitake were both anxious to strike a positive tone, while at the same time emphasizing that their goals would likely remain out of reach so long as funding remained at its current paltry levels.[3]

Asked by a legislator to name the locations in Japan they thought most at risk of a major earthquake, Hagiwara answered with what some might have seen as evasiveness and others as an accurate assessment of the situation they were in. The danger was everywhere, he said. There was almost no place in Japan he could think of that was not at risk of an earthquake, and so he hesitated to pick just one or two locations. Rikitake was less circumspect. He told legislators that researchers were focused on two high-risk sites. One was offshore from the Nemuro peninsula on Hokkaido's east coast. Powerful earthquakes in that region were not uncommon, relatively speaking, so Rikitake wouldn't have been going out on much of a limb to suggest that one might happen there soon, but, in this instance, he was drawing on recent work by Yochiren member Utsu Tokuji, then at Hokkaido University.

Utsu had identified six or so earthquake source regions off Hokkaido's coast, which he mapped as ellipses running in sequence along a southwest-by-northeast axis, beginning in the bay between Aomori and Hokkaido and extending well into the Kurile island chain, far to the north. Utsu associated each of these source regions with clusters of M7.9–M8.1 earthquakes going back many years in the historical record. Since 1952, Utsu noted, earthquakes of that same magnitude had occurred in five of his six source regions. The only one where there hadn't been an event of that magnitude in the past few decades—Utsu's region "C" on the map—was adjacent to the Nemuro district. There had been an M7.9 event in that zone in 1894, but nothing since. Utsu was also able to map the distribution of shallow M5 or larger earthquakes offshore Hokkaido and northeastern Honshu for the period 1961–1970. When the map was done, it showed some sectors with a great deal of activity and a few with noticeably fewer earthquakes, but there was only one—region "C"—where there essentially had been none at all.[4]

One of the conclusions that Utsu drew from his study, and which Rikitake echoed in the Diet in April 1973, was that the next M8 earthquake in Hokkaido's vicinity would probably be near Nemuro. There was no certainty about the timing, and actually Utsu expected that they would have some warning in the form of a spike in the number of M6 events in region "C" before the more powerful earthquake struck. When Rikitake identified Nemuro as a likely location of one of Japan's next big earthquakes, like Utsu, he avoided implying that they knew when to expect it, exactly.[5]

The Tōkai district was the second location that Rikitake told legislators should expect an M8 earthquake sometime soon. Here he was drawing on Mogi's observations from a few years before and on measurements since then offering more evidence that strain was continuing to build along segments offshore of the Tōkai coast. In a paper Rikitake finished around the time of his Diet testimony in April 1973, he calculated the probabilities for a series of time intervals of a major earthquake occurring in the southern Kantō region, off the eastern Hokkaido coast, and off the Tōkai coast, based on his own estimates of current crustal strain and the upper limits of the sites' capacities to absorb it.[6] For the southern Kantō area, where measurements of crustal deformation were relatively easy to obtain, Rikitake concluded the probability of a major earthquake in the short term was low, at present. For Hokkaido and the Tōkai district, there was no way to measure crustal deformation at the focal points of the earthquakes, themselves, and Rikitake was quick to acknowledge that there was still a great deal of uncertainty about how stress accumulated during subduction or how faulting occurred. That said, the conclusion he reached for both locations was that "the probabilities of recurrence of a large earthquake in these two areas are shockingly high."[7] "How to evaluate the probabilities thus estimated," he continued, "depends on one's subjective mind. But the writer cannot help feeling that the area off eastern Hokkaido and the Tōkai district are in danger of being affected by a large earthquake in the near future."[8] Rikitake conveyed those same feelings to legislators when he spoke with them in April.

A little less than two months later, shortly before one in the afternoon on June 17, people in communities across eastern Hokkaido were shaken by a powerful earthquake. Reports of damage to homes and buildings in the Nemuro district confirmed what the JMA's instruments had recorded, which was that the earthquake's focal point was southeast of the Nemuro peninsula, or just about where Utsu and later Rikitake had said that the next big quake would start. At M7.4, the earthquake was considerably less powerful than the M8 event the scientists had earlier warned of (that gap was a source of concern going forward, since it raised the possibility that not all of the built-up energy in the vicinity had been released), and although the damage to the built environment was significant, injuries, fortunately, were few.

The Yochiren's scientists would probably have gained the support they needed for a third five-year plan for earthquake prediction, regardless, but Hagiwara and the others didn't hesitate to promote their cause in the Diet hearings and press coverage that followed the Nemuro quake. It might seem paradoxical to suggest that the prediction campaign had more credibility after the disaster in Hokkaido—which it had, after all, failed to predict—than it could have claimed before. There were other reasons coming into focus around the same time for scientists and policy makers to be optimistic about prediction's capabilities, as we'll see, but the circumstances of the Nemuro quake were also understood to have affirmed prediction's potential in Japan. The Yochiren had designated the region off Nemuro's coast for special observation, years earlier. Had a powerful quake occurred instead somewhere well outside one of the committee's designated zones, the prediction initiative would have had to defend its judgement, not least about whether they could be sure they were looking in the right places for the evidence they sought. Since the Nemuro earthquake happened more or less right where the committee had said it would, its judgement on those grounds was not in doubt. It was reasonable to assume too that if the committee's scientists had been correct about the risks to Nemuro, they were probably right about the risks to the other designated districts.[9]

After the earthquake in June, Rikitake made sure to remind everyone about the warning he had given in the Diet back in April. Meeting with reporters within hours of the Nemuro earthquake, Rikitake said that his first thought upon hearing the news was, "So it's happened, as I expected it would."[10] He was careful not to imply that he had known with any certainty when the quake was coming, but it was clear he was not at all shocked that it had. As Rikitake explained it, borrowing from Utsu's argument as he did, the absence of a major earthquake off Nemuro for so many years, along with reports of continued declension along the shoreline there, had convinced him that it was really only a matter of time before the overdue quake hit.[11] And, he went on, in the aftermath of the Nemuro quake, he found himself even more worried about what might happen along the Enshū Sea (Enshūnada) coast in the Tōkai district. As had been the case in Hokkaido, the situation there included a lengthy period of inactivity in an area with a long history of powerful, M8 quakes, as well as signs that the pressure exerted along that segment of shoreline was building at an alarming rate. In light of those findings, in Rikitake's opinion, it was imperative that the Enshū Sea region be subject to intense scrutiny going forward.[12] The *Yomiuri* helped convey both points with its choice of headlines for the article on Rikitake—the first of which read, "Warning Right on the Money—Nemuro," and a second, quoting the professor himself, "'I'm scared that Enshūnada will be next.'"[13]

For most Japanese, the significance of a threat to that part of the country would have been immediately obvious. The Enshū Sea area was one sector of the much larger Tōkai district, which includes Aiichi, Shizuoka, and Mie and Gifu prefectures, and which was itself part of the vast urban conglomeration stretching from Tokyo to Kobe, a densely populated and heavily industrialized belt along Japan's Pacific coast.[14] The Tōkai district had been key to the functioning of early modern Japan's political structures and economy (the Tokugawa regime's main trunk road connecting Edo with points west, such as Kyoto and Osaka, ran through it) and was at the center of many of the networks and industries that shaped twentieth-century Japan. It would be difficult to exaggerate the district's economic importance. In 1970, around two-thirds of Japan's entire manufacturing output was concentrated along the Pacific Belt, of which the Tōkai region was a large part. Even during an era of especially rapid urbanization in Japan, population growth in the Tōkai district had outpaced that in the rest of the country throughout the 1950s and 1960s, so that roughly 13 million people were living there by the mid-1970s.[15]

The Enshū Sea Earthquake

At just about the same time that Rikitake was calling attention to the potential for an Enshū Sea Earthquake, Andō Masataka, a graduate student in geophysics at the University of Tokyo, was adding an important layer to the field's understanding of seismicity along the Nankai Trough. Andō's work showed that the fault region along the trough appeared to be made up of "four parts, A, B, C, and D, from west to east, which are more or less decoupled mechanically from each other." Andō's dissertation research suggested that each of these four segments could rupture independently of the others and that they could also slip together in blocks.[16] In 1707, according to Andō, all four segments moved more or less simultaneously, resulting in the massive Hōei earthquake. In 1854 the A-B and C-D segments ruptured separately from each other but, as in 1707, all the energy that had accumulated along those four segments appears to have been released. In 1944 and 1946, however, Andō showed that only three of the four segments had behaved as expected. Where segment D's earthquake should have been, assuming that the previous patterns were still relevant, there was only a worrying "seismic gap."[17] If previous patterns held, then at some point in the relatively near term Japan could expect an earthquake somewhere along segment D. Andō suggested that it would be a powerful and potentially destructive one, when it happened.[18] Based on the effects of the 1854 Ansei and 1944 Tōnankai earthquakes, Andō calculated that the one they were anticipating would generate tsunami with waves as high as five meters at some locations along the coast. A large swath of the region, starting

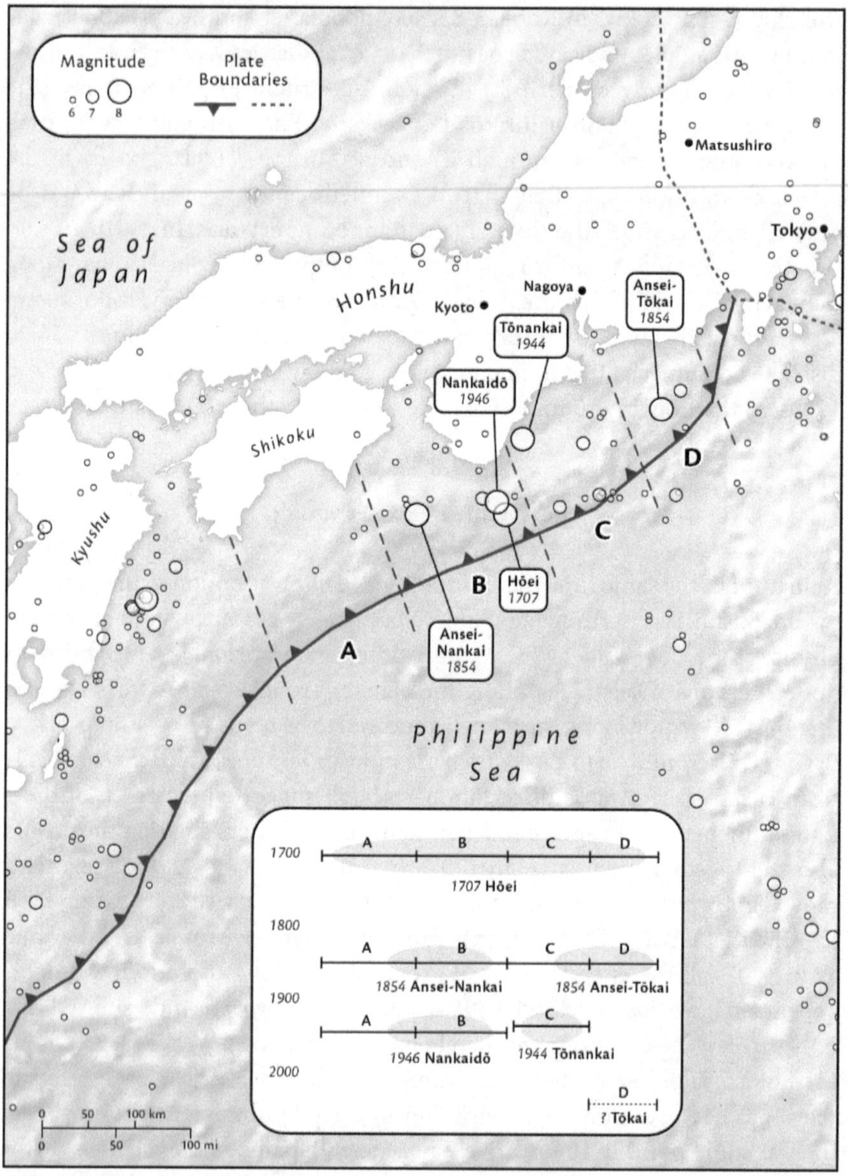

Figure 10. Major earthquakes and fault segments along the Nankai Trough, 1700–2022, as described by Andō Masataka in the early 1970s. Map by Erin Greb Cartography.

with the Kantō plain and Tokyo in the north and extending to parts of eastern Mie prefecture to the southwest, would experience shaking of JMA Intensity Level 5; communities along the Enshū Sea coastline, along the edges of Ise Bay, and north into the Nōbi Plain would be harder hit; they would experience the earthquake as an Intensity Level 6 event.[19]

Andō wasn't the first, by any means, to draw attention to "missing" earthquakes and their implications for the communities that had so far benefitted from their absence. Utsu had done so to good effect in his work leading up to the Nemuro earthquake, for example, as had Rikitake. Writing in a post-Nemuro context of growing concern about a potentially destructive earthquake in the Tōkai region, Andō affirmed those earlier warnings about the risk ahead for the Enshū Sea district. At the same time, Andō also helped change the terminology and basic frameworks for describing seismicity in the district to reflect the field's embrace of plate tectonics. From the moment that move was complete, warnings like the one Andō had issued about the Tōkai district would be validated by how well plate tectonics was working to explain so many of the other phenomena that Japanese seismologists encountered on a regular basis.

The analyses and concerns that Rikitake and Andō brought forward in 1973 were in keeping with what the Yochiren had been saying for some time about which parts of Japan were most at risk and with the committee's strategies for addressing those threats. Hagiwara announced in early 1974 that the committee had decided to step up monitoring efforts in the Tōkai zone, which would join the capital district as one of Japan's two "enhanced observation regions," effective immediately. Hagiwara explained that the decision to change the district's status had been based less on any single piece of new information than on the gradual accumulation of reasons for the committee to take a closer look.[20] The Yochiren's measured response as new information about a potential Tōkai earthquake came to light implied that more drastic interventions would be unnecessary.[21] Acknowledging that the district merited more attention also had the effect of assuaging some of the public's post-Nemuro anxieties about an Enshū Sea earthquake. The media's interest in a possible event in the Tōkai district was most evident in the months following the Nemuro quake, with a few bursts of coverage again when the Yochiren began its discussions of the issue later in 1973. After the new year and the committee's announcement in early 1974 that it was stepping up monitoring in the area, the Enshū Sea earthquake dropped off the media's radar. Within the scientific community, interest in the topic seems to have diminished as well, both as a subject for debate and as a potential research focus. One scholar described the atmosphere among earth scientists at the time as generally skeptical of the idea that a Tōkai earthquake would happen anytime soon.[22]

Ishibashi Katsuhiko and the Making of the Tōkai Earthquake

Ishibashi Katsuhiko remembers that his theory about a Suruga Bay earthquake started out as a hunch. It came to him a few days after he presented the results of his dissertation research at the Seismological Society's 1975 spring conference.[23] Ishibashi's work up to that point had focused on seismicity associated with the collision of the Philippine Sea and Eurasian plates beneath eastern Yamanashi Prefecture. Those two plates interacted differently there than they did offshore and along the Nankai Trough. Beneath Yamanashi, the plates were colliding; along the Nankai Trough, the Philippine Sea plate subducted beneath the other. The subducting sectors of the plates regularly produced very powerful earthquakes, including ones like the Enshū Sea event that Andō was concerned about. The colliding sectors of the plates, as under Yamanashi, were also seismically active but the earthquakes that resulted from movement at those boundaries typically released much less energy than those offshore.

Ishibashi's "hunch" had to do with how the same two plates were interacting beneath Suruga Bay. The consensus in 1975 was that the Philippine and Eurasian plates were colliding under the bay, just as they were to the northeast far below Yamanashi. That was an important determination; it meant that the plate boundary beneath the bay would not be the source of a significant release of energy in the event of an Enshū Sea earthquake. Experts could focus just on what would be released by offshore, subducting sectors of the plates when estimating how much damage that event was likely to do. Andō was among the scientists who had looked for geomorphological evidence that subduction—not collision—was taking place beneath the bay, but had found none. "It is assumed," Andō wrote about the situation under Suruga Bay, "that no major earthquakes related to the plate boundaries will occur there in the future."[24]

Ishibashi wasn't so sure. He hadn't been paying that much attention to the discussions that Andō's research had triggered but, as he started to, it occurred to him that his colleagues perhaps had been too quick to discount the possibility that the Philippine plate was indeed subducting along the

Suruga Trough.[25] Signs that he might be right surfaced soon after Ishibashi began reading through local accounts and other historical records from around the time of the 1854 Ansei-Tōkai earthquake. He was looking for reports of damage or other evidence consistent with the effects of an underthrust event along the plate boundary and found several. One was a description of a section of the coastline between Okitsu and Yui, at the northwestern edge of the bay, that was suddenly much higher relative to the waterline after the 1854 earthquake than it had been before.[26] At Yui, Ishibashi put the vertical displacement at just over a meter and, at Okitsu, at around three.[27] Eventually, he found more than a half dozen references to locations with vertical displacements of between one and two meters at sites all along the bay's western edge, from Omaezaki to Fuji. He also uncovered reports of tsunami with waves that were too big to have originated anywhere else but within the bay itself. Ishibashi was soon confident that the plate underthrust associated with the 1854 earthquake had extended past the Enshū Sea coast all the way to the northern end of Suruga Bay.

His model of the 1854 Ansei earthquake differed from Andō's in important ways, but the two were not so far apart when it came to descriptions of what had and hadn't happened along the fault zone during the more recent 1944 Tōnankai earthquake. Both were of the opinion that in that event, the fault zone began in the Kumano Sea, off the Mie coast, and terminated somewhere near the western edge of the Enshū Sea.[28] That assessment had been the basis for Andō's warning that the rest of the plate to the east and up to the mouth of Suruga Bay hadn't moved since 1854 and that it might soon. He expected such an earthquake to have a powerful impact along the Enshū Sea coast and to the southwest. The earthquake that Ishibashi's theory made possible, however, would have a much larger underthrusting fault zone than the one in Andō's scenario and would be considerably more powerful, in absolute terms. Size matters in earthquakes and Ishibashi expected this one to top out somewhere between M8.2 and M8.3. What mattered even more—and this is the aspect of his discovery that likely alarmed Ishibashi most—was that a Suruga Bay earthquake had the potential to inflict much more damage on many more communities than Andō and others had imagined. The disaster that he had conjured up was a different animal entirely than the ones his colleagues had said to expect.

Ishibashi studied the data and refined his theory for about a year before he was ready to come forward with his findings. In May 1976, Ishibashi was thirty-one years old and had been done with graduate school for all of two months. He was well aware that his youth and status would influence the field's response to his Suruga Bay earthquake. It would also have occurred to Ishibashi that coming forward with an argument as provocative as his was sure to influence the field's response to him—not necessarily for the better. If it turned out that he had read the data incorrectly or made other errors along the way, his career prospects could take a sudden and public turn for the worse. It was hard to know what the odds of that happening might be, but Andō had worked for years on his analyses of the tectonics along the Nankai Trough. Ishibashi had mapped out his new theory in much less time. Even if his findings held up to the scrutiny they were sure to get, there was no way to know ahead of time how senior figures in the earth sciences would react once they realized what agreeing with Ishibashi would require of them.

So, Ishibashi was careful and he was patient. He brought the Suruga Bay earthquake theory to Asada Toshi first, then to the Yochiren in May, and finally to the public and the field as a whole, a process that took roughly six months from start to finish. He didn't have to make any significant adjustments to the theory along the way; it held up well to his colleagues' scrutiny. Ishibashi's views on what the scientific community or policy makers ought to do in response to the theory, on the other hand, did change over time. He started out with calls for the Yochiren to take immediate action and designate the Suruga Bay coastline as an "intensive observation district," a category that the Yochiren had yet to use even once.[29] Ishibashi also made it clear that he didn't think the prediction initiative as it was currently constituted was up to the task of dealing with a Tōkai earthquake. "There are many obstacles to contend with," he wrote, "but in the end we are going to have to get to the point where we can successfully issue a warning just prior [to an earthquake]. Unless we want to miss this once in a lifetime chance and let one hundred years of seismology go to waste, then we had better come up with way to make this work for real [*jissen taisei o totonoeru*]." "The main event," he told them, "has already begun."[30]

A little more than two weeks went by between the first time the Yochiren learned about Ishibashi's concerns and his initial discussions of his theory with groups of senior scientists. He met on June 10 with representatives of the Geographical Survey Institute and then the following month in quick succession with different working groups in the Yochiren, including the one Asada Toshi was responsible for and that focused on prediction in the Tōkai district.[31] Ishibashi had little to say about what those meetings were like or if he felt that his colleagues were generally sympathetic to his ideas or intent on finding flaws in them (and him), but he was clearly not discouraged by whatever critiques might have been coming his way. He continued to try to pin down exactly where the plate boundaries were in the district and to revise his fault zone models accordingly. Ishibashi had given some more thought to what it was the scientific community and the state needed to do be doing to prepare for a Suruga Bay earthquake and was ready to link those recommendations to his theory. He wanted to be ready to make the best case he could in front of the entire Yochiren at its next full meeting, scheduled for late August.

Ishibashi faced difficult choices that summer about when and how to reveal what he knew to the world outside his small circle of colleagues. How long should he wait to warn the communities that didn't know they were at risk? Or the fire and police agencies that were going to have to rethink their disaster planning—when should they be told? Ishibashi was confident that he understood what the evidence was telling him and that the threat of a major, destructive earthquake in the Tōkai district was quite real. He was also reasonably sure that it wouldn't happen right away, relatively speaking. There had been an M6.4 earthquake in Shizuoka thirteen years before the 1854 Ansei quake and another slightly more powerful one near Odawara roughly two years prior. Assuming that a pattern similar to that one would repeat before the next very large event in the area, there was mixed news to consider. Shizuoka had been shaken by an M6.1 earthquake in 1965 and, for the last several years, crustal deformation in the Izu district had been happening at an accelerated rate. One way to interpret those results was that they could expect another quake in the M6 range sometime soon and the much more destructive M8 event a year or two after that, if the previous pattern repeated itself. The conclusion that Ishibashi appears to have drawn under those circumstances was that if he had to, he could wait a few months before going public and it wouldn't make a significant difference in how much time people would have to prepare.

Ishibashi also made a deal with Tamura Kazuko, a reporter from the Kyōdō Tsushin's (one of Japan's largest news agencies) science desk, who had reached out to him earlier in the summer to do a story on his new theory. (The journalist had heard about it from Rikitake Tsuneji after the Yochiren's meeting in May.) Ishibashi was at first hesitant about cooperating with the reporter, since it wasn't exactly a given that the Japanese press could be counted on to take the time to get the details right, especially when the topic at hand was a potentially catastrophic earthquake in a densely populated part of the country. It was only after Asada assured him that Tamura could be trusted that

Ishibashi agreed to confide in her.[32] The arrangement Ishibashi and Tamura came to was that Ishibashi would share what he knew with her but only if Tamura promised not to go public with the story until after Ishibashi had made his presentation to the Yochiren at its next meeting, in late August. It was a lot to ask. Two months was a long time for Kyōdō to have to sit on the piece, especially when there were no guarantees that one of their competitors wouldn't break the story before they did. Ishibashi insisted—it was important that he had a chance to brief the Yochiren on his theory and the warnings that accompanied it before either were shared with the public. Even though Ishibashi had discussed his work with colleagues and in meetings of the Yochiren's subcommittees, he hadn't presented it to the entire committee and couldn't do so until it met in August. He was under no formal or legal obligation to share his work with the committee before he made it available to the press, but Ishibashi wanted to make sure that that was the order in which events unfolded—Yochiren first, then the press and the public.

It wasn't hard to imagine what would happen if the Suruga Bay earthquake theory showed up in the headlines before Ishibashi had had a chance to talk to members of the Yochiren. In that scenario, when reporters reached out to the best-known scientists in earthquake prediction, as they were sure to do, the best Ishibashi could hope for was that Hagiwara and the others wouldn't dismiss his ideas out of hand. Asada and Rikitake would likely be supportive on the record but other scientists the press contacted would say that they had never heard of Ishibashi's theory or that they had but either disagreed with it or that they were withholding judgement until all the facts were in. The media would end up with a story about how Ishibashi's theory was the product of something other than "normal" scientific practices—that it was stirring up controversy in the field. All it would take would be a few quotes from panicked residents of one of the communities that Ishibashi claimed a Suruga Bay earthquake could damage and whatever chance he might still have had of being taken seriously would be as good as gone.

The arrangement with the reporters from Kyodo was clever in the way it kept the Yochiren's options open at the same time that it helped guarantee that the committee would have to do something. So long as the news report came out after the Yochiren had met and formally engaged with the Suruga Bay earthquake problem, the ball started out in the committee's court, not the public's and not the media's. Had the order been reversed and publication preceded the meeting, then the committee would have had no choice but to respond to the most problematic or alarmist reports by the press. Its chance to control the narrative would be gone. Historically speaking, neither the Yochiren nor its predecessors had shown much of a willingness to support new approaches to prediction that their members hadn't had a hand in making. If the committee had been forced to react to Ishibashi's ideas under circumstances such as those, the odds of the new theory being taken seriously, much less acted on, would not be good. Arranging the theory's public debut so that the Yochiren got to it before the public did, on the other hand, improved those odds considerably.

When the Yochiren's August meeting finally took place, Ishibashi's was one of half a dozen presentations that dealt with the Tōkai district that day, but it was the only one to suggest that there was any reason to be concerned about the possibility of a major earthquake in the area. His report reiterated the reasons he had offered back in May explaining why he believed a Suruga Bay earthquake was possible. Because they were fortunate enough to have records going back many years that documented crustal deformation and changes in the distances between different sites along the bay, he had been able to use methods developed by Rikitake and others to estimate how much strain had built up in those fault zones since the Ansei earthquake. His calculations revealed, he told the Yochiren, that "the amount of crustal strain accumulated in the Suruga Bay zone is estimated to be roughly equal to the level at which release will occur, based on the fault model. It is also believed to be equivalent to the upper limit on strain as per Mogi [Kiyō]."[33] Ishibashi conveyed the rest of what he wanted the Yochiren to hear in language that was as straightforward and confident as his matter-of-fact observation that the necessary conditions for a Suruga Bay earthquake appeared to have been met.

His report in August made few direct demands on the committee. The closest he came to suggesting that there was something to be done followed this warning: "There is a risk that a 'Suruga Bay earthquake' is imminent. More specifically," he went on, "and based on the results of long-term forecasts, the concern is that we are now at the point where it would come as no surprise if precursory phenomena (assuming there are to be any) were to begin at any time."[34] He had concluded that the Tōkai earthquake would announce itself in one of three ways: precursory phenomena could begin and continue for some time before the primary event; precursory phenomena could begin but last for only a short time before the primary event; and finally, there might not be any detectable precursory signs at all before the Tōkai earthquake struck. All he asked of the Yochiren at this point was that they move quickly to begin "intensive observation" (*shūchū kansoku*) in the district, so that they stood a better chance of noticing any precursory signs, if they started to appear. Ishibashi otherwise seemed willing to let the implications of his findings sink in and for the committee to decide on its own what its next steps should be.[35]

At the press conference after the meeting, no one said a word about Ishibashi's new theory. Hagiwara didn't mention it or concerns about a Tōkai earthquake when he summarized the meeting for reporters. Nor did any of the journalists bring it up. Almost all of the questions that day had to do with recent earthquake activity in the vicinity of the Izu peninsula, and the committee's assessment of how long it was likely to continue. Reporter Kawabata Nobumasa, who was at the press conference that day on assignment from Shizuoka Broadcasting, remembers filing his copy on exactly that topic as soon as the conference wrapped up, fully expecting it to be the lead story on that night's local news. Instead, the headline story that evening was that a young University of Tokyo researcher named Ishibashi had warned at that day's Yochiren meeting of the possibility of a major earthquake centered on Suruga Bay. Kawabata was stunned, both by the implications of what he was hearing for Shizuoka and because he didn't understand where the story had come from. Journalists weren't allowed to observe Yochiren meetings when they were in session. All he and his colleagues normally knew about what had gone on in them was what they were told at the press conferences afterward by Hagiwara or other spokesmen. But Kawabata was sure that nothing at all had been said that afternoon about Ishibashi, much less about a Suruga Bay earthquake.[36]

The answer, of course, was that they had all been scooped by Tamura and her colleague at Kyōdō. The two had timed the release of their story so that it came right on the heels of the Yochiren meeting, which explained how it had preempted Kawabata's report so effectively. It was a clever move. It made news of Ishibashi's theory available to broadcast media in time for their evening reports and gave Kyōdō's affiliate newspapers and possibly others, too, plenty of time to decide whether to include it in the next day's morning edition. Ishibashi remembers that the media did pick up the story right away, and that it was soon receiving coverage in local and regional papers throughout Japan. As we'll see, the interest in a possible Suruga Bay earthquake was intense from day one in Shizuoka, for obvious reasons. Interestingly, though, the fact that the story started out as a Kyōdō scoop made a difference in how the large, national media outlets reacted to it at the time. Kawabata put it this way: when a competitor delivered a scoop, other newspapers would almost never cover it themselves. They'd normally do so only if editors thought the story was important enough to assign their own reporters to it, in the hope that by taking a different approach they could turn it into an even bigger, but unique, news item.[37] Most of the major dailies chose option number one and just didn't mention Ishibashi or the Suruga Bay earthquake, at least not right away.[38]

The story did receive some early coverage in major markets. The *Asahi* summarized Ishibashi's findings in an article nestled between an interview with Hagiwara about earthquake prediction (in which the Tōkai quake is not mentioned) and a discussion of Chinese methods of earthquake prediction, for example. It isn't clear whether anyone at the paper actually spoke with Ishibashi before running the article, which reads as if it had been cribbed from someone else's summary of what he had shared with the Yochiren. (The *Asahi* does deserve credit for its artwork showing the extent of the area that would be affected by a Suruga Bay earthquake; it may have been the first major paper in the country to share that with readers.) The article covers the basic components of the theory, from Ishibashi's evidence that the Suruga Bay was part of the Ansei Tōkai earthquake, to his claim that the Tōkai district could be at risk for another powerful event in the near future. Its tone is detached rather than alarmed. "Ishibashi's idea has not been accepted as an established theory by any academic society," the article concludes, "but flexible countermeasures that reflect developments in research are to be hoped for."[39]

The reaction to "Ishibashi's idea" in the Tōkai district itself, on the other hand, was immediate and intense. The front page headline in the *Shizuoka shinbun* on the morning of August 24 read, "Massive Suruga Bay Earthquake Predicted—Jishin Yochiren—Shocking Report by a Young Researcher."[40] Other local media outlets were equally enthusiastic, if that is the right word, and news spread quickly about the theory and its potential impact on communities in the area. The Numazu City–based Suruga Bank had asked Ishibashi months earlier to come speak as part of the company's public lecture series and had settled on September 3 as the best date for the event. The invitation predated Ishibashi's new notoriety and bank officials reportedly had no idea when they first reached out to him that his work was about to attract so much attention.[41] Ishibashi appeared before a packed auditorium that Friday evening. Some in the audience took careful notes, while others had portable tape recorders and were using those to capture as much as they could of the lecture and the questions and answers that followed.[42] Kawabata, who was in the audience and reported on the lecture afterward, thought that the talk went well and that Ishibashi did a good job of explaining why he had come to the conclusions he had about an earthquake in Suruga Bay.[43] He didn't talk down to his audience, but it wasn't easy to find a balance between making sure his listeners understood what a Suruga Bay earthquake might mean for them (and encourage them to prepare for it) and giving them honest answers to the question on everyone's mind: when was it going to happen?

What he told them at the lecture in Numazu was that "It wouldn't be a surprise if it were to happen within the next two to three years, but it could just as well be twenty or thirty years from now."[44] As he would in many other appearances after this one, Ishibashi stressed how hard it would be for anyone to offer a narrower time frame than that with any degree of confidence. He didn't say that it would be impossible but, by describing the production of such a prediction as "exceptionally difficult," he hoped that his audiences understood that it wasn't lack of interest or effort that had brought them to this point. Careful studies had shown that the crust associated with the fault zone was already at or very close to its upper limit in terms of the amount of strain it could withstand before it gave way. Even with that level of understanding of what was happening at the local level, "could be three years, could be thirty," was the best he and his colleagues could do.

Who Ishibashi was and how he situated himself in relation to the earthquake prediction establishment shaped how the Tōkai earthquake was introduced to the nation in 1976. Ishibashi was young—he had turned thirty-two in August—in a field in which its spokesmen tended to be much older. Hagiwara was in his late sixties and Asada and Rikitake in their mid-to-late fifties, at that point. Photographs of Ishibashi that accompanied articles about the earthquake in October and November could only have reinforced an image of him as youthful and dynamic. In one, he is seen in the middle of a presentation at a city hall in Shizuoka, microphone in one hand while the other draws his audience's attention to a feature on the map of Suruga Bay he has brought along to show them.[45] In this photo and in other settings, too, Ishibashi was engaging directly with the public, communicating what he knew in ways that were meant to be accessible and useful to his listeners. None of the other scientists who took part in discussions about the Tōkai earthquake reached out to the public as he did.

The media gave Ishibashi plenty of opportunities to share his conclusions with their audiences. In just the first month after his presentation at the Seismological Society meeting in early October, he had been on television six times and sent in half a dozen magazine and newspaper articles for publication.[46] On the day the photograph mentioned above was taken, Ishibashi spent the morning on an aerial tour of southern Izu and Suruga Bay in a helicopter belonging to one of the major newspapers, before giving his presentation in Numazu City to an audience of around five hundred. Later that day, he met with a smaller group of local residents for a discussion on the topic of "the discovery of old records documenting the Ansei earthquake." Reporters later squeezed in an interview with him on the train trip back home to Hiratsuka that evening. He confessed that he might be feeling a bit overwhelmed, but explained why he was so committed to getting the word out. "As a scientist," Ishibashi said, "it would be irresponsible to just stay in the lab and debate seismology. . . . From a disaster-prevention perspective, it is vital that the earthquake's dangers be accurately understood."[47]

The Yochiren and the Tōkai Earthquake

There is an interesting tension between Ishibashi's image as someone trying to challenge the field almost as an outsider to it—he described what he was doing at one point as "Don Quixote-like"—and his adherence to the practices through which the Yochiren and the field would assess his arguments.[48] Hagiwara offered this comment to a reporter in October: "It is correct that if a massive earthquake were to strike the Japan archipelago, an earthquake in the Tōkai district would be the number one candidate," he said. "Ishibashi's thesis contains material that merits close examination and is indeed of deep interest. My own hope going forward is that we will be able to monitor the entire Tōkai district, while incorporating the results of new research, including Ishibashi's thesis."[49]

In a series of Diet appearances that followed over the next few weeks, Hagiwara continued to express interest in what Ishibashi was suggesting, while stopping short of formally endorsing all his conclusions. Most of the other scientists asked to share their expertise with legislators at that point, on the other hand, were more explicit in their support for Ishibashi's work and openly expressed their alarm at its implications. Diet committees took up the Tōkai earthquake problem at length on at least five separate occasions in October 1976. Hagiwara was an invited expert at three of them. The first two times, he was the only earth scientist asked to speak and, on the third, he was joined at the Upper House Special Committee on Disaster Countermeasures by Rikitake and former ERI director Tsubokawa Ietsune. Asada's second appearance that month, after a memorable performance on October 4 (discussed below), was on October 22 before the Upper House Special Committee on the Advancement of Science and Technology. Only one of the four scientists to testify, Tsubokawa, openly challenged the idea that Ishibashi's work was cause to worry about a major Tōkai earthquake in the near term, and his doubts were immediately dismissed by the others.[50]

Asada, Hagiwara, Rikitake, and Tsubokawa each offered slightly different perspectives on Ishibashi's theory and how it related to the Yochiren's assumptions about what was happening in the Tōkai district. Legislators (and a concerned public), not surprisingly, were hoping to hear more definitive answers from the experts than most of them were willing to provide in October. It was frustrating not to have clear responses to questions like "Could a Suruga Bay earthquake as Ishibashi had described it happen?" and "If so, when can we expect it?" One of the ways that Hagiwara helped temper hopes for a quick, clear consensus to emerge around Ishibashi's arguments was to remind his audiences that the Yochiren was the proper forum for those debates to play out. When he was asked about Ishibashi's ideas a few days after the younger researcher had presented them at a meeting of the Seismological Society, Hagiwara replied that the Yochiren itself had not yet issued a final opinion on the subject. "I suspect that it will formally do so in the statement that will follow its next general meeting on November 29." Whenever Hagiwara or officials from the different agencies involved in some way with earthquake prediction projects had to respond that fall to questions about what the government's position was on the Tōkai earthquake, their standard response was to defer to an opinion the Yochiren didn't plan to share until almost the end of the year.

Hagiwara sought to reassure legislators that he and his colleagues had no interest in stifling the free expression of ideas. In one of his appearances in October, he offered this defense of academic freedom:

> On the subject of this research, Mr. Ishibashi is one of the researchers working zealously on the Suruga Bay, Tōkai earthquake. It is absolutely essential that these individuals present their work in academic conferences and debate their work freely in those settings. To hold back from speaking, or to refrain from asking certain questions because to do otherwise would mean that it would somehow get out into the public would pose huge impediments to the progress of scholarship. As a matter of freedom of expression, we absolutely cannot as a society say that we'll give those things up.[51]

Hagiwara was even willing to put a positive spin on the press's tendency to transmit the content of such discussions, unfiltered, straight to the public. "I'm optimistic that, after who knows how many times, an understanding of earthquakes will have spread little by little amongst scholars, those in the media, and the public too, and at that point things will get better."[52] In other words, it was fine that Ishibashi was speaking directly to the public, even if in the short term it was likely to raise anxiety levels. Hagiwara might have been a little disingenuous with remarks like these, since it is hard to imagine him being all that happy having to deal with the fallout from Ishibashi's warnings.[53] He and the Yochiren had at least temporarily lost control of the agenda and the narrative surrounding earthquake prediction.

Hagiwara also pushed back hard against the suggestions that he and the Yochiren might try to downplay Ishibashi's concerns and that they had not been taking the Tōkai issue seriously enough. Ōmi Mikio, a Lower House legislator representing Osaka, suggested as much in this exchange with Hagiwara:

> I would respectfully urge the government's [committee], as well as Professor Hagiwara and the others, to move quickly to examine Ishibashi's statements. I realize that Ishibashi is a young and up-and-coming scholar, but what he has said is so important, isn't there a need for authorities in the field such as yourselves to be proactive in taking this up and analyzing it, however young the professor is who makes the statement?[54]

Hagiwara was having none of that. "The Yochiren had been focused on that area long before Ishibashi said anything about it," he insisted. They'd had a great many dedicated researchers monitoring it quite closely for some time. "As for Ishibashi," Hagiwara said, "the Yochiren itself has not issued a statement regarding the Suruga Bay earthquake and the like. I suspect there will be one after the meeting in November." With that out of the way, Hagiwara then went ahead and offered a summary of the issues on which he believed there was already broad agreement between Ishibashi's work and the Yochiren's understanding of the situation in Suruga Bay.

It was clear, for example, that in the 1854 Ansei-Tōkai earthquake the fault had been active all the way into Suruga Bay and that in 1944 it had not. Evidence from multiple sources, including recently uncovered historical documents and the results of ongoing surveys in the area, showed that stresses had been building for some time now in the vicinity of the faulting area. This was not in doubt. Presumably what was left, then, for the Yochiren to weigh in on was whether Ishibashi's contributions to their understanding of the situation required a different set of responses than the ones they had relied on up to this point. As Hagiwara pointed out, the Yochiren had already set up a special section under the direction of Asada Toshi to focus more attention on the Tōkai district and had been calling for close monitoring of the region for years. Heightened monitoring of the district was merited, in his opinion, but they had reached that conclusion without Ishibashi's prompting.[55]

In addition to Hagiwara's defense of academic freedom and the open exchange of ideas, he also inoculated the Yochiren against suspicions that it might dismiss Ishibashi's warning by drawing attention to its own longstanding concerns with the district. Although Hagiwara made it clear that his younger colleague's contributions built on work the Yochiren had already done and would have continued to pursue with or without Ishibashi's intervention, this was also a powerful statement of how closely aligned the "establishment" and the up-and-coming young professor seemed to be.[56] When the Yochiren's statement, "On the Tōkai Earthquake," finally appeared in late November, it contained no surprises. It reiterated Hagiwara's earlier comments in the Diet almost point-by-point, thus agreeing with Ishibashi's theory that one plate was indeed underthrusting the other beneath Suruga Bay and with his argument that the boundary between the two plates had ruptured during the 1854 Ansei earthquake but had not done so since. The Yochiren also acknowledged that ground elevations along the western coast of Suruga Bay had been decreasing since the Meiji era, even as instruments also recorded ongoing horizontal compression along a north-by-northwest/south-by-southeast axis relative to the coast. That data aligned with Ishibashi's observation that the processes responsible for the district's past earthquakes could well trigger another, and soon.[57]

As for what should happen next, in its report, the Yochiren simply repeated the same points it had been making for some time, which were essentially these: it might be possible for scientists to spot precursory phenomena that would tell them when a Tōkai earthquake was imminent, but they would never know without access to more and better monitoring capabilities in the district. The best way to prepare for the threat that Ishibashi had helped make real, according to the Yochiren, was to make sure that the prediction project had the resources it needed to succeed.[58] Hagiwara and his colleagues on the committee eventually would have more to say on the topic of next steps before the year was out.

Confronting the Threat

In Shizuoka and elsewhere in the Tōkai region, Ishibashi's announcement that a Suruga Bay earthquake was likely to be much more destructive than scientists had previously thought was taken seriously from the start. Skeptics of Ishibashi's approach or local officials anxious to downplay the risk a Tōkai earthquake posed to their communities were soon outnumbered by those persuaded that the threat was in all likelihood real. A key figure in efforts to respond to the news and protect Shizuoka was its governor, Yamamoto Keizaburō.

That summer, Yamamoto (1913–2006) was two years into his first term as governor but already a seasoned politician. His path to the governorship had started early and gone smoothly since. Yamamoto, a Tokyo Imperial University graduate, was elected to Shizuoka's prefectural legislature in 1960 and from his seat in that chamber to one in the Upper House of the Diet in 1968. Six years later, the LDP and the outgoing governor both backed his bid to lead the prefecture.[59] It was a close election.

The Tōkai earthquake, in one form or another, defined the rest of Yamamoto's political career. Local media references to him as the "earthquake governor" (*jishin chiji*) reflect some of the ambiguity in the public's views of his response to Ishibashi's warnings. On the one hand, his efforts to convince the government in Tokyo to do all that it could to protect Shizuoka from the next Tōkai earthquake eventually bore fruit. The most ambitious legislative responses to fears of a Tōkai quake were mostly Yamamoto's ideas, as we'll see. By spring 1978, the government's commitment to anti-earthquake disaster measures in Shizuoka (and nearby prefectures) was of a type and on a scale few would have imagined possible in 1976. In the next two gubernatorial elections in the prefecture, in 1978 and 1982, Yamamoto trounced his opponents. On the other hand, there were downsides to having a governor who kept drawing attention to the catastrophe about to engulf the prefecture, as one might imagine. For those hopeful that Yamamoto might move on to some new concern, and perhaps stop frightening off the tourists, the "earthquake governor" moniker was more a lament than an expression of gratitude.

Yamamoto's first reaction to the news of Ishibashi's announcement in late August 1976, indeed, was to try to minimize the fallout from it. The following month he mobilized a small "earthquake countermeasures squad" (*jishin taisakudan*) in the prefecture's disaster prevention office and ordered them to begin learning everything they could about the Suruga Bay earthquake and its potential effects on Shizuoka. Much later, setting up the squad was seen as an important first step toward the eventual passage of legislation to respond to the Tōkai earthquake and as evidence of just how committed the governor was to addressing this new threat. As an astute reporter pointed out at the time, however, one of Yamamoto's main goals in creating the squad was to help craft the prefecture's public relations strategy against the onslaught of bad news Ishibashi's announcement had unleashed. It was only after scientists other than Ishibashi began discussing the theory and its implications in public venues that the governor and his squad began to see the earthquake itself as their primary problem.

The timing of the squad's creation coincided with major developments in the theory's legibility to the public and to policy makers. One was foreseeable, the other less so. Ishibashi was expected to present his work on the Suruga Bay earthquake at the Seismological Society's annual fall meeting, hosted that year by Kyushu University. The conference would begin on October 7. This would be the largest public venue yet in which he had shared his work. Interest in what Ishibashi would say and how his peers would react was growing and, as a result, the conference had attracted much more attention from the media and the public than was normal. Yamamoto dispatched two members of his "earthquake countermeasures squad" to Fukuoka to sit in on Ishibashi's session.

The less foreseeable development unfolded in the Diet and came a few days before the conference in Kyushu got underway. The Upper House Budget Committee had invited Asada Toshi to appear before them at their October 4 session, ostensibly for his input on next year's spending plan and its impact on earthquake science. Asada had appeared before other Diet committees—most recently in November 1974, when he answered questions about earthquake prediction put to him by members of the Special Committee on Disaster Countermeasures—but this was the first time the Budget Committee had sought his input.[60]

Asada's interlocutor was the LDP's Hatano Akira, who had several points he wanted the professor to comment on, including the state of earthquake prediction in Japan.[61] But before he got to that topic, Hatano invited Asada to provide the committee with his thoughts as to which areas in Japan were most at risk from a major earthquake and why. Asada's answers were embedded in a very lengthy response, interrupted just once by a question from Hatano. The professor spoke in language that most of his audience would have had little difficulty understanding.

Asada began by noting that while there really wasn't any place in Japan that could be said to be entirely safe from earthquakes, some locations were more

Figure 11. Asada Toshi (facing the camera) meets with Shizuoka Governor Yamamoto at the Shizuoka Prefectural Government Office, October 6, 1976. Photo by Yomiuri Newspaper/AFLO.

associated with potentially catastrophic earthquakes than others. In the capital district, for example, the Great Kantō Earthquake had been both exceptionally powerful and massively destructive. Earthquakes originating directly beneath Tokyo—of which accounts were plentiful—tended to be less powerful by comparison but still destructive by virtue of their proximity to the city. It was unlikely that another event like the one that had struck the city in 1923 would happen again soon, Asada noted, since researchers had concluded that it would be some time still before enough stress had accumulated to make that likely. As far as the near-field earthquakes were concerned, there really was no way to tell when one might strike. The geology under the city made it exceptionally difficult to monitor conditions there and the city's constant background noise interfered with what instrumentation they did have in place.

Which brought Asada to what he really wanted to talk about. "Compared to these other earthquakes," Asada said, "the so-called Tōkai earthquake is an exceptionally easy earthquake to understand. In fact," he continued, "I've been saying for years now in pieces I've written in Japan and for foreign audiences that it probably poses a danger." Still, he hadn't been certain about it, at least not until this year. Over the past few months, he told the assembled legislators and officials, evidence had come to light that had convinced him that the earthquake was likely to occur. Asada listed some of the factors that had gone into reaching that conclusion, including the emergence of a better understanding of how far into Suruga Bay the 1854 earthquake had reached. It was clear to him now that the Tōkai earthquake would occur along those sections of the faulting zone that had not ruptured in 1944.[62]

It was at this point in his response to Hatano's question that Asada offered the following observation about the Tōkai earthquake: "It's been said quite often [lately] that it wouldn't be strange at all if it were to happen tomorrow, and in a sense that is true. Which is to say, if it had ruptured in 1944, then we would expect that everything would be okay. But it did not and, therefore, it is reasonable to think that it has the capacity to do so." "However," Asada concluded, "can such a thing be said definitively? Given the state of the field at present, a statement such as that cannot be made definitively."[63]

Asada was trying to make a point by highlighting the uncertainty around when the Tōkai quake might occur, which was that there were steps they could take to try to narrow its scope. Using techniques and equipment that earthquake prediction researchers had already embraced, Asada went on to say it was possible that, in time, he and his fellow scientists would be able to identify signs of an imminent earthquake in the Tōkai district. Of all the places in Japan where major earthquakes were expected, Asada believed that the Tōkai district was where earthquake prediction stood the best chance of success.

Asada had said quite a bit, but what the nation focused on was his comment that, "It wouldn't be at all strange if it happened tomorrow." That phrase and variations on it entered the public lexicon for discussing a Tōkai earthquake almost as soon as it left Asada's lips and was soon on everyone else's too. Papers the next morning reported Asada's warning, some at length. The *Asahi* called it a "shocking assessment."[64] Not all managed to get across the idea that he had been repeating a widely held view in order to comment on it, as opposed to making a claim of his own about the likelihood of the quake happening very soon.[65] It wasn't long before the press just did away with whatever nuance there might have been in Asada's original comment and simply credited him with saying, for example, that "It would come as no surprise if an earthquake with an epicenter in Suruga Bay were to happen tomorrow."[66]

Asada's juxtaposition of this potentially monstrous earthquake with a memorable expression of the uncertainty surrounding it resonated with legislators too. The LDP's Hatano used the phrase to make a point of his own before Asada was even done speaking that morning and, within a few days, other legislators in different committees in both Houses were deploying it freely as they turned their attention to prediction and to the Suruga Bay earthquake. While there were other factors at work that helped make those issues ones the Diet would have to address that fall, the questions that legislators put to ministry officials, to expert witnesses (including Hagiwara, Rikitake, and Asada, invited back later that month), and their own observations all suggest that it had become almost second nature to think of the Tōkai earthquake as an event that could happen at any time. The setting in which Asada made the comment—in the Diet—together with his status as a University of Tokyo professor, lent credibility to an idea that likely would not have carried nearly as much weight had it come up in a different venue, or been expressed by a less prominent scientist. The timing of his appearance before the Budget Committee was also a factor in how an understanding of the Tōkai earthquake took shape that fall. Asada spoke to legislators on a Monday morning; Ishibashi was scheduled to present his own work on the Suruga Bay earthquake at the Seismological Society meeting in Fukuoka later that same week, but the *Asahi* published detailed accounts of his conclusions as early as Wednesday, before any presentation had taken place. Other papers were not far behind in their coverage of Ishibashi's theory.[67]

Since the major dailies had not touched much on Ishibashi's work when news of it first broke in August, the spate of articles triggered by Asada's comment and the conference in Fukuoka mark the beginning of much wider public engagement with the Tōkai earthquake. Those articles tend not to vary much in how they present the arguments at the center of Ishibashi's claim, which (as we've already encountered) point to historical documents as proof that the subducting fault zone in the 1854 earthquake in the region had extended well into Suruga Bay and that it hadn't ruptured since. Recent observations in the area by the JMA and other researchers, meanwhile, had shown that strain was continuing to accumulate in the vicinity. Therefore, Ishibashi's theory went, when that strain next exceeded whatever the fault's capacity for it happened to be, then the earthquake that followed would be much more powerful and its impact more widely felt than scientists and the public had planned for.

A fuller picture of just how powerful and how destructive an earthquake the Tōkai district could expect also began to emerge around the time of the conference. Descriptions of it as likely to be an "M8 class" earthquake, for example, are common from this point on, as are reminders when that figure comes up that the earthquake that caused the disaster in 1923 had been an M7.9 event.[68] The *Asahi* explored the implications of such a massive earthquake in the not-so- imaginatively titled "Magnitude 8" article series it launched later in October, the first of which featured this exchange:

> Q: Is a big earthquake really coming?
>
> A: At some point, absolutely. It is the Kantō's and the Tōkai's fate.[69]

CHAPTER 8

The Tōkai Earthquake and the Turn to Prediction

As scientists like Hagiwara, Asada, and others began signaling their support for Ishibashi's findings in the fall of 1976, questions about whether the threat was real or not gave way to anxiety about what could be done about it. People expressed their unease in different ways. Stores in Shizuoka reported runs on disaster goods even before Ishibashi's September lecture there.[1] Department stores and other retailers were quick to respond to growing concern over the earthquake by setting up special displays of "emergency supplies" to make it easier for consumers to stock up on necessities. The sixth-floor display at the Seibu department store in Shizuoka City, for example, featured preassembled "Disaster Preparedness Backpacks" (*Bōsai ryukku*) for the princely sum of 9,800 yen. While the store typically only sold two or three of those a day, it was not uncommon for it go through fifty plastic containers for storing drinking water daily; those were only 550 yen apiece. Biscuits, preserved foods, rope, and protective outerwear were among the better selling items, store managers reported.

This was, of course, a clever bit of marketing, too, since the sudden appearance of these goods and the department store aisle displays that they were part of would not have escaped shoppers' notice. Even if they hadn't come to the store intending to purchase extra water containers or first aid supplies, it isn't hard to imagine that many would have felt the need to do so. Store managers reported feeling torn about whether or how much to advertise their emergency goods and rescue kits, out of worries that they'd be accused of trying to exploit the public's fears. A manager at a Seibu department store remarked that they'd refrained from any advertising of their emergency preparedness goods but that sales were fantastic anyway. He was having trouble keeping items in stock. His counterpart at the Yaohan department store in Shimoda City repeated a similar story. Customers had been asking him to get in a supply of dried biscuits, a basic emergency food item, so he placed what, to him, seemed like an absurdly large order. He must have worried that he'd overestimated demand and would be stuck trying to sell boxes of unappetizing biscuits for months to come but told reporters he was stunned to see every last one of the packages fly off the shelves by the afternoon. The manager called his wholesaler back to order another delivery. The wholesaler was sold out too.[2]

Worries about the earthquake were acted on in other ways as well. A number of elementary schools (and some preschools and kindergartens) began regular earthquake drills, intrusions on the school day that had not been thought necessary before. As part of the preparation for the exercise, mothers were asked to provide their children with a "*Bōsai zukin*" (disaster-prevention hood) to keep at school. Usually made of a piece of quilted, lightly padded cloth that lies flat when not in use, when worn as a hood it provides some protection to one's head and neck. Grandparents and perhaps some parents, too, would be more familiar with the wartime version of this same article, when it was called a *Bōkū zukin* (air-defense hood). Its reappearance in this new context was, for some, quite jarring. The drills themselves were a combination of duck-and-cover routines that sent students into their hoods and under their desks and closely supervised evacuations of school buildings.[3] The principal at the Jitōgata Elementary School, where a two-lane road was all that stood between it and Suruga Bay, favored the latter. He insisted that the almost-four hundred students in his charge practice the drill again and again. The school had lost three students to the 1944 Tōnankai earthquake and, if Ishibashi was right, its location near the tip of the Omaezaki Peninsula meant that it was in harm's way again. Images of cheerful, hooded children performing these new routines appeared in the major national papers.[4]

Anecdotal reports of local reactions to the earthquake reflected the range of impacts it was having. In Izu, where communities were heavily dependent on the tourist trade, local businesses and leaders were struggling to strike a balance between assuring potential visitors that their towns were safe and trying to avoid drawing even more of the public's attention to warnings about the earthquake.[5] The mayor of Kawazu, a town on Izu's coast, reportedly approached a University of Tokyo scientist doing research nearby to ask whether the town could buy its own seismometer, since it was thinking about setting up its own "prediction system." The mayor was disappointed to learn that a single seismometer could easily cost a million yen and that making sense of what it recorded was not at all straightforward. Elsewhere, builders and contractors reported that orders for new homes and renovations to existing ones had fallen far below normal levels. The earthquake was a likely culprit. As more and more clients chose to postpone plans to build, reported one contractor, carpenters and other tradesmen were having to leave the prefecture to find work, "just like farmers pushed into migrant labor by cold weather damage [to their crops]."[6]

The central government responded slowly to the rising anxiety over a potential earthquake. In October, the Cabinet's announcement that it would establish a new Earthquake Prediction Promotion Headquarters (Jishin yochi suishin honbu) made it sound as if it had achieved something it hadn't. The creation of the headquarters was more bureaucratic sleight-of-hand than a meaningful intervention; it was little more than an upgraded version of an existing committee within the Science and Technology Agency.[7] It took almost two more years for policy makers in Japan to decide how the country was going to prepare for the next Tōkai earthquake and what role earthquake prediction would play in those preparations. Their solution to the problems posed by Ishibashi's discovery was a remarkable piece of legislation, the Large-Scale Earthquake Countermeasures Act of 1978. Part of what set the act apart from the government's approach to planning for catastrophes up to that point was the scope of the interventions it authorized the state to undertake—mandatory evacuations, the pre-deployment of the SDF, and the closure of road and rail networks in and out of the Tōkai district among them. What was really remarkable about the new legislation, however, was that many of the most important steps it proposed the country take to reduce the harm a potential Tōkai earthquake might cause were predicated on the assumption that scientists would know ahead of time when that earthquake was about to occur. A central premise of the Japanese government's response to what was potentially one of the gravest threats the country had faced in recent memory, in other words, was that short-term earthquake prediction worked.

Ishibashi's warnings about a Tōkai earthquake surfaced during a period of unprecedented confidence among seismologists in Japan and elsewhere that they were on the cusp of discovering how to perform successful short-term earthquake predictions. There were by then many reasons to believe that this might be the case; scientists in the US and the USSR had recently offered up a series of monitoring practices that could, they believed, detect changes in the qualities of the earth's crust that signaled an imminent earthquake.

Researchers in Japan and the United States started public discussions about the social and economic implications of short-term earthquake predictions at about the same time, for example, and for similar reasons. Interest in the "Palmdale bulge" and the Parkfield experiment are other markers of the consideration that scientists in the United States gave to earthquake prediction in the 1970s. Scientists in the People's Republic of China went one better and pointed to the evacuation warning officials there had issued shortly before the 1975 Haicheng earthquake—no doubt saving many lives—as an example of prediction in actual practice. Enthusiasm for short-term earthquake prediction among scientists worldwide was arguably at its peak just as the implications of the next Tōkai earthquake were coming into focus in Japan.

It was never a given, however, that the choices policy makers made about how to mitigate the harm a Tōkai earthquake was expected to do would depend quite so much on scientists' ability to predict it. The process that culminated in the passage of the Countermeasures Act was shaped to a considerable degree by bureaucratic infighting within the government, the messy day-to-day politics of the legislature, and many contingencies—earthquakes, election outcomes, and so on. The pressure to "do something" about the coming disaster was felt by different groups of actors at different times. Those with a voice in the process usually approached the problems posed by the earthquake through the lens of their own areas of expertise or leverage and their own interests. For most of that almost-two-year period, policy makers and scientists declined to take any dramatic steps at all and instead exhibited a preference for incremental adjustments to the status quo, where preparations for the Tōkai earthquake were concerned. The legislative campaign that culminated in the passage of the Countermeasures Act didn't begin until after the January 1978 Izu-Ōshima earthquake, or, sixteen months after Ishibashi's announcement. The change in trajectory was abrupt. There had been no scientific discoveries or warnings to account for the government's decision to launch an all-out defense of the district at that point, much less one that relied on earthquake prediction.

This chapter explains how Japan went from point A to point B, where "A" is the stage at which earthquake prediction in Japan was primarily an interesting project for seismologists (1976) and "B" the moment at which it became an essential part of a plan to save tens of thousands of lives (1978). This was the first time the Japanese government explicitly adopted earthquake prediction as one of the tools it used to protect its citizens, which raises interesting questions about, among other things, how different scientists communicated their understandings of prediction's capabilities at the time, how well policy makers understood them, and what significance those cohorts and others attached to prediction's new role. Understanding the Tōkai earthquake's and the Countermeasures Act's origins may also help explain why the Japanese government maintained for four decades after the passage of the Daishinhō that the Tōkai earthquake was the country's most worrisome seismic threat even as a lengthening list of earthquake disasters elsewhere in Japan argued otherwise, and that prediction was a useful tool against that threat, despite a rising tide of expressions of doubt by experts at home and abroad. What I show below is that how the Countermeasures Act came to be and how earthquake prediction became such an important part of it in the late 1970s, made it very difficult to disentangle the two or provide compelling alternatives to either.

Peak Prediction

In the 1970s in the People's Republic of China, earthquake prediction and preparedness were work that the public was encouraged to pursue, right alongside the seismologists and other experts whose professional duties focused on those goals. "Collective monitoring" performed by citizen-scientists targeted anomalous natural phenomena associated with seismicity, be it unusual animal behaviors, changes in well water, or unexpected results from locally operated seismometers and other instrumentation.[8] It was just such a sudden upturn in nearby earthquake activity that convinced local officials in the city of Haicheng, in China's northeastern Liaonang Province, to urge residents to leave their homes late in the day on February 3, 1975, and be ready for the worst. When an M7.3 earthquake struck early the next morning, the loss of life was considerably less than it would have been had many families not already been outside. Officials in the PRC announced the news of their successful, life-saving earthquake prediction a few weeks later. International press coverage and visits by delegations of scientists from the United States, Canada, and Japan soon followed.[9]

Japanese scientists had been interested in China's earthquake prediction program before the Haicheng earthquake.[10] In mid-1974, Asada Toshi and several colleagues spent three weeks in China travelling to different research centers, meeting with seismologists, and discussing their methods and plans. They were impressed by the sheer scale of the undertaking, by the government's strong support for it, and by the considerable optimism of their Chinese colleagues.[11] Several months after the earthquake, a delegation of seismologists from China arrived in Japan on a reciprocal visit and shared at least some of what they knew about the Haicheng prediction during their stay.[12] Later, at the height of the arguments in Japan over Tōkai earthquake policies, both Shizuoka's governor (Yamamoto Keizaburo) and Rikitake Tsuneji went to China for briefings on the prediction work being done there.

The earthquake and the reports that the Chinese had successfully predicted it were mixed blessings for those involved in the Japanese program. The good news was that what happened in Haicheng seemed to show that short-term earthquake prediction was possible and that it could be put to effective use; the less-good news was that it was the Chinese who had done the showing. Questions about why the Chinese scientific establishment had been able to accomplish something that had, so far, eluded Japanese scientists surfaced first in 1975 and then returned with a vengeance during the debates over how to prepare for the Tōkai earthquake. Legislators demanded to know how it was possible that Japanese seismologists had been working on the earthquake prediction problem since the early 1960s, using far more sophisticated equipment than the Chinese and with access to a great deal more money, but had still fallen behind. Were seismologists in Japan not working hard enough? Were they less effectively organized than their counterparts in the PRC? There had to be some explanation for why prediction wasn't working in Japan. China's success was certainly a factor in how Japanese scientists situated their prediction work relative to the expectations that policy makers had for them after 1975.

In the United States, mainstream seismologists in the mid-1970s were as willing as they had ever been to consider earthquake prediction's possibilities.[13] What had happened in San Fernando in 1971 brought earthquake hazards to the public's attention in a dramatic fashion (much as it had in Japan). The 1971 event was the most powerful earthquake in California since 1952 and the deadliest since the Long Beach Earthquake of 1933, so it was no surprise that finding ways to be better prepared for the next one was a priority for many. New research also made a difference. In 1973, researchers at Columbia's Lamont-Doherty Geological Observatory published the results of their recent work on dilatancy, which refers to the minute expansion in the volume of a body of rock that will occur when it is subject to great stress. The particular contribution of their work was to relate observations of dilatancy in rock samples in the lab to potentially observable changes underway in the earth's crust prior to an earthquake.[14] The "dilatancy model" was exciting in part because it not only identified specific precursory changes to watch for but also because it explained the physical processes behind those changes and how they were directly related to the earthquake event itself.[15] The discovery of the so-called "Palmdale bulge" in 1976—named for the community about thirty miles north of Los Angeles where geodetic surveys had recorded a surface uplift of more than a foot, extending over several thousand square miles—meshed well with what the dilatancy model suggested would be observable in areas of the crust on the pathway to fracture. Needless to say, this threat was proximate enough to invite speculation about what might be done to protect that part of Southern California.

Finally, like their Japanese counterparts, seismologists in the United States were intrigued by the reports from China after the Haicheng earthquake. Afterward, MIT's Frank Press was among the more visible figures in the field to endorse the idea that earthquake prediction might be possible, given the proper application of resources and effort. Other seismologists, including some at the Department of Interior and United States Geological Survey, had also begun to think about prediction in a new light.[16] As Susan Hough points out, the shift was pronounced enough that even *People* and *Popular Science* remarked on it; writing for the latter magazine in late 1976, journalist George Alexander observed that "earthquake prediction, long treated as the seismological family's weird uncle, has in the last few years become everyone's favorite nephew."[17]

Efforts to steer this enthusiasm in practical directions took several forms. In late 1975, officials and scientists from the western United States gathered for discussions about the practical, real-life implications of a publicly-announced prediction.[18] That same year and then again the next, the National Academy of Sciences released the results of two studies addressing the same topic.[19] In early 1976, Frank Press called on Ford administration officials to back funding for earthquake prediction and, later that same year, a panel convened by the administration included support for prediction in its recommendation that the government more than quadruple the budget for earthquake hazard studies over the next few years. The National Science Foundation and US Geological Survey also called for significant investments in prediction research to be part of proposed legislation for a federal earthquake hazards reduction program that same year.[20]

Many of these recommendations ended up being reflected in the Earthquake Hazards Reduction Act when it was signed into law by President Carter in 1977. Although they were a relatively small part of the total package, the provisions in the act relating to prediction were what most captivated the public and the media. One of the stated objectives of the act was, for example, "the implementation, to the greatest extent practicable, in all areas of high or moderate seismic risk, of a system (including personnel, technology, and procedures) for predicting damaging earthquakes and for identifying, evaluating, and accurately characterizing seismic hazards."[21] The law created the National Earthquake Hazards Reduction Program to coordinate amongst the many different agencies tasked with different pieces of this new undertaking and led to the establishment of a National Earthquake Prediction Evaluation Council, in 1980.

One thing to say about the emerging support for earthquake prediction in the United States is that seismologists and some policy makers in Japan were paying close attention to it. The rise in concern over the Palmdale bulge developed at roughly the same time that Ishibashi was starting to raise the alarm about a Tōkai earthquake and the debates over how to move forward highlighted similar issues in both countries. Whose responsibility would it be to decide if the data at hand merited announcing when and where a major earthquake would strike? On what would they base their decision? If there were clear indications that an earthquake would soon strike, what then? What policies had to be in place for a prediction to do the most good for the most people?

The Assessment Committee

It had come up frequently in the media and in the Diet in the fall of 1976 that, even if scientists somehow learned that an earthquake was imminent, it wasn't clear what they could do with that information.[22] Earthquakes were different from tsunami, typhoons, and floods, which the JMA was empowered to forecast and to warn people about well before their arrival. It had no legal standing to make predictions about earthquakes that hadn't happened yet. Nor, for that matter, did the Yochiren.[23] As Hagiwara told legislators, the Yochiren had no authority of its own to issue an official "forecast," much less a warning. He could hold a press conference and share information with reporters that way but, so far as he knew, there were as yet no processes in place that would allow him or anyone else to issue an actual "earthquake warning."[24]

In December, Hagiwara and the Yochiren used a few short paragraphs at the end of an otherwise routine proposal to revise the latest five-year plan for earthquake prediction to define an important new role for themselves in the lead-up to a Tōkai earthquake. By then, Hagiwara and the others had brokered an agreement to allow the JMA to collect data—in real time—from a collection of different micro-seismicity monitoring networks in the region.[25] The scientists needed the JMA to take that step so that they could move forward with another, which was the establishment of a new earthquake prediction "decision-making structure" (*hantei taisei*), distinct from the Yochiren. "So far," the proposal noted, "the Yochiren has provided overall assessments based on information exchanged among all the different institutions involved in earthquake prediction. However," it continued, with the new procedures for gathering and consolidating data from ongoing observation efforts that were "accompanying the development of methods for short-term prediction . . . an improvement in our capacity to make decisions based on that data with a high probability is essential."[26] How such a function might fit within the overall structure of earthquake prediction in Japan was an important question, they acknowledged, but, in the near term, "it should be tentatively appropriate to set up an organ that works on evaluation and judgment of data taken by continuous observations in the Tōkai area."[27] Hagiwara, as careful as ever with his language, promised little but sounded moderately upbeat about what these new arrangements might lead to. "If this system is put in place," he told reporters, "might it not be possible to make short-term predictions several weeks or several days before [an earthquake]? For the moment, given the Tōkai earthquake problem, I would hope that the budgetary provisions necessary to make these arrangements are implemented as soon as possible."[28]

The government didn't act on the Yochiren's recommendations until the following April, when it announced the creation of a new "Tōkai District Assessment Committee" (Tōkai chiiki hanteikai, called the Hanteikai for short).[29] The new council's function as it was described that spring was in keeping with what the Yochiren's scientists had imagined it would be, back in December. The JMA had already started remodeling some of its offices so it would have enough room to accommodate new monitoring equipment for the Tōkai district. As that equipment came online and the institutions with instrumentation in the Tōkai region that weren't already sending their real-time data directly to the JMA began to do so, its technicians would begin their night-and-day lookout for the unusual and inexplicable. If an anomaly did appear and it was deemed sufficiently worrisome by the committee chair, he would issue an "emergency summons" to his fellow panelists. Once gathered at JMA headquarters, the group could decide to do nothing, in which case no announcement would be necessary. Or, if they were convinced based on the information available to them that an earthquake was imminent, the panel would pass that judgement on to the JMA.[30]

There had been some evolution since December in the government's thinking about how to constitute the council. The April announcement stipulated that it would function as a subcommittee of the Yochiren, an arrangement that was fairly easy to implement and that sidestepped for the moment any debate about how to assign responsibility for making predictions and announcing them to the public. That approach also meant that the government didn't have to go very far to find experts to appoint to it. Six scientists were chosen to serve: Hagiwara, who would chair the council, Asada, Rikitake, Usami Tatsuo (ERI), Kurihara Keiichi (ERI), and Mogi Kiyō (ERI).[31] All were Tokyo residents, all were employed at public universities (except Hagiwara, who had retired from one) and all were well-credentialed, established scientists. (Mogi, in his late forties at that point, was the youngest in the group.) Each was also already a member of the Yochiren when they agreed to be on the panel.

At no point leading up to the creation of this new body had the Yochiren or any of the six scientists on the committee stipulated that they knew how to make short-term earthquake predictions using the tools currently at their disposal. The underlying assumption was that, with better access to better data, anomalous observations would stand out and, with time and practice, it would eventually be possible to identify correlations between those signals and subsequent powerful seismic events. The promise of continuous supervision of the Tōkai district was an important prerequisite to that process but each step along the way raised difficult questions about how to get to the next. Committee members did eventually come up with guidelines to help JMA technicians decide how to differentiate between normal and anomalous results in the information that would soon be flowing in from the Tōkai district. It was agreed that Hagiwara would be contacted if: (a) instruments recorded ten or more micro-earthquakes in a one-hour span for at least two hours straight, or (b) one of the seven or so strain gauges in the area reported a change of at least five parts per ten million (the equivalent of a change of five millimeters over ten kilometers) in a short period and if there was at least one other strain gauge in the vicinity producing similar results.[32]

The committee members, for their part, had given some thought to what they themselves would be able to determine once summoned to review the data. If together they concluded that there was a good possibility that the anomalies were precursory signs of a coming Tōkai earthquake—the exact criteria for that decision remained up in the air—the panelists planned to choose one from their list of three predefined statements. In the first, they would tell the public that an earthquake was expected within the next several hours. Option two expanded the time frame to "within a few days and up to a week," while the last was much less specific and suggested only that the earthquake could happen "at some point in the not-very-distant future."[33] They did not offer opinions as to what steps officials or the public might take in response to one of these announcements.

As of the middle of 1977, the Yochiren and its scientists had taken a tentative step toward linking the pursuit of short-term prediction to preparedness for a potential Tōkai earthquake. This development was a unique product of the Yochiren's history but it is telling that earth scientists in the United States were also in the midst of discussions about how to handle the prediction-making process. Both communities were grappling with questions about what to do now that their attempts at reducing uncertainty appeared about to bear fruit.

Lobbying on Shizuoka's Behalf

On June 8, 1977—a weekday—prefectures throughout the Tōkai district, along with Tokyo and other major cities in the area, held the largest disaster preparedness drills ever attempted in Japan. In the scenario planners had designed for the exercise, participants were told to imagine that, at 6:30 that morning, an M8.4 earthquake had struck off the Shizuoka coastline. Local police agencies mobilized all available personnel—some 75,000 officers—within an hour. With phone lines cut, police and rescue officials relied on portable and shortwave radios (and sometimes Morse code) to coordinate their efforts and collect damage reports. Communities on the coast reported violent Shindo 7 shaking and the almost complete destruction of buildings and roads. Further inland, in Shizuoka, the tremors were only slightly less destructive and news of major damage and injuries there and in nearby areas were also coming in to local disaster headquarters and the police.

By 9:30 a.m., mock evacuations were underway, ultimately involving 170,000 of the region's residents destined for one hundred fifty pre-designated evacuation sites. In Shizuoka, officials staged broken down cars and fallen telephone poles along the evacuation routes to simulate obstacles that residents might have to deal with. Thousands of kindergartners and elementary school students, disaster-prevention-hoods in place, followed police and their teachers to safety as helicopters hovered overhead. Each class held tight to a long rope that served to keep them together and in line—don't talk, don't push, and don't run, their teachers told them. Elsewhere, volunteers pretended to be injured or trapped, so that rescue workers would have a chance to perform their roles as well.[34] Planners noted that the previous year's exercises had involved only 30,000 people. This year's drills were on a much larger scale than ever before, due, in large part, officials said, to what everyone had been hearing about a Tōkai earthquake.[35]

Spectacles like this show that the Tōkai earthquake had not faded from public consciousness once the first few months of intense legislative interest had passed. It clearly hadn't and, the closer one's home was to Suruga Bay, the more heavily the earthquake must have weighed on residents' minds. What does seem to have been the case, however, is that the arguments over what the government ought to be doing about the possibility of a Tōkai earthquake dropped out of public sight once the Cabinet signed off on the creation of the Hanteikai. References in the press to Ishibashi and his theory dropped off sharply that spring, for example, and in the Diet neither the earthquake nor how to respond to it attracted much attention after April 1977. With the Assessment Committee in place, the scope of the state's response and the perceived severity of the threat reached a temporary equilibrium.

The lull lasted for several months. The Tōkai earthquake next resurfaced as a priority for policy makers late in 1977 and it did so because of a concerted lobbying effort by Shizuoka's governor, Yamamoto.[36] Yamamoto had been trying since 1976 to enlist his fellow governors in the region in plans to solicit the national government's help preparing for the Tōkai earthquake.[37] The pace of those efforts picked up in June 1977, after Yamamoto turned to Nose (pronounced noh-say) Niyuki, a career bureaucrat on assignment to Shizuoka from the Ministry of Home Affairs, for help. Yamamoto asked Nose to come up with some language for a new earthquake disaster law, the details of which he seemed willing to let Nose work out on his own.

Nose's draft—he said later that he wrote the entire document in a single evening—was unusual in several respects, but nowhere more so than in its treatment of earthquake prediction.[38] Nose's proposal took it as a given that scientists would be able to predict when the Tōkai earthquake was about to strike and that there would be time after they did so for the government to intervene in ways that would save lives. Many of the scientists Nose had met by then, he said, had assured him that prediction could work and he could see that the government was moving ahead with prediction research. His draft legislation therefore focused almost entirely on the interventions the prime minister would be empowered to take after scientists had determined that a Tōkai earthquake was imminent. "As I was writing it, it occurred even to me that it was something novel," Nose told an interviewer many years later.[39] He was right.

Over the next several months, Yamamoto, Nose, and prefectural government staff went back and forth between Shizuoka and Tokyo as they lobbied legislators, former colleagues, and ministry and agency officials to support this new approach to dealing with the Tōkai earthquake threat. They had a hard time getting anyone to take them seriously. Police, fire, and defense officials were generally sympathetic but decision makers in other parts of the government were not. Many asked what would happen if the prime minister issued an "emergency warning," based on what the scientists had said, but it turned out to be a false alarm? How could they possibly recover from that? The prime minister himself was said to be opposed to what Yamamoto was proposing.[40]

Undaunted, Yamamoto's team went ahead and circulated a full draft of their bill in early December. The Major Earthquake Countermeasures Special Emergency Facilities Bill (Daijishin taisaku tokubetsu kinkyū setsubi hōan) included what were, by then, some familiar ideas, such as support for real-time monitoring in parts of the country at high risk for a destructive earthquake. But most of what was in the draft legislation had not been up for discussion before and most of it focused on what would happen *after* the prime minister learned that an earthquake was imminent and issued a warning to the public. The draft legislation said that he could, for example, direct schools, hospitals, and department stores to evacuate; he was also empowered to bring rail and automobile traffic within the district and all traffic headed into it to a complete halt. Large factories, oil refineries, and other potentially hazardous industrial sites were similarly subject to directives from Tokyo requiring them to shut down. If necessary, the proposal stated, the prime minister could freeze prices for basic goods and take steps to stabilize local financial institutions. The draft legislation opened the door for the dispatch of medical personnel, police, and the SDF to threatened areas once the government's warning went out.[41] Since none of these preemptive interventions were available to the prime minister at the time, Yamamoto's team was proposing a radical shift not just in that office's authority but also in the temporal scope of the government's disaster countermeasures.

Government officials were, at first, no more enthusiastic in December than they had been earlier. Particularly vocal resistance to the new legislation came from the National Land Agency; its critiques described the bill as flawed and as unworkable in practice. Agency spokesmen argued that any mechanism for the prime minister to launch emergency measures before an earthquake ought to be managed through revisions to the existing Disaster Relief Basic Law. The Land Agency also challenged the proposal's assumptions about earthquake prediction. The whole premise of Yamamoto's approach was that scientists would be able to warn the public that the Tōkai earthquake was about to occur. The Land Agency targeted that claim and argued that the governor's bill had failed to take into account the fact that earthquake prediction was not yet sophisticated enough to consistently yield accurate information.[42] Prime Minister Fukuda himself was at best noncommittal when Yamamoto met with him in person to discuss the proposal. In one encounter, after the governor had explained how much of an impact the uncertainty surrounding the Tōkai earthquake was having on the prefecture, Fukuda coldly acknowledged only that, "Yes, that must be a difficult situation to be in."[43] It wasn't what the governor was hoping to hear.

The Izu-Ōshima Earthquake

The JMA began observing a sharp spike in the number of small earthquakes near Izu-Ōshima island beginning around 9:30 a.m. on January 14, 1978. A short while later, the agency took the unusual step of issuing a warning statement: "A moderately large earthquake, one that could give rise to slight damage, may occur shortly."[44] NHK and local television and radio broadcasters in Shizuoka reported the announcement, which was featured in noon news broadcasts, as well.[45] The residents of Izu-Ōshima began to take precautions but most people living in other areas nearby, including those on the Izu peninsula, did not. The island was almost 25 kilometers east of the peninsula itself; there seemed little reason to be concerned.

The earthquake that struck about an hour and a half after the JMA's announcement was not at all like the one that officials had said might happen. At M7.0, it was many times more powerful than the JMA's "moderately large" description implied would be the case, its epicenter was closer to the Izu peninsula itself, and the combination of those two factors meant that the earthquake did far more damage than the experts had implied was likely.[46]

More than two dozen people lost their lives as the tremors damaged hundreds of buildings, triggered landslides, and sparked numerous fires (almost all of which were quickly extinguished, fortunately).[47] Several major roadways, rail lines, and tunnels on the Izu peninsula were blocked by debris or otherwise rendered impassable, in some cases for weeks. The earthquake was powerful enough that its tremors were felt throughout Shizuoka and in Yokohama and Tokyo as well, where Shinjuku skyscrapers swayed from side to side and subway cars far underground stopped dead in their tracks.[48]

The misalignment of the warning with the severity of the actual earthquake was a topic of discussion in the press and the Diet in the days following the disaster. One obvious question was why the JMA, if it understood what was about to happen well enough to issue a warning, had underestimated the severity of the coming earthquake by such a large factor. The M6 event the JMA had said might be coming in its statement and the M7 earthquake ninety minutes later were so far apart in scale that it was as if the agency had said to watch out for a few rain showers offshore and a typhoon had made landfall right next door. According to Hagiwara and the other scientists who weighed in on this issue, the evidence the field had been working with up until January 14 all pointed toward the same conclusion the JMA had ultimately come to that day, namely that an offshore M6 event was the most powerful earthquake likely to occur in that vicinity.[49] There was a broad consensus on this point. As Hagiwara told legislators in early February, he and his colleagues had been convinced that if the section of the fault zone associated with the earthquake clusters ruptured on a large scale, it would do so well away from land. Even if the earthquake itself turned out to be more powerful than the ones they had seen there in the past (meaning, much larger than M6), its location meant that it was unlikely to do any significant damage inland or on the Izu peninsula.[50] Because they had misjudged both the extent of the fault zone and its proximity to the peninsula, Hagiwara acknowledged, they had also badly misjudged how hazardous the earthquake could be for Izu's residents, "who ended up being taken by surprise. This has been an extremely valuable experience for us," he said, "and we hope to make the best possible use of it going forward."[51]

Shizuoka's residents had another encounter with an earthquake-related warning just a few days after the first. The Yochiren released statements on January 15 and 17, alerting the public that aftershocks were likely to continue for some time. "The maximum magnitude expected for the largest aftershock," read the first, "would be 6 or thereabouts."[52] On January 18, Governor Yamamoto took the unusual step of issuing his own statement about the aftershock

Figure 12. Jishin Yochi Renrakukai meeting after the Izu-Ōshima earthquake, January 15, 1978. From the left: Rikitake Tsuneji (Tokyo Institute of Technology), Fujita Naomi (Geographical Survey Institute), Tsumura Kenshirō (ERI), Kasahara Keiichi (ERI), Mogi Kiyō (ERI), and Hagiwara Takahirō. Photo by Mainichi Newspaper/AFLO.

threat. People were used to receiving advisories from the JMA and occasionally the Yochiren about aftershocks, but not from the prefectural government. Yamamoto's decision to take on that role himself was unexpected, and that break from routine practice was one factor that led many in the prefecture to conclude that something extraordinary was about to happen. The statement alone wouldn't have given anyone that impression, since all it did was repeat what the Yochiren had said the day before and listed some precautions that residents in Izu in particular should be thinking about, just in case, such as maintaining a supply of emergency food and water, or being careful with fire. "In the worst case," Yamamoto announced, "an M6 earthquake may possibly occur during the aftershock activity of the Izu-Ōshima Kinkai earthquake." "Moderate to severe damage may well be expected" in the event of a shallow, M6 earthquake in the vicinity of the Izu peninsula.[53]

The governor released his statement at a press conference around 1:30 in the afternoon. Local radio stations began reporting on it a little after 2:00, and local television stations summarized it in chyrons that ran with regular programming not long after.[54] The news also reached communities via official channels, as the prefectural government instructed cities and towns to let residents know what was in the announcement. Local officials had firetrucks and cars equipped with loudspeakers drive around broadcasting summaries of what the governor had said.

What happened next was afterward labelled a "panic," although it was at worst a mild one. Starting a little after three, police stations and prefectural and town offices were inundated with phone calls from residents, most of whom wanted confirmation that what they had heard about an imminent earthquake threat was true.[55] The rumors they had heard were all very different from what Yamamoto had actually said and were something of a smorgasbord of alarming bits of news. According to one of the prefectural police units working on the disaster, it received three hundred calls between 2:00, around the time the first television coverage of the "aftershock information" started, and 5:00 that evening. Among the things callers wanted to know were, "Is it true that there will be an earthquake between 4:30 and 5:30 today, as the prefectural police have announced?" (from a call made at 3:15 p.m.); "I heard that there is going to be Shindo 6 or 7 earthquake in the Yokohama area within the next two hours. Is that true?" (4:47 p.m.); "I heard that both Kanaya and Shimada Elementary Schools have been evacuated because of the earthquake information. Is that true?"[56] In some communities, spontaneous evacuations did happen, as frightened residents fled their homes and gathered in parking lots and open fields, bundled against the cold. Supermarket shelves fully stocked that morning were said to have been stripped bare by nightfall by shoppers anxious to lay in supplies before whatever was going to happen, happened.

Because these reactions were so unexpected and so widespread, prefectural officials realized quickly that they had a problem on their hands. The governor issued a statement at 4:30 assuring citizens that there was no immediate danger of an earthquake and, soon afterward, both NHK and local television stations interrupted regular broadcasting several times to make that same point and to let people know that the rumors of an imminent earthquake were simply misunderstandings. By the time prefectural officials made several hastily arranged appearances on the early evening television news shows to further calm the waters, the volume of panicked telephone calls to the police and other offices was already dropping off.[57] Those who had evacuated made their ways back home.

The "Aftershock Information Panic" was frightening but thankfully short-lived for those who were in the middle of it. For those watching it unfold from afar, the search for "lessons" from what had happened began almost immediately. Much as Hagiwara had promised to learn from the mistakes he and his colleagues had made leading up to the Izu-Ōshima earthquake, the panic offered opportunities to understand how a seemingly simple message could be corrupted as quickly and thoroughly as the governor's had been. The Science and Technology Agency conducted phone interviews with hundreds of Izu residents just a week after the event to find out what they had heard, from whom, and what actions, if any, they had taken in response.[58] Researchers from the University of Tokyo later interviewed officials and hundreds of people affected by the panic and determined that none of the routes the announcement traveled to reach the public had guaranteed an accurate reception. In addition, most people seem to have believed that what they were hearing, however unlikely, was probably true. Roughly a fifth of the people the researchers talked to were thoroughly convinced that a powerful earthquake was imminent; another 60 percent were more sure than not that one was coming. Only one in six told interviewers that they thought such reports must be a mistake.[59]

One of the other questions that researchers asked was how people felt after the fact about the government's decision to issue the information about the aftershocks. Were they angry? Did they wish the governor had kept the news to himself? According to a Science and Technology Agency spokesman, more than 70 percent of respondents had positive things to say about the decision to make the announcement. They either understood why it had to be done or suggested that it was for the best if information like that was released.[60] This was all potentially useful information; the government's intense interest in understanding precisely what had gone wrong and how to prevent it from happening again had been fueled, at least in part, by changes in its thinking about the Tōkai earthquake. The events that led up to the panic were similar to those that one could expect to see if officials were ever in a position to warn the public that an earthquake was expected.

Before the Izu-Ōshima earthquake and the aftershock information panic, preparing for the Tōkai earthquake had not been one of the government's priorities. Afterward, it was. A similarly damaging earthquake anywhere in the country might also have helped build support for new legislation just as well, but it did matter that the one in January had been in Shizuoka. People were reminded that a much more powerful earthquake still threatened the entire district. It also meant that Governor Yamamoto would be invited to the capital to meet with legislators and appear before Diet committees to describe the damage to Shizuoka. Yamamoto was quick to turn those sessions into opportunities to remind his audiences of the larger threat still waiting in the wings and to lobby once again for the new bill.

The Daishinhō

The earthquake on January 14 and the aftershock information panic a few days later put Shizuoka's earthquake problem in a new light. The administration moved quickly to signal that it was embracing at least some of the elements of Yamamoto's proposal. As early as January 17, Prime Minister Fukuda said that the Cabinet was considering new "major earthquake policy" legislation. That same day, the LDP's Sakurauchi Yoshio (1912–2003), then serving simultaneously as Minister of Construction and as director of the Land Agency, let it be known that he expected to submit such a bill during the current Diet session and that it would be based on the one that Yamamoto had been working on.[61] The Land Agency's earlier objections to Yamamoto's proposal, including its criticism that the bill's assumptions about earthquake prediction were flawed, were set aside.

Five months passed between Fukuda's about-face in January and the Diet's final vote to approve the new law, by then renamed the Large-Scale Earthquake Countermeasures Act (Daikibō jishin taisaku tokubetsu sōchihō, or Daishinhō for short). Politicians in both chambers showed a surprising degree of interest in the draft legislation. In the Lower House's Special Committee on Disaster Countermeasures hearings alone, thirty-nine different legislators raised questions about the proposal between April 13 and May 1, an unusually high number for a typical bill.[62] The press also tracked the legislation closely as it made its way through committee meetings and general sessions in both houses. The *Asahi* took the unusual step of publishing the entire text of the bill in March, so readers could follow along with the paper's coverage and see for themselves what the government was proposing to do.[63]

Governor Yamamoto appeared before the Lower House's Special Committee on Disaster Countermeasures twice before it voted on the new legislation and used both opportunities to appeal to policy makers to support the act.[64] He reminded them of the damage caused by the recent quake in Izu-Ōshima and then shared experts' recent estimates of how much worse the outcome would be if the Tōkai earthquake were to strike. According to them, Yamamoto stated, one hundred thousand homes in Shizuoka would be flattened and a fifth of the prefecture's population (or around 600,000 people) would be harmed in one way or another as a direct result of the earthquake. Sixteen thousand citizens would be badly injured, another ten thousand killed. Those estimates only reflected the initial effects of the earthquake, he pointed out—none of the secondary hazards had been factored in. "Fires were not included and neither was tsunami damage. The Shinkansen, expressways, and the like—none of that was included. . . . Please, I ask you to fix in your minds that the damage will be widespread and severe."[65] In our pursuit of rapid growth and economic efficiency, Yamamoto argued, "we've ended up creating a country (*kokudo*) that is truly fragile in the face of earthquakes and other disasters."[66] "Underground shopping centers, skyscrapers, *kombinato*, nuclear power plants, giant dams—there are so many things that have never been through a massive earthquake," he observed.[67]

The Daishinhō did not realign the country's priorities to favor safety over growth (it didn't even realign Shizuoka's) but instead proposed a series of novel interventions that could, in theory, preempt some of the worst possible effects of a Tōkai earthquake.[68] One of the issues that had come up quite a bit in the aftermath of Ishibashi's announcement was that there were no legal or administrative frameworks through which warnings of an imminent earthquake could be transmitted. The JMA couldn't, the Yochiren wasn't designed to, and even if the Hanteikai's members decided that an earthquake was coming, it wasn't clear who they should tell or to what end. The Countermeasures Act addressed parts of that puzzle in two ways. First, passage of the new law would simultaneously amend the Meteorological Service Act to broaden the definition of the information the JMA was permitted to share. Before the amendment went into effect, the JMA could issue warnings and forecasts for floods, tsunami, meteorological phenomena, and so on, but not for earthquakes. The amendment added language to the Meteorological Service Act giving the director-general the authority to tell the prime minister that "the occurrence of a large-scale earthquake pertaining to the Areas for Intensified Measures against Earthquake Disasters" was possible, "based on the results of observations and research with regard to terrestrial phenomena, tremors and ground deformations, terrestrial magnetism, terrestrial electricity, hydrological phenomena, and surveys of land and water areas concerning earthquakes."[69]

The second way the new Countermeasures Act clarified how information about predictions would be managed was by stipulating that, ultimately, it would be up to the prime minister to decide, in consultation with the Cabinet, whether or not to issue an "earthquake warning statement" (*keikai sengen*) based on what he had learned from the director-general. A warning would not automatically follow the receipt of "earthquake prediction information" from the JMA, in other words. If the prime minister rejected the JMA's findings outright or simply chose not to decide, no action would be taken, no measures announced. There was nothing in the law that required either of them to immediately share the director-general's thinking or what he had passed on to the prime minister with the public.[70]

If the prime minister ultimately decided to go ahead and issue an earthquake warning, however, its contents would include instructions to prefectural and other officials to begin putting emergency measures into operation. The Daishinhō piggybacked on the existing provisions of the Disasters Countermeasures Basic Law by requiring communities in zones designated by the prime minister as "Areas Under Intensified Measures Against Earthquake Disaster" to add interventions to their existing disaster plans that would go into effect after an earthquake warning was issued.[71] (In 1978, it was assumed that the Tōkai district would be the first to be named as an "Area Under Intensified Measures" and that the southern Kantō district might not be far behind.) The Daishinhō also empowered central and local governments in ways that the Disaster Countermeasures Basic Law had not. Many of the provisions that had been in Yamamoto's proposal of the previous year, such as language authorizing the prime minister to preemptively dispatch the SDF to threatened districts or to enforce limits on travel in or out of the region, had become part of the new draft act. Provisional scripts drafted by the NHK for broadcast after a warning announcement included directives to viewers (or listeners) in the affected area to stay off the roads, to extinguish any open flames, and to prepare for evacuation and firefighting.[72] Those outside the area were asked not to attempt entry, to obey traffic control measures in the surrounding districts, and to help keep emergency lines of communication open by not making telephone calls to anyone inside.

Locally, public spaces such as schools, hospitals, large stores and the like were expected to either evacuate or take other steps to reduce the risk to those who hadn't left. Factories and businesses would have to shut down or come close to doing so, depending on how vulnerable their facilities were and how hazardous the materials they handled might be. Some roads or highways would be closed entirely, and travel significantly restricted within the affected zone. The new law imposed hefty punishments and fines on drivers who failed to follow orders. It also specified that all the costs associated with any of these measures—lost wages, lost sales, or any other expenses incurred because of the slowdown—were to be borne entirely by the business owners or private citizens who carried them out. The state wasn't going to compensate anyone, even if the earthquake warning turned out to be a false alarm.

Together these measures would mark a sharp break between "normal" time and whatever this new post-warning status might be called. They would also highlight the boundaries between "intensified areas" and the zones around them that were not implicated by the Daishinhō. The prime minister's warning would transform boundaries that, up until that moment, would have been visible only on maps into borders of another kind. On one side of the line, people in an "intensified area" would enter a time and space of forced expectation, diminished mobility, and slowness. On the other, no such transformation was mandated and the pace of daily life could proceed normally. There would of course be spillover effects but there would be no mistaking the differences between the parts of the country affected by the new law and those that were not.

How long they would have to wait was unknown. What was clear is that they would be waiting for one of two things to happen—either there would be an earthquake, as the warning had implied was likely, or the prime minister would announce that the warning statement had been cancelled. Either outcome was understood to be possible, since the Daishinhō included language explaining that a warning could be withdrawn "when the prime minister recognizes that the possibility of the occurrence of the earthquake which has been warned had disappeared upon receipt of additional earthquake prediction information from the director-general of the Meteorological Agency."[73] As was true for the process that led to the release of the original warning statement, the prime minister had to rely on input from the director-general before a warning could be rescinded, but it could be done.

Earthquake Prediction and the Countermeasures Act

The section that follows examines how legislators, agency and ministry officials, and the by-now-familiar scientist triumvirate of Asada, Hagiwara, and Rikitake saw earthquake prediction in relation to the new law. The latter two groups—officials and scientists—were regularly challenged by the first to explain prediction's capabilities. From February 1978 on, when a legislator raised a question about the law's goals or functions, more often than not the government's reply would come from either the Land Agency director,

Sakurauchi Yoshio (a career LDP politician), or the JMA's Suehiro Shigeji (a seismologist).[74] If the question had anything to do with earthquake prediction's capabilities and its roles in the new legislation, it was even more likely that Sakurauchi or Suehiro would be the ones to answer. Sakurauchi appeared before the two Disaster Countermeasures committees eighteen times between February and June; Suehiro was present for just under a dozen. The two grew into their roles as enthusiastic spokesmen and advocates for the law.

Asada, Hagiwara, and Rikitake also appeared before legislators but much less often than their government colleagues. Hagiwara made two appearances in February, when he spoke to the Disaster Countermeasures committees in both chambers.[75] Asada and Rikitake were guests of the Lower House's Committee for the Promotion of Science and Technology in March and Asada spoke to that same chamber's Disaster Countermeasures Committee again in April. Legislators pressed Hagiwara and the others for their thoughts on a range of prediction-related topics as part of their conversations about the Daishinhō. Some asked the scientists to comment directly on the proposed law, while others focused on long-standing issues with the prediction effort, including centralization, funding shortfalls, and comparisons with China's prediction program, to name but a few of the directions that those discussions took. The scientists differed with each other on some issues but on the big questions they were like-minded. When asked to reflect on the current state of earthquake prediction in Japan, they were mostly in agreement on where things stood, as this exchange between a legislator and Asada suggests:

> *Legislator:* When the Yochiren was created and when the Hanteikai was set up, both of those were widely reported in the papers and the like, and the impression I get is that many Japanese have come to expect that with prediction techniques arising from all these models it must mean that [earthquakes] can be predicted with a lot of accuracy. But based on what I've learned from you, my sense is that—whether it's prediction theory or prediction technique—that research is just getting started.
>
> *Asada:* As you said, I think it is exactly right to view this as just getting underway. It is just getting started. I should add this about the Tōkai district, which is that what we're doing right now is putting

> a lot of things in place (and trying them out). Because we don't know how many instruments will be enough in the Tōkai area, we'd therefore like to put as many in place as possible, watch what's happening, and do our best to make it possible to make a prediction before [an earthquake happens].[76]

Rikitake agreed. "The realization that earthquake prediction is just getting started," he said, "is little-by-little becoming the prevailing opinion."[77]

Later that same day, a different legislator quoted a sentence back to Asada from an article the professor had just published in *Shizen* (*Nature*). "Is Earthquake Prediction 'Big Science'?" was Asada's attempt to grapple with questions of scale and the debates over centralization in the field, and to work through some of the implications of prediction research becoming more like "big science" than it already was.[78] What the legislator was interested in was Asada's observation in the article that "Earthquake prediction stands at the threshold of success." So far, there had been lots of discussion about when and how long-term and short-term predictions might be possible, he said, "but from a layman's perspective, when I read this word 'threshold' [*tobaguchi*], I take it to mean something that is very close. How should I be interpreting it?"[79]

"It's a very tricky expression," Asada said, "but what I meant by '*tobaguchi*' was that, if things go well, then that threshold will be crossed and, if they don't go well, then we won't make it across. That was the feeling I was trying to capture. . . . We can't say whether we can make an earthquake prediction, but we can't say for sure that we can't either. What that implies is that there is a very good chance that earthquake prediction is possible."[80]

Hagiwara, Rikitake, and others made similar observations about the distance between prediction's capabilities in 1978 and what they hoped to achieve some day. Hagiwara was asked by legislators what he thought about the law in his roles as head of the Yochiren and of the Hanteikai. It was an unfair question, to some extent, since Hagiwara attended both Disaster Countermeasures Committee meetings before a full draft of the act would have been in wide circulation. Some of the act's priorities had been discussed in the press at that point, but details about its provisions were just beginning to emerge. Even so, Kodaira Kohei of the Kōmeitō was blunt in his own assessment of the law, which he offered before asking Hagiwara what he thought of it. Kodaira put it this way: "We find ourselves in a situation in which neither the capability to respond to earthquake information if it were to be produced, nor the technology for prediction itself have been developed, but we're being compelled to consider this special law. In such a case . . . what are your thoughts with regard to the creation of such a law, from your perspective, as it were?"[81]

Hagiwara avoided having to agree or disagree with what Kodaira had said about the law's weaknesses. He instead moved the discussion to a framework that he was more comfortable with—one in which he could discuss prediction as encompassing not just one method or outcome, as Kodaira was doing, but several. The biggest problem that the new law was going to face, Hagiwara suggested, was the accuracy of any predictions associated with it. Objective, consistent methods like the ones that produced reliable weather forecasts didn't exist yet for earthquake prediction. But, he pointed out, that was not to say that it was impossible for predictions to be right. There had been many instances in which useful forecasts of earthquakes had been possible, even if they didn't meet the formal definition of a prediction. It was unlikely that they'd get to the point any time soon where all their predictions would be correct, Hagiwara noted, but he reminded legislators to consider how different the situation was in 1978 compared to what it had been just ten years earlier. The technology had improved and would likely continue to do so; the Yochiren was able to put out long-term forecasts, a Hanteikai had been created as a temporary measure just for the Tōkai district, and real-time monitoring of a wide array of instruments was also underway.[82] While he never directly addressed Kodaira's claim that prediction wasn't ready to do the work the law needed it to do, Hagiwara's response implies that even if its capabilities were limited at the moment they were unlikely to remain so forever.[83]

By the time Asada and Rikitake appeared before the Disaster Countermeasures Committee in March, the role that scientists such as themselves would play in the events leading up to the act's implementation was coming into focus. The exact processes leading up to a decision by the JMA director to share "earthquake prediction information" with the prime minister were not spelled out in the act, but it wasn't a mystery where warnings about an imminent Tōkai earthquake would be coming from. Asada, Hagiwara, and Rikitake had already agreed the previous April to serve on the Tōkai District Assessment Committee. That the Hanteikai existed at all and that they were all part of it implied a collective acceptance of the idea that a Tōkai earthquake might be predictable at some point in the near future. The act itself therefore didn't have to directly address questions about whether prediction was possible or not, or come up with a mechanism for producing one.

Legislators were nevertheless curious about how the scientists felt about the responsibilities they were being asked to carry. Some of the slipperiness of their situation is evident in an exchange between Rikitake and the Japan Communist Party's Sesaki Hiroshi, in March. The legislator had prefaced his question by reading excerpts from a draft of the act, including sections that referred to the release of "earthquake prediction information" and subsequent declarations of an "emergency alert" by the prime minister. What, Sesaki wondered out loud, did the professor make of all this?

"My assessment is that it is a big step forward for a law like this, a special law for earthquakes, to come about," Rikitake responded. He went on:

> My impression though, and maybe this is to be expected, is that its main focus is on disaster preparedness [*bōsai*]. Prediction was touched on in just a summary fashion. I wonder if the reason for that isn't that prediction has not yet left the developmental stage. As has been brought up a couple of times already, any debate over whether a decision [*hantei*] is possible or not has been omitted, and the outline—maybe you would say the draft—of the law has been created on the assumption that a judgement is possible. I think that there are at present cases where a judgment might be possible, but I also think it would be okay to go so far as to say that it could only be in the Tōkai district. For example, if you were to say that we should make a decision about an earthquake in Kyushu, there's isn't a thing we could do, since no [preparations] at all have been attempted there.[84]

Rikitake's comment is worth quoting at length both because it offers some insight into how scientists like him saw the law and because it captures some of the ambiguities of the discussions surrounding it. Here, Rikitake seems to be implying several contradictory things at the same time. He characterizes earthquake prediction as still in the early stages of its development or, in other words, as not yet ready to be counted on in practice. He also points out that the law is based on an assumption about judgements and that it deliberately avoids the debate over whether they are possible or not. On the other hand, Rikitake does not come out and say that the assumption is wrong. In fact, he offers up that in at least one location—the Tōkai district—a "judgement" may be possible. The way forward, Rikitake ends up suggesting, would be for them to begin rethinking how they go about making those judgements and perhaps remaking the way that process is organized. It almost seems as if Rikitake is coming to grips with this new law's implications for the Hanteikai's work for the first time, as he is talking about it with Sesaki.[85]

If legislators had been listening carefully, the overall message that Hagiwara, Asada, and Rikitake ended up sending about predicting a Tōkai earthquake was that it might be possible to do so but that success would be highly contingent on a great many things all falling into place in exactly the right way at just the right time. Their answers didn't undermine the premise behind the Hanteikai, since if they did get lucky and the pieces did start to fall into place, it would be important to have a structure like the Assessment Committee there to review the data. Along the same lines, their answers also didn't directly challenge the premise that an earthquake prediction could be used to trigger interventions like those described in the Daishinhō. If they were lucky and if the Hanteikai agreed an earthquake was imminent, then perhaps it did make sense to have disaster prevention measures ready to go, just in case.

What their answers did do was make the case that earthquake prediction even in the Tōkai district was going to be exceptionally difficult. None of them were confident they would be able to tell which anomalous phenomena were precursors to an earthquake and which were not. It might take many false alarms before they could be sure. "If we were to live for several hundred years," Asada speculated, "and experienced a Tōkai earthquake many times, and we had the appropriate equipment in the right numbers in the right places, then maybe we would be able to provide very accurate predictions."[86] But, he went on, the next Tōkai earthquake would be their very first, they didn't yet have all the equipment in place that they'd like, and, because they'd never done this before, they didn't even really know how much equipment was enough.[87] Asada added to this assessment, a few weeks later, in response to a question about the strengths and weaknesses of the current monitoring system, and whether predictions were possible based on the information it would provide. "For the Tōkai area, it depends entirely on how it unfolds," Asada replied. "If, through chance, a sufficiently large precursory sign appears over a sufficiently wide area, then I think it might be possible to say that something would happen in ten hours or so with a good degree of accuracy."[88] "But this is our first time doing this," he went on to say, "and we don't know what to expect, so we have to make every possible preparation."[89]

Magnitude Matters

What government spokesmen told legislators about earthquake prediction during discussions of the draft Countermeasures Act differed from what the university scientists said. Where Asada, Rikitake, and the others emphasized how difficult and imprecise short-term earthquake prediction was and said that they would recognize the indicators of a coming Tōkai earthquake before it happened only if they were very lucky, Sakurauchi, Suehiro, and their colleagues in other agencies insisted to legislators again and again that they were all but certain that the next Tōkai earthquake could be predicted. These men talked about the nation's ability to predict an M8-class Tōkai earthquake as a given.

The disconnect between the scientists and the government spokesmen is the latter's doing. As the act was taking shape, Hagiwara and his colleagues continued to hold the same positions about prediction's viability as they had the previous year and the year before that. They were generally optimistic but unwilling to guarantee that short-term prediction—even in the Tōkai district—would be within reach soon. Prior to the start of the Diet's discussion of the Daishinhō, there hadn't been much daylight between the positions the scientists were taking on prediction and the answers that Sakurauchi, Suehiro, and others in their cohort shared publicly. If anything, agency officials like Suehiro had tended to defer to Hagiwara and the others on prediction related issues. Once the Lower House began to engage in earnest with the draft legislation, however, these same officials began providing assessments of earthquake prediction that diverged from what Hagiwara and the others had been saying up to that point.

Here, for example, is how Sonoyama Shigemichi (Science and Technology Research Coordination Bureau) responded in late March 1978 to one legislator's questions about prediction's capabilities:

> Because of the energy believed to be stored in the Tōkai district, a wide array of monitoring methods has been deployed there. If these observation methods are focused, then the professors of seismology

> are now confident that if a large earthquake were to occur, then without fail its precursory phenomena would be captured, whether several hours or a day or two ahead of time. In that sense, it is definitely possible to capture precursory phenomena immediately before [an earthquake].[90]

A week later, Sonoyama gave an almost identical answer when asked in a committee session in the Upper House whether it was true that the JMA would absolutely not fail to predict an M8 or larger earthquake. "At present, if an M8 level event were to happen," he said, referring to the Tōkai district, "the precursory phenomena would certainly be captured. We have not gone through this before, so it cannot be said clearly when they will be detected but, in the opinion of the majority of scientists, unusual phenomena will certainly be evident several hours or a day or two before."[91] "The way the scientists often put it," Sonoyama continued, "is that they may swing and miss [raise a false alarm] but they will never let a good ball go by, and I trust that such is the case."[92]

The next day, Suehiro offered legislators in the Upper House a slightly more cautious version of the same argument. "When it comes to an M8-class, very large earthquake," he said, "if we deploy a densely concentrated observation net in the area where it is expected to occur, and if we subject the results arising out of that observation net to constant monitoring, day and night, it is our assessment that we are now at a stage where we can detect the precursory phenomena that will lead to the earthquake prediction information referred to in the draft law."[93] Suehiro's path to prediction had more contingencies in it than Sonoyama's but several of the qualifiers that he mentioned were already in place, as we've seen, and the others were not out of reach.

One of the inflections that Sonoyama and Suehiro introduced, at this point, was that they began to refer not just to "the Tōkai earthquake" as the focus of prediction efforts central to the act's success but specifically to an M8 Tōkai earthquake as the key to the whole process. It may seem like a distinction without much of a difference. The reason everyone was worried about a Tōkai earthquake in the first place was because it was expected to be very powerful. Still, this was not a qualification that Asada or the others had employed in their discussions of when prediction might or might not work; they were more likely to distinguish between earthquakes below and above M7 when discussing prediction's best chances for success. Asada had mentioned M8 events in the Diet in March when talking about how far in advance of a major earthquake precursory phenomena might be evident, but he hadn't been making a point about prediction's capacity to detect signals in the near term and it hadn't been in reference to the Tōkai district. And, while many estimates of how powerful the next Tōkai earthquake would be did put it in the M8 range, "*magunichūdo hachi*" really wasn't part of the conversation around the Countermeasures Act until Suehiro and his colleagues made sure that it was.

This exchange in early April between a legislator and the director of the Geographical Survey Institute is illuminating. The legislator began by acknowledging the work the Yochiren had been doing over the past eight or nine years in pursuit of putting prediction into practice and its recent focus on predicting events in the M8 range. "However," he continued, "so far there hasn't been even a single instance of this happening. There hasn't been a single instance of a prediction of when and where [an earthquake would occur]. My understanding, based on the replies that [Suehiro] and now you have provided is that from now on we will be able to do this. Can it be done from here on out, prediction? Are you saying that in the case of a Tōkai magnitude 8 [earthquake] [prediction] is possible?"

The GSI director's reply—"That is correct" (*sō de gozaimasu*)—was unambiguous.[94] Sakurauchi offered a more convoluted but similarly upbeat assessment later that month in response to a question about shortfalls in the prediction program: "Now, as for the law we've brought before you, its reliance on predicting a magnitude 8 [earthquake] and the question of whether or not that can be accomplished, given that data is even now being collected centrally [*ichigenteki ni*] by the JMA and a 24-hour monitoring system is in place, what is there to worry about?" [Literally: "isn't it that there isn't anything to worry about?"][95]

One of the reasons that officials like Sakurauchi began emphasizing that magnitude was the determining factor in whether an earthquake could be predicted or not was that it helped them respond to legislators' questions about why Tokyo and other areas of the country were not being offered the same protections that the Daishinhō would provide to the Tōkai district. Once they began arguing that prediction was possible, Suehiro and the other officials were pressed to explain why it wouldn't also work for Tokyo, or the northeast, wherever damaging earthquakes also threatened. Although one response might be that prediction was dependent on a certain density and sophistication of monitoring equipment that so far existed only in the Tōkai district, the next question was sure to be why wasn't the government rushing to replicate that capacity elsewhere? If the only barrier to a successful prediction was whether or not the state was willing to invest in the infrastructure necessary to do so, shouldn't the JMA already be building new monitoring networks all over Japan? The Yochiren had already designated multiple zones for close observation, so, clearly, the experts already had a good idea of where they ought to watch. Why weren't they prepared to predict the next major earthquake in those locations too?

Reframing the discussion about the Daishinhō and prediction specifically around an M8 Tōkai earthquake made it fairly easy to explain why the rest of the country wasn't getting the same treatment. If the only earthquakes that generated precursory phenomena that scientists were confident they could detect were ones in the M8-class—and this was the argument Suehiro and the others were using—then almost all of the other potentially damaging events that the Yochiren and the public knew about were not, by that definition, predictable, at least not yet. An earthquake directly beneath Tokyo would likely be in the M6–M7 range. The next "Great Kantō Earthquake" might well be in the right magnitude category but there seemed to be a solid consensus that it wasn't going to happen anytime soon. Of the other zones under close surveillance by the Yochiren, few were thought to be capable of producing an M8-class earthquake under circumstances that would allow precursory phenomena to be observed. The faulting mechanisms that had historically produced massive M8-range earthquakes off the coasts of northeastern Japan and Hokkaido were expected to do so again, at some point, but there was simply no way for researchers to monitor those processes with any precision.[96] So far as anyone knew, in other words, the only M8 earthquake threatening Japan in 1978 that would produce observable precursory signals was the one that Ishibashi had told them to expect in the Tōkai district.

In mid-April, Shizuoka's Harada Shōzō gave Sakurauchi an opportunity to explain why they were limiting themselves to M8-and-above events. He also asked why they weren't going to address threats like an inland earthquake in Tokyo, "which even though it might be [just] a magnitude 7 [earthquake], can be expected to bring about a great deal of damage."[97] To take the last question first, the director said that, "with regard to earthquakes of magnitude 7 and below, the current state of the art is such that the relevant prediction techniques are still in the research phase. We have not yet reached the stage at which such predictions can be made with any degree of confidence."[98] On the other hand, Sakurauchi went on to say this: "We are at the point where we can capture the precursory phenomena associated with large-scale earthquakes of around magnitude 8 in the zone of enhanced observation, as a result of earlier planning that resulted in concentrated, enhanced observation there. This is the premise for the 'prediction information' [*yochi jōhō*] in the special facilities law."[99] Later that same day, the director noted that their ability to "satisfactorily predict in the event of a magnitude 8" earthquake wasn't a product of changes in organization prompted by the new law but, instead, reflected results of work that had been under way for some time.[100]

Sakurauchi and Suehiro returned to these themes again and again during interactions with legislators in both chambers for the rest of April and on into June. They used the same terms and often the same phrases to assert that M8 earthquakes could be predicted and that less powerful earthquakes could not. As Suehiro explained it to members of the Lower House Disaster Countermeasures Committee after they raised questions about why the smaller events were out of bounds, "As you've pointed out, magnitude 8 and 7 differ by only one step but, in fact, their energy as natural phenomena differ by a factor of thirty. Thus, there is a remarkable difference between magnitude 8 and 7 in terms of the degree of articulation of the precursory phenomena, the range over which they will appear, and so on. Given our present monitoring network and scientific capacities, unfortunately, magnitude 7 earthquakes cannot be predicted with the certainty necessary for the purposes of disaster prevention."[101]

One of the reasons that the arguments Suehiro, Sakurauchi, and the others made are so fully fleshed out in the record is that they kept having to repeat them. Agency heads and their spokesmen were called on again and again to explain why they thought prediction would work. Suehiro, Sakurauchi, and other officials together offered more than a dozen separate statements about the predictability of an M8 Tōkai earthquake to Lower House legislators on April 18 alone. Three days later they were back in front of the same committee, responding to another round of questioning on the same topics. Their answers didn't vary much from one response or meeting to the next and they faced fewer questions about prediction in the Lower House as a vote on the law approached in late April, but they never went away altogether. Once the discussion shifted to the Upper House, legislators there were generally more interested in having Sakurauchi and Suehiro confirm that prediction was going to work than they were in raising questions about its limitations but, even then, it was clear that legislators were making the connection between the act's viability and prediction's capabilities.

Second Thoughts and Last Steps

Is it possible that scientists like Hagiwara, Asada, and the others were just unaware of the claims that Sakurauchi and others were making about earthquake prediction and the Tōkai earthquake? Suehiro and his colleagues were present on several of the occasions when Asada, Hagiwara, and other scientists gave their analyses of prediction's capabilities that spring. Suehiro was there with Hagiwara in the Upper House on February 9 and present on April 9 when Asada appeared before the Lower House's Disaster Countermeasures Committee. It seems reasonable to think that agency spokesmen in general, and people like Suehiro in particular, were familiar with what Asada and the others thought about prediction and how they talked about its capabilities. (Suehiro had been a member of the Yochiren since 1969; he was a diligent attender of its meetings.) At no point did the government spokesmen lead legislators to believe that they disagreed with the more cautious assessments that Asada, Hagiwara, and the others had offered, and that their reading of the situation differed from that of the scientists. Quite the contrary—when making a case for the Countermeasures Act, officials like Suehiro represented their argument about the predictability of the Tōkai earthquake as if it were the scientific consensus.

It would have been harder for the university scientists to keep track of what Suehiro, Sakurauchi, and the others were telling legislators. With a few exceptions, their comments during Diet committee meetings seldom made the papers. Press coverage of the debates over the Daishinhō at that point focused more on questions about the SDF's role and the scope of pre-quake interventions by the state than on the nuances of prediction. None of the scientists were present in person—at least not as official participants—when government spokesmen told legislators that they would almost certainly know ahead of time if an M8 Tōkai quake was coming, for example. Neither Asada nor any of his colleagues publicly challenged that argument at the time.

Mogi Kiyō, who himself testified about prediction's capabilities in the Diet in 1978, suggested many years later that he and others in the earth sciences community were not aware of the extent to which Suehiro and the others were making such optimistic and definitive claims about prediction's capabilities.[102] It was only after he took over from Asada as chair of the Yochiren and the Hanteikai in 1991 that Mogi says his curiosity led him to start reading through Diet transcripts from the sessions leading up to passage of the Daishinhō. Mogi thought he might have been among the first in the earthquake sciences community to have gone through those records.[103]

Mogi wrote that he was shocked by what he found. His reaction was prompted in part by the disconnect he saw between what he and his colleagues had said about prediction in 1978 and what government spokesmen had insisted would be possible under the act. Some of the exchanges that stood out for Mogi as clear illustrations of a gap between the two camps are ones we've already encountered in this chapter; others make similar arguments. Mogi points to Asada and Rikitake's cautious assessments in front of one of the Lower House committees in March and to a number of the more optimistic assertions made by Suehiro and others in April and May to help readers see his point about how different the scientists' perspectives on prediction were from the positions the government spokesmen ended up taking.[104] Mogi's effort to draw attention to the gap between the experts' cautious approach to prediction and the government's more enthusiastic embrace of it came just a few years after the Hanshin-Awaji Earthquake, at a time when neither earthquake prediction in general nor the Daishinhō were held in high regard. While it is clear from Mogi's analysis that he thinks the gap is significant in some way to our understanding of the act's passage in 1978, he doesn't tell us much more than that. He does not say, for example, that legislators were somehow tricked into voting for the law by government officials who knowingly misrepresented the science to them (although that is a conclusion one might draw). Nor does he argue that the act would not have passed if he and other Hanteikai scientists had directly challenged Sakurauchi, Suehiro, and the others' claims. His point rather seems to be that what emerged at that point as a widely shared narrative of prediction's capabilities was significantly different from how scientists themselves understood the problem.[105]

Closing statements and a vote in each of the two Disaster Countermeasures Committees preceded the final votes in both houses and it was in those settings that legislators signaled whether or not their party planned to support the law and why. Representatives of the Komeitō and the Minshatō (Democratic Socialist Party) had both indicated in those sessions that their members would be voting in favor of the act. Since, together with the LDP—whose support for the law was a given—the three parties held about two-thirds of the seats in both houses of the Diet that year, their support meant that it was all but certain that the law would have the votes it needed. Party spokesmen justified their support for the act by pointing out how vulnerable the country had become to the dangers earthquakes posed. Urbanization, the growth of complex transportation and communication networks, and the concentration of industry and businesses in a handful of densely populated areas were among the changes they said had left Japan more at risk from a devastating earthquake than ever before. Several juxtaposed those vulnerabilities against the threat of the magnitude 8 earthquake that scientists expected to strike the Tōkai district soon and noted that they could hardly ignore the appeals from elected officials and the residents of those areas for the state's help. The Daishinhō's emphasis on coordinated planning and interventions to address the shortfalls in existing disaster prevention efforts, they felt, was the right response.[106]

Only the Japan Socialist Party (JSP) and the Japan Communist Party came out strongly against the act. Concerns about the role of the SDF and the new powers granted to the prime minister under the law led the lists of reasons for the parties' opposition. As a spokesman for the JSP put it, in their view, the law's primary purposes were to concentrate power in the hands of the prime minister and to make it easier to deploy the SDF. Disaster prevention and initiatives that would actually protect communities and save lives had, in their opinion, been given far too little attention. Legislators from the JCP generally agreed with that assessment.[107] Both parties made it clear that they were very much in favor of addressing the problems that the law had claimed to concern itself with but, in their opinion, actually hadn't. Where was the financial help that communities needed in order to prepare for the next big earthquake? Why hadn't the new law provided meaningful support for research on earthquake prediction? "There is nothing more important than accurate prediction for reducing the nation's unease and [creating] the right disaster policies," a JSP spokesman told his colleagues, "and yet all the government's proposal has in it [to address this] are the most unpromising things possible."[108] The government had put the cart before the horse, the JSP argued.

The JCP spokesmen also cited their party's longstanding commitment to protecting Japan's citizens from earthquake hazards and the legislation it had proposed to help large urban areas prepare for disasters. They wanted to make it clear that they, too, supported some of the ideas behind the act. "The most important thing right now," as one put it, "is to move ahead with a law (or laws) that will help strengthen communities and the nation against earthquakes, create disaster preparedness policies that prioritize citizens' safety, and radically improve the earthquake prediction system, even as we anticipate the dangers of the next big earthquake."[109] The problem, of course, was that the new law didn't do any of these things well enough to merit moving forward with it. The JCP was, if anything, even more skeptical than the JSP about earthquake prediction's role in the new law. As one JCP legislator put it, "It is obvious that you can't put your trust in a law that relies on emergency measures based on [acts of] prediction that are as yet unproven and unreliable."[110] The Lower House nevertheless voted to approve the new law on May 9. The Upper House did the same about a month later and the Large-Scale Earthquake Countermeasures Act went into effect in December 1978.

CHAPTER 9

Uncertainty and the Hanshin-Awaji Earthquake

Of all the different parts of the new apparatus created in 1978 to prepare the country for a Tōkai earthquake, the Hanteikai was the element that seemed to fascinate the public the most. Pop culture references to the Assessment Committee surfaced right away, as in television dramas like *Great Tokyo Earthquake: Magnitude 8.1* (*Tokyo Daijishin: Magunichūdo 8.1*), first broadcast on Yomiuri Television in April 1980, and in the Tōei feature film *Earthquake Archipelago* (*Jisshin rettō*) released later that same year. Both films take it as a given that short-term earthquake prediction works and both make the committee's assessment of the earthquake threat a key element of their respective plots. In the former, the panel is portrayed in a favorable light; its members do their duty and warn the government that an earthquake will soon strike Tokyo, only to have the feckless prime minister and his cabinet refuse to heed the committee's advice. In the latter, panel members are depicted as narrow-minded and timid. In both, Tokyo ends up being destroyed.[1]

Media coverage of the committee's activities (in real life) may have lacked the dramatic tension of the era's "panic films," but its members and their responsibilities were the frequent focus of news and television reports nevertheless. "The Six Samurai"—one of the *Yomiuri*'s monikers for the panel members—were queried by journalists about their hobbies, how they felt about having to carry a government-issued pager day and night, and whether they were worried about the prospects of an earthquake in the near future. Readers learned, for example, that Hagiwara, who led the group until 1981, liked to work on old cars and was a fan of the Yomiuri Giants baseball team; Asada, who succeeded Hagiwara and stayed on as committee chair until 1991, enjoyed photography and long drives out of the city.[2] Both are portrayed as taking their responsibilities quite seriously and as burdened but not overwhelmed by them. Hagiwara joked in late summer 1979 that his pager had already gone off four or five times in the few months that he had it but that all the calls had been wrong numbers. His larger point, building off of the "wrong number" story, was that he didn't think they were at the stage yet where they should expect to be able to spot anomalous signals, although that day might come eventually. A few years later, Asada was confident enough to tell a reporter that because the instruments hadn't detected anything unusual lately, the Tōkai district was therefore probably fine for the next month or so. The same could not be said, however, about what to expect over the next five years. Asada's own opinion, which he realized would not be a popular one, was that the sooner the earthquake happened, the better. The more time passed without one, he said, the worse the destruction was going to be.[3]

Japan's annual Disaster Preparedness Day exercises were another venue in which the Assessment Committee and the new methods for dealing with a Tōkai earthquake came to play prominent roles. The exercises designed to prepare vulnerable communities for a Tōkai earthquake were supposed to simulate a coordinated response across as many as ten different prefectures and metropolitan areas and, as such, followed a carefully scripted timeline. The schedule officials adopted in the early 1980s and stuck with for years worked like this: at 7:00 a.m. on September 1, the JMA would send an alert to the Assessment Committee members' pagers, summoning them to duty. Police patrol cars dispatched to the members' homes would collect the scientists and rush them to the JMA headquarters. (Those dramatic scenes were sometimes broadcast live.) There the scientists pretended to review whatever unusual developments had supposedly prompted the summons and, as per the plan, quickly determined that the threat was real. At 9:00 a.m., the prime minister would issue his warning announcement, setting in motion some of the preemptory steps that the Countermeasures Act had made possible. Traffic on preselected stretches of highway would temporarily come to a halt or be diverted to allow emergency vehicles to pass; trains on some rail lines would reduce speed or, in some cases, come to a complete (if brief) standstill. Communities in the Tōkai district were encouraged to use this time to perform mock evacuations and make other preparations prior the next event in the day's program, the earthquake itself. Participants knew to expect it at 11:00 a.m. sharp.[4] By the mid-1980s, government estimates put the number of people participating in coordinated exercises in preparation for the Tōkai earthquake at several million.[5]

Public opinion surveys in the 1970s and 1980s suggest that people were aware of the threats posed by future earthquakes and knew about the government's efforts to put earthquake prediction into practice. Nine out of ten Japanese polled in 1980 had heard that a Tōkai earthquake might happen soon, more than four out of five were aware of the government's efforts to provide advance warning of an earthquake, and just over 70 percent claimed to be familiar with at least some of the provisions of the Daishinhō.[6] Roughly two out of five residents of the Kantō and Tōkai districts that year believed that a major earthquake would likely occur near where they lived within the next decade.[7] Confidence in prediction's capabilities was high. "Given current levels of science and technology," pollsters asked in 1978, "to what extent do you think it is possible to know ahead of time when a large earthquake will occur?" While only 6 percent of those responding were convinced that earthquakes could already be predicted with absolute certainty, another 53 percent believed that prediction already worked "to some extent."[8]

The effects of the continuing concerns over Japan's vulnerability were evident in other registers as well. The budget for the 1974–1978 Earthquake Prediction Plan had come in at slightly more than 12.8 billion yen; plans to expand the monitoring networks in the Tōkai district (and elsewhere) and to create the infrastructure needed to monitor the data they produced grew the cost of the next five-year plan (1979–1983) to an impressive 31 billion yen.[9] The new funding was put to use, in part, to install new seismometers and a variety of other tools to track a wide array of potential indicators of seismic activity. By the early 1980s, universities and the JMA were responsible for close to fifty different instrument installations in the Tōkai district alone, each tied into at least one regional data-collection network. Many transmitted data in real-time directly to the JMA, where technicians and automated systems watched day and night for signs of the next earthquake.[10] The technicians' salaries and those of the staff who kept the monitoring equipment in the field up and running were also part of the fixed cost of keeping the lookout for precursors going.

Finally, the region's communities and residents were also affected by the new policies as they waited for the earthquake to arrive. One of the Daishinhō's stipulations was that its provisions only applied to those jurisdictions that had been duly designated as part of the "Area under Intensified Measures against Earthquake Disaster" (Jishin bōsai taisaku kyōka chiiki). The task of deciding which parts of the Tōkai district belonged and which did not fell to a committee headed up by Hagiwara. By late in the summer of 1979, he and his colleagues had produced a list of one hundred seventy communities spread over six prefectures that calculations suggested would be subject to high-intensity tremors in the event of an M8 Tōkai earthquake.[11] There was almost no pushback when the list of designated communities was first made public earlier in the year, despite concerns that had been raised about the label's potential detrimental effects on land prices, tourism, and so on.[12] Every city, town, and village in Shizuoka was on the list, as were communities in each of the prefectures on its borders: Aiichi, Gifu, Kanagawa, Nagano, and Yamanashi. All were compelled to produce new disaster-reduction plans that took into account the scenarios the Countermeasures Act suggested were now possible, ones in which scientists (via the prime minister) would warn them ahead of time that an earthquake was coming. They were also expected, as part of that planning process, to begin reinforcing public buildings and to encourage private property owners to do the same. Altogether, the *Yomiuri* estimated that almost 5.7 million people lived in the newly designated "danger zone."[13]

It wasn't all bad news for the communities inside the new district's boundaries. Governor Yamamoto had said from the start that any new disaster-reduction initiatives in the Tōkai district would be expensive and that the government was going to have to help the prefectures shoulder that financial burden. After the Daishinhō went into effect, Yamamoto quickly launched another lobbying campaign, one that led to the passage of the Special Earthquake Fiscal Measures Act (Jishinzai tokuhō) in May 1980.[14] The new law did several things for Yamamoto and the governors of the other prefectures that fell under the Daishinhō's jurisdiction.[15] It expanded the categories of projects that governors could include in plans to repair or upgrade "emergency facilities" and it committed the central government to covering a larger share of their costs than before.[16] Lawmakers, aware that communities in the newly designated zone had been told to finish their disaster prevention construction and retrofitting projects within five years, stipulated that the new financial support law would also be in effect for five years, until 1985. That meant that the funding program could run in parallel with new construction and it also reflected the sense at the time that the Tōkai quake was likely to happen sooner rather than later, in any case. Once it did, obviously, there would no longer be a need for money to prepare for it. Legislators left themselves the option of extending the law for another five years, if they made it to 1985 without an earthquake and if there were projects that still needed funding.

Five years later, many communities and prefectures were grateful when the Diet agreed to renew the funding program. There were projects still undone and, even if awareness of the threat of a Tōkai earthquake had faded a bit, scientists hadn't signaled that the threat itself was gone or significantly diminished. Nor would they. Thirty-five years later, in 2015, when the funding law was renewed for the seventh time, the brilliance of Yamamoto's plan was hard to deny. Ministry of Finance efforts to pull the plug on the funding program had been consistently rebuffed. As one Land Agency official reportedly pointed out, the government would be in an impossible position if communities that had been eligible for help were badly damaged in an earthquake after the government had cut off support for projects that might have provided them with some protection. "We're going to have to keep this system going," he acknowledged, "until the Tōkai earthquake strikes."[17]

Prediction's Critics

Not everyone in the Japanese earth sciences community had been convinced from the start that the nation's investments of time and resources into short-term earthquake prediction were a good idea, but the number of researchers who were openly critical of the field's priorities was not large. University of Tokyo geophysicist Takeuchi Hitoshi (1920–2004) had gone public with his doubts about the prediction project in 1973 and was, for years, the only prominent scientist to have done so.[18] Takeuchi's credentials as a researcher were impeccable. He had also made a name for himself from early in his career as the author or coauthor of dozens of popular works on physics, the earth sciences, and earthquakes. He was of roughly the same generation as Asada Toshi and Rikitake Tsuneji at the University of Tokyo but, unlike them, had not been drawn into prediction's orbit. Takeuchi had not been part of the group that produced the Blueprint, was never on the Yochiren's roster, and was not called on to share his expertise with legislators during any of their discussions in the 1970s about prediction's merits, or lack thereof.

Takeuchi's concerns with the prediction project had several layers but most of them are variations on a single argument, which was that the physical processes that produced earthquakes were fundamentally unpredictable, aleatory phenomena. Takeuchi described them as stochastic events (*kakuritsu tōkeiteki genshō*) and, to illustrate what he meant, asked his readers to imagine a piece of thread as someone slowly pulled it taut and then continued to stretch it for as long as it took until the thread broke. The thread's rupture, Takeuchi said, was the equivalent of the first fracture along a fault or elsewhere in the earth's crust that marked an earthquake's first moments. And, just as no one could possibly have told you precisely when and precisely where that piece of thread was going to break, no one could say with any certainly when and where an earthquake would occur.[19] In Takeuchi's assessment, short-term earthquake prediction—which his colleagues had been pursuing since the mid-1960s—was simply impossible.

Takeuchi made the case against the prediction project in additional detail a few years later, in articles intended for a general readership. In them, he again tried to draw attention to what he saw as the prediction project's fundamental scientific flaw, while also making it clear to readers why he thought it mattered that his colleagues remained so committed to it. Takeuchi shared their concerns about the risk of a powerful earthquake striking the capital, or the Enshū Sea district, or off the coast of northeastern Japan, he wrote, but was convinced that the focus on prediction was getting in the way of interventions that might actually save lives and mitigate the worst of the hazards to come. As his titles suggest—"Preparing for the Great Japan Earthquake" (1976), "Earthquake Prediction is not an Earthquake Policy" (1977), and "Wells Rather than Earthquake Prediction" (1978), Takeuchi had given some thought to what policy makers ought to be focused on instead; he had a ten-item list of suggestions just for Tokyo.[20]

Rikitake Tsuneji thought enough of Takeuchi's arguments that he published a rejoinder, of sorts, to them: "Earthquake Prediction is not a Pipe-dream" (Jishin yochi wa yumemonogatari de wa nai) appeared in *Kagaku Asahi* a few issues after Takeuchi's 1978 article.[21] The "of sorts" qualifier stems from Rikitake's decision not to engage directly with Takeuchi's critiques and to instead rely on a set of rhetorical strategies that later defenders of prediction would find useful, as well. Rikitake reminded readers, for example, of what was at stake, given the threat of a Tōkai earthquake, and pointed out that whatever prediction might cost it was still far less than the price the nation would have to pay if that disaster occurred without warning. "Of course," he acknowledged, "there is no proof that the Tōkai earthquake can be predicted," but this did not stop Rikitake from pointing out to his readers that all the new measuring devices and communications equipment procured by the prediction project were generating so much more information about earthquakes than anyone had been able to access before—the implication being that even if they hadn't done so yet, the odds were good that he and his colleagues would soon spot the precursory signals they had been looking for.[22] No matter how much those opposed to moving forward with prediction might criticize them for having jumped the gun, Rikitake concluded, the Japanese public's desire for it was so absolute that expanding the nation's earthquake prediction monitoring capabilities was an inevitability at this point.[23]

Rikitake was right, in that the prediction project's momentum showed no signs of being slowed by Takeuchi's arguments.[24] For more than a decade after that first burst of criticism, there were few if any overt challenges to the project's underlying premises or complaints that funding and other resources were increasingly going to it and not to other research areas. As historian Tomari Jirō notes, there was some grumbling about these and other issues at annual symposia and in other professional settings, but these occasions were easily dismissed, if not encouraged, as a way for disgruntled researchers to "let off steam."[25]

A cross section of the sorts of concerns that these scientists shared can be seen in the debates that surfaced during a two-day Earthquake Prediction Research Symposium in June 1994. There had been other symposia dedicated to prediction research in the early 1990s, but this one differed from those earlier meetings in at least two ways. One was that its organizers allowed specialists from outside the usual earthquake prediction research community to submit paper proposals; the other difference was that the organizers acknowledged that, after thirty years, it was time to "fundamentally rethink how earthquake prediction research is conducted." The symposium, they suggested, was an opportunity to craft "a new blueprint, based on the most recent findings in earthquake science."[26]

Of the symposium's two dozen or so papers and comments, only a handful actually called for fundamental changes to the status quo. These came from scientists who had not been directly involved in the earthquake prediction program. Ishibashi Katsuhiko, for example, offered a carefully constructed critique of Japan's "empirical and phenomenological" focus on the detection of earthquake precursors and suggested that continuing to employ those same methods was unlikely to produce any useful outcomes. "In order to drastically improve the present irrational situation," he wrote, "I propose to disintegrate the current Earthquake Prediction Program and to establish [a] more scientific Earthquake Prediction Research Program as well as [a] comprehensive Earthquake Hazards Reduction Program."[27] This restructuring would provide the support for research on earthquake source physics and other topics that were missing in the current program; in so doing, it was Ishibashi's hope that they would eventually be in a position to evaluate earthquake potential on a national scale and perhaps to provide "generalized forecast[s] of specific events," as well as what he called "'first-aid earthquake prediction' like the Tōkai project."[28] The proposal managed to validate the work that had gone into the prediction program to date, while also making it clear that it was time to bring that chapter to a close.[29]

Ishibashi hadn't been in the public eye much since 1976; the year after he raised the alarm about the Tōkai earthquake, he started work at the International Institute of Seismology and Earthquake Engineering, where he helped develop training programs for engineers and scientists from earthquake-prone developing countries.[30] Ishibashi continued to do research in historical seismology, publishing a number of papers in that area, but it would be fair to say that, for many years, his role in the field and in the public discussions around earthquake disasters in Japan was mostly defined by what he had set in motion in 1976.

That began to change in 1994, with the Earthquake Prediction Research Symposium and with the publication of his new book, *An Era of the Earth's Unrest: A Seismologist Warns* (*Daichi dōran no jidai: jishin gakusha wa keikokusuru*). In it, Ishibashi again raised an alarm, this time about earthquake threats to not just one part of the country but to many.[31] He described what he saw as the troubling vulnerabilities of Japan's largest cities and their built environments, and drew particular attention to the harm that a "directly-beneath" earthquake would do to the capital district.[32] The context for raising the alarm in 1994 was his belief that what had been a period of relative seismic quiet for the country was ending and that people should expect to experience more frequent and more powerful earthquakes than recent history had prepared them for. It was also his assessment, in contrast with the confidence regularly expressed by other experts, that Japan's buildings and its infrastructure were not nearly as safe from earthquakes as people had been led to believe. The Hanshin-Awaji earthquake struck just a few months after the book's publication and the disaster that followed worked to confirm many of Ishibashi's arguments. His warnings certainly seemed prescient. Many who had not been paying much attention to the threat to Tokyo, for example, began to do so. *An Era of the Earth's Unrest* became a bestseller; bookstores had trouble keeping copies on the shelves, and once again Ishibashi was part of a national discussion about earthquake risks and how to respond to them.[33]

A second call for major changes in the status quo came from someone even further outside of the earthquake prediction camp than Ishibashi. Robert J. Geller did his undergraduate and graduate work in geophysics at the California Institute of Technology before becoming, in 1984, the first foreign national to be appointed to a permanent faculty position at the University of Tokyo.[34] His move to Japan had been set in motion years earlier by Hiroo Kanamori, who joined the faculty at Caltech in Geller's senior year and was an important influence on him thereafter. Kanamori had started his own career as a seismologist at the ERI before moving to Pasadena in 1972 and retained close ties with colleagues in Japan, who arrived at Caltech as a steady stream of visiting researchers. Geller attributes his initial willingness to consider a job in Japan to his close relationship with Kanamori and his Japanese friends and told himself he could always come back to the United States if it didn't work out.[35]

Geller ultimately found the position at Tōdai and the work to be a good fit.[36] His research in the Department of Earth and Planetary Sciences focused on computational and 3D modeling of the earth's interior, subjects on which he published extensively. That work did not bear directly on earthquake prediction and there was nothing about his job that required him to engage in any way with that topic. By 1994, however, Geller had already begun to establish himself as the most outspoken critic of earthquake prediction in Japan, which explains why he, too, was a participant in the symposium that year.

Geller hadn't started small, or quietly. His "Shake-up for Earthquake Prediction" appeared as a two-page commentary in *Nature* in July 1991.[37] Part of what the article did was review the history of the prediction program from the Blueprint to the Daishinhō and establish for readers that the search for earthquake precursors had been underway in Japan since the 1960s. Geller coupled that overview to blunt assessments of what he described as the deep flaws in the assumptions behind the prediction project. About the search for precursors, he wrote, "There might arguably have been reasons for supposing, in 1962, that such precursors existed," Geller wrote, "but there is no longer room for such a belief." "Empirical earthquake prediction," he concluded, "seems to have come to a dead end. If there were reliable and obvious precursors they probably would have been found a long time ago."[38] Geller attributed the fact that work on earthquake prediction in Japan continued to be well funded to entrenched interests who benefitted from the "locked-in" quality of the spending on observation and other prediction-related work. He also cited the absence of an external review process for the funding and research plans the Geodesy Council issued every five years as an additional reason for prediction's persistence.[39]

Geller's recommendations for fixing what was wrong with the prediction project were in keeping with how broken he thought it was. He called for the data collected by the monitoring systems to be made freely and quickly available, not sequestered for long periods before being released to other researchers. Geller also suggested that earthquake prediction work and basic research in the earth sciences be linked more closely than the current arrangements allowed. It was also time, he said, for "a full and open review of earthquake prediction research." Finally, Geller insisted that "the public and government should be clearly informed that earthquake prediction is presently not possible."[40]

A few months after Geller's piece first appeared, *Nature* published a response defending the prediction project. Hamada Kazuo, who worked for the Science and Technology Agency at the National Research Center for Disaster Prevention, agreed that an external review was a reasonable request and one that he supported. As for Geller's other points, however, he offered his own observations about what the prediction program had already achieved and insisted that more and better observation work was still essential. Hamada argued that researchers had, in fact, identified more than one "precursory phenomenon." "There is a clear consensus," he went on, "that the state of the art in predicting earthquakes has progressed to the level that short-term prediction for a great interplate earthquake, the so-called Tōkai earthquake, can probably be made using the present instrumental networks, although neither the current state of knowledge nor the distribution of instrumentation permits useful prediction on a routine basis except for this earthquake."[41]

Geller's reply to Hamada appeared in the same issue and rather handily refuted each of the author's claims, including his assertion that "a clear consensus" existed among Japanese scientists with regard to short-term prediction's efficacy. Geller cited the "spirited discussion and standing-room attendance" at a recent session on earthquake precursors at the joint Japan Earth and Planetary Science meeting in April as evidence that no such consensus was in place. He also pointed out that none of Hamada's references, save for Geller's commentary itself, had appeared in peer-reviewed publications. "This epitomizes," Geller wrote, "the closed nature of Japan's present earthquake prediction programme and further underscores the need for an open review."[42]

Both the Japanese media and journalists writing about science in Japan for *Nature* and *Science* reported on Geller's critiques and on how prediction's supporters were responding to them. The *Asahi*'s Tomari Jirō wrote one of the first articles to appear in the national press in Japan about the controversy and Geller's place in it.[43] Tomari made it clear that he took Geller's criticisms of earthquake prediction seriously and as representative of what others in the field were thinking but were unwilling to voice publicly about Japan's long-term commitment to earthquake prediction. "A debate is now under way among seismologists," Tomari wrote, "about whether or not the time has come to rethink that plan."[44]

The English-language science media reiterated many of Geller's points in the early nineties, emphasizing the prediction project's lack of transparency and drawing attention, as well, to the tensions between scientists who insisted that the current research agenda was fine the way it was and those who wanted to take the field in different directions.[45] David Swinbanks's description in *Nature* of the divisions within the scientific community pitted "a handful of senior seismologists, who have been in the earthquake prediction business since the 1960s" not only against their younger colleagues in Japan but also against the norms of scientific practices elsewhere. "Ten to fifteen years ago US researchers put a lot of effort into an empirical approach to earthquake prediction, just like Japan," Caltech's Kanamori is quoted as saying, before noting that the Americans had since moved on to more productive problems. That "Japan is practically the only country in which earthquake mitigation and hazard reduction takes a back seat to predicting the next big one," Swinbanks implied, was due more to the selfish behavior of a handful of senior scientists than to any scientific insight on their part.[46]

Writing for *Science* in 1994, Dennis Normile also emphasized the growing tension between those scientists in Japan who sought to challenge prediction's grip on funding and research agendas, more generally, and those who preferred the status quo. Normile seemed reluctant to characterize the prediction program's supporters as operating out of self-interest, instead portraying their motivation as a commitment to the safety of their countrymen. The article featured photos of Mogi Kiyō and Geller, and allowed both to stake out their respective positions on the prediction problem. The differences between the two were clear: Mogi, at the time chairman of the Assessment Committee, asserted that "there is a possibility of prediction" and that "prediction would minimize the disaster."[47] Geller, for his part, is quoted as stating that "prediction is impossible" and describing the program's supporters' claims as "irresponsible" and its premises and procedures as "specious."[48] And while none of his Japanese colleagues were quite as blunt as Geller, Normile did find several who were willing to go on the record expressing their own problems with the prediction program. Ishibashi Katsuhiko and others shared their frustrations over how paying for the prediction program had, for years, diverted funds from basic research in the earth sciences, for example. Some staked out more provocative positions. "As of this moment," shared Yokohama City University physicist Kikuchi Masayuki, "the majority of seismologists in Japan believe earthquake forecasting is either impossible or very difficult," suggesting that Geller was giving voice to what many other scientists also believed to be true and, therefore, that the field as a whole in Japan had lost confidence in the program.[49]

Geller's 1991 article in *Nature* was just the first of many in which he called attention to the prediction project's flaws. His contribution to the 1994 symposium on prediction also called for a restructuring of the whole project, along lines similar to the ones Ishibashi had laid out; Geller also hoped for "an orderly transition to a new program for earthquake hazard mitigation that would include strong support for seismological observations and basic research."[50] Whether it was describing the gap between the media's generous portrayal of prediction's capabilities and what the field's practitioners knew to be its limitations, or highlighting the faulty logic behind some of his colleagues' explanations for why the program's search for precursors should be allowed to go forward, Geller's criticism was unambiguous and wide-ranging. He quoted directly from colleagues' published work and from their contributions to the symposium, from the 1962 Blueprint, and from an array of other documents to highlight contradictions in their reasoning and arguments that did not hold up well to scrutiny. This blunt (by Japanese academic standards) style was characteristic of Geller's approach to critiquing the earthquake prediction project.

Geller also suggested that prediction's supporters were avoiding a debate over its merits by insisting that the burden was on skeptics like him to provide proof that prediction didn't work, and not on them to show that it did.[51] Since the only event that the prediction program expected to be able to provide a short-term warning for was the Tōkai earthquake and because that earthquake hadn't yet occurred, the logic went, it was impossible to say for certain that prediction was impossible. The ERI's Hirata Naoshi, for example, argued in 1994 that, "We won't know whether the current research structure is effective or not for several hundred years, or after two or three Tōkai earthquakes." "Earthquake prediction has not yet had a success since research into prediction began," Hirata continued, "but, at the same time, nor has it ever failed."[52]

The Hanshin-Awaji Earthquake

Afterward, it was not uncommon for survivors of the Hanshin-Awaji earthquake of January 17, 1995, to remark that they simply hadn't known that such an event was even possible in the region of Japan where they lived. It was certainly true that even long-time residents of Kobe (population 1.4 million in 1995) or nearby Nishinomiya (population 412,000) would have had a hard time remembering the last time either of their cities had experienced a significant seismic event. One person had been killed, five injured, and three buildings destroyed by a powerful earthquake near Kobe in 1916, but nothing on that scale or even close to it had happened nearby since. (The last earthquake-related fatality in the entire Kinki district was in 1952.) Like it was everywhere else in Japan, the crust beneath Kobe and its environs was seismically active, but relatively few of the earthquakes it produced were powerful enough to be felt by anyone at the surface. Compared to the inhabitants of Tokyo, where feeling the ground shake was a regular enough experience, residents of Kobe experienced few such direct reminders of any seismicity in their vicinity.[53]

There was a wide gap between the worst disaster that residents had thought possible and the one that actually occurred on January 17. The earthquake struck very early on a Tuesday morning. It caught most people at home; many were still in bed. The rupture took place over a large area at a relatively shallow depth, with an epicenter just 20 kilometers or so offshore from downtown Kobe. The shaking it caused there, in Nishinomiya, and elsewhere was exceptionally violent; in some sectors it even met the JMA's criterion for "Shindo 7" designation. "Shindo 7" had been added to the JMA's intensity scale after the 1948 Fukui earthquake to provide a new upper register for the intensity of the shaking at ground level at a given location; this was the first time since its inception that tremors powerful enough to justify the label had ever been recorded.[54] In Kobe, the most violent shaking was distributed along a ribbon of land that ran through densely inhabited residential neighborhoods. Many of the homes and other buildings in those neighborhoods were older, built before more stringent construction guidelines went into effect, and often failed catastrophically when the earthquake hit. Some "pancaked" as heavy roofs bore down on floors below, while others failed in different ways but to similar effect. Most of those killed by the earthquake are thought to have died in the first few minutes of the event or soon thereafter, due to blunt force trauma or suffocation as the buildings they were in collapsed around them.[55] The disaster took 6,434 lives.

Japan's deadliest disaster since 1923 was broadcast live, or very nearly so. Everything from grainy surveillance camera video, recorded in shops and offices as the tremors first hit, to semi-apocalyptic scenes of towering fires burning out of control in the city center and of long stretches of twisted and fallen spans of the Hanshin Expressway filled television screens and newspaper front pages in Japan and abroad. Like the iconic images from the Niigata earthquake more than thirty years earlier, the ones coming out of Kobe and Nishinomiya in 1995 were shocking both because they showed how much harm had been done to those places and because of what they implied about how vulnerable Japan's other cities probably were to a similar event. The rigor and trustworthiness of Japanese engineering and its building standards were themes that had long run through public-facing rhetoric about Japan's defenses against seismicity. Images of the aftermath of the 1994 Northridge earthquake had circulated widely in Japan and commentators had made a point of insisting that Japanese-built structures and highways would have fared much better than their American counterparts under similar circumstances.[56] The scenes and stories coming out of Kobe certainly cast doubt on the presumption that Japanese scientists, architects, or engineers really knew what they were doing. The slow and sometimes inept rescue and recovery effort that followed only made matters worse.

Earthquake science and scientists came under fire from several directions. For the most part, the earth sciences community had gone along with the new status quo after 1978. Keeping the existing monitoring systems going and upgrading as better technologies became available had been the backbone of the Geodesy Council's "Earthquake Prediction Research Plans" from the start, and those goals remained front and center as new plans were rolled out in five-year intervals after 1978. The Geodesy Council, the Yochiren, and the other institutions and agencies entangled with the prediction program simultaneously deflected the questions that arose every time a powerful earthquake other than the one they were waiting for in the Tōkai district caused damage somewhere in Japan. In May 1983, for example, a towering tsunami triggered by a powerful earthquake in the Japan Sea off the coast of Akita reached a number of communities on the coast before the official tsunami warning did. More than one hundred people were killed. The following year, an earthquake in western Nagano prefecture left twenty-nine people dead or missing and, in 1987, two people died in Chiba as the result of an M6.7 event near its coast. One thing these earthquakes had in common was that they were all within areas designated by the Yochiren for ongoing observation; Chiba even fell within the boundaries of the south Kantō district's "intensive observation" zone. The overlap between where the Yochiren had said to expect earthquakes and where they were actually happening was so close that the scientists' judgment seemed sound.

Still, none of these events yielded any evidence of the existence of earthquake precursors, despite the availability of plenty of data after the fact. It was almost always possible to go back and identify some form of seismic activity or a change in the makeup of the local groundwater that had occurred before the earthquake and flag them as potential precursors, but what researchers and the Yochiren were discovering was that it was much harder than they had anticipated to be sure that what they were looking at was linked to the earthquake mechanism and not just part of the background "noise" in a seismically active sector of the lithosphere.[57] After the Chiba quake, a spokesman for the Yochiren attributed the absence of clear precursors to the event's modest size and its relatively deep focal point; purposefully or not, statements like that implied that under other circumstances—had the earthquake been more powerful or its epicenter shallower—the program's instruments would have spotted the missing precursors.[58]

A slightly different dynamic was evident a few years later, when northern Japan was hit by a series of powerful quakes, all of which occurred outside of the Yochiren's observation zones. The Hokkaido Nansei-oki earthquake in July 1993 killed 229 people; in October of the following year, an M8.2 event in the Kuril Islands took ten lives and in December two more died following an M7.6 earthquake offshore of Aomori prefecture. The Yochiren turned up no clear precursors for these events, either. The discussion in the Diet following these events, however, suggested that legislators were in favor of expanding the existing monitoring network to include those parts of the country not already under surveillance. What this step was expected to achieve is not entirely clear since at no point did the JMA or the Yochiren suggest that additional monitoring would make it possible to predict the next earthquake in any of those areas.[59] What it shows, though, is that legislators and, presumably, the public were willing to equate scientific watchfulness for earthquake phenomena with some measure of defense against their hazards.

The apparent failure of earthquake prediction to offer any warning prior to what happened in 1995 was not so easily dismissed. Kobe was poorly prepared for the disaster, some argued, because no one had warned the city that such a powerful earthquake was possible. While it was true that scientists had not specifically urged Kobe or other cities in the area to be ready for an M7 event, they had been drawing attention to the presence of a large number of active faults in the vicinity and the risks associated with them since the early 1970s. The Yochiren had designated the Hanshin district for special observation (*tokutei kansoku*) in 1970 and again in 1978, citing the faults and the Nagoya-Osaka-Kobe region's history of producing M7 earthquakes.[60] As Tomari Jirō has pointed out, the local press kept the public up-to-date on these developments and continued to do so as scientists learned more about the region's seismicity.[61] Even so, some survivors told researchers that the seemingly constant attention to the earthquakes threatening Tokyo and the Tōkai district had led them to assume that Kobe must be safe; if it wasn't, they reasoned, presumably experts would have been as vocal about their concerns for the Hanshin district's safety as they had about the risks facing the capital. That they had not raised the alarm was reassuring.[62]

Japan's investments in short-term earthquake prediction technologies also came under scrutiny after the disaster. Some residents in the areas hardest hit by the earthquake questioned why there had been no prior warning; they had been under the impression that the protocols that applied to the Tōkai district also applied to them and that scientists would tell them if an earthquake was imminent. After the disaster, when it became clear that those assumptions had been misplaced, those same scientists were called on to explain why it was that the tools they were counting on to protect people in the Tōkai district had not been deployed anywhere else. Why were they focused so closely on those communities and on the capital district and ignoring the rest of the country? Questions about prediction's usefulness and status in Japan surfaced much more openly and often after the 1995 Hanshin-Awaji earthquake than they had before it, prompted by the tragedy but fueled as well by critiques predating the disaster from within the earth sciences community itself.

Prediction Holds On

The government's post-disaster push for a new approach to earthquake science's role in protecting the nation was led by Science and Technology Agency Director Tanaka Makiko. In her initial public comments after the disaster, Tanaka had expressed confidence in the earthquake prediction research project. By early February, her attitude toward it had changed.[63] By then, she had met with Mogi Kiyō, then chair of both the Yochiren and the Hanteikai, and a number of other scientists involved in different parts of the prediction initiative. They disabused Tanaka of some of the more optimistic assumptions she had held about how much progress had actually been made toward putting short-term earthquake prediction into practice.[64]

Not long after her meeting with Mogi and the others, and working closely with a group of legislators calling themselves the "Association of Diet Members for the Protection of Japan from Earthquakes," Tanaka called for the creation of a new government body to oversee and coordinate the collection and dissemination of information about earthquake risks and hazards throughout Japan. Its roles would differ from those of the Yochiren and the Headquarters for the Promotion of Earthquake Prediction (which dated from 1976) in several ways; one was evident in the announcement that the new body would oversee detailed, comprehensive surveys to identify and map as many faults as could be found, wherever they were in Japan, something the Yochiren had never attempted. Another was that Tanaka's plans for a new research agenda downplayed the importance of short-term prediction and the search for reliable earthquake precursors in favor of a commitment to mid- to long-term earthquake forecasting and to studies of the processes leading up to the ruptures themselves. Finally, she insisted that the new Headquarters for Earthquake Research Promotion (Jishin chōsa kenkyū suishin honbu), or HERP, also make communication between scientists and the people in charge of earthquake disaster-reduction programs a priority.

The Diet endorsed Tanaka's call for a shift in priorities and the creation of the HERP in July 1995 with the passage of the Special Measures Law on Earthquake Disaster Countermeasures (Jishin bōsai taisaku tokubetsu sōchihō). The new legislation was an important milestone in the Diet's and the government's efforts to demonstrate to the public that they were taking steps to learn from the most recent disaster. Roughly half of its provisions outlined new roles for prefectural governments as architects of five-year earthquake disaster preparedness plans, and for the central government as a source of funding and support for putting the plans into effect. The other half focused on the new Headquarters and its responsibilities. Legislators made sure that HERP was under the direct administrative umbrella of the prime minister's office, which gave it and its members a legal standing and authority that the all-volunteer Yochiren lacked. In so doing, and in other gestures, the government implied that the earthquake threat was too important a problem to be left wholly in the hands of scientists any longer. As Yochiren chairman Mogi noted at the time, neither he nor his colleagues had been consulted as the decisions about the new Headquarters and the administrative scaffolding around it were being made.[65]

The new law and the new Headquarters were achievements that officials could point to as proof that the government was taking the lessons learned from the most recent earthquake disaster to heart. From the public's perspective, it certainly would have looked like a major break with past practices was underway, as the old Headquarters (for the Promotion of Earthquake Prediction) was abolished and a new Headquarters (for Earthquake Research Promotion) took its place. Newspaper articles described the post-disaster changes as the realization of long-sought-after "consolidation" (*ichigenka*) of earthquake disaster mitigation efforts. For the first time, earthquake disaster planning and scientific work that might help the nation prepare for earthquakes would be housed under the same roof, so to speak, their efforts coordinated like never before.

The post-disaster changes also seemed to signal a move away from "prediction" (*yochi*) as the most important goal for the nation's new earthquake research initiative. In this context, "prediction" was shorthand for the efforts that had been underway since the 1960s to spot the changes in seismicity, stresses in the crust, and so on, that were believed to precede at least some earthquakes. Legislators had made sure that "prediction" wasn't part of the name they assigned to the new headquarters and, in keeping with the apparent shift in priorities, a number of research centers and government labs with "earthquake prediction" or even just "prediction" in their titles quickly dropped that language and renamed themselves using less problematic terminology.[66]

The government moved quickly, after the disaster in 1995, to change earthquake science's priorities, at least as they related to the measures that policy makers believed to be the most effective defenses against future disasters. The scientific agencies and institutions that had been doing the work of earthquake prediction research took longer to decide how they planned to move forward after the disaster. The Science Council of Japan, the Seismological Society of Japan, the Geodesy Council, and a number of ad hoc committees all launched reviews of prediction's progress to date, several of which ended up recommending sweeping changes in how the earth sciences community engaged with earthquake prediction research and attempts to put it into practice. They did so at least in part as a defensive measure against the possibility that government technocrats in the newly created Headquarters would seize control of the research agendas (and funding) if the scientists themselves did not take bold steps, but it was also the case that at least some earthquake scientists had already expressed serious misgivings about where the field was headed, as we've seen. For them, the disaster in Kobe created an opening to push for the changes they sought.[67]

There were few signs that funding agencies or research institutions took Geller or Ishibashi's criticisms all that seriously before the Hanshin-Awaji earthquake, but there are echoes of some of their ideas in changes that were pursued afterwards. The first wave of post-disaster reports and conference proceedings in 1997, for example, included a number of calls to set aside the search for earthquake precursors, on the grounds that the more scientists had learned about earthquakes the clearer it had become that everything about them was more complex than had been anticipated back when the prediction initiative was first getting under way.[68] Ishibashi's recommendation that researchers shift their efforts away from prediction toward trying to better understand the forces and physical processes that actually produced earthquakes also surfaced again. The Japanese media followed the discussions in 1997 fairly closely and reported the gist of the reports and meetings over the course of the year. The story the press conveyed was one in which the earth sciences in Japan were at a fork in the road, with dramatic changes just over the horizon. The media paid particular attention to the Geodesy Council's report that year, which included language acknowledging that short-term earthquake prediction—even for the Tōkai earthquake—was, at present, "*konnan*," a usefully ambiguous term that describes everything from tasks that are quite difficult but might still get done to those that are simply impossible, with a bias toward the latter.[69] Although Hagiwara, Asada, and others in their circle had used similar language when discussing prediction's prospects in the 1970s, journalists treated its appearance in 1997 as a remarkable admission on the part of the scientists involved.[70] Descriptions of the outcomes of the reviews and debates as expressions of "self-criticism" (*jikohihan*) or even as "declarations of surrender" (*haiboku sengen*) gave the impression that short-term prediction had been discredited, its supporters in retreat.[71] There was an element of truth in what the press was suggesting but, of course, the reality wasn't so straightforward.

In 1997, scientists like Ishibashi and other advocates of the new research agenda were better positioned to influence the planning process than they had been before and, as a result, many of their suggestions were reflected in the ambitious new ten-year plan for seismic research that the new Headquarters (HERP) released in 1999.[72] The Headquarters' plan described a research and funding agenda that was substantially different from those that had come before it. It included a commitment to the "preparation of maps showing predicted earthquake activity which consolidate surveys of active faults, long-range assessment of the probability of earthquake occurrence, and prediction of strong tremors," each element of which was itself a new initiative, most of them adapted from the proposals developed in 1997.[73] It called, too, for investments in the sort of basic scientific inquiry that Ishibashi had asked for in 1994, explaining that "a succession of new discoveries made possible by a deepening of the understanding of earthquakes, has shown how complicated the subject is, and indicates a necessity to return to basic research."[74] The authors of the new documents broke with prior practice by announcing steps that scientists and research institutions would use from then on to disseminate "information deemed necessary to gain public understanding in the application [of] earthquake research results."[75] The maps mentioned earlier were one manifestation of this desire to communicate with the public using language and in a format they were likely to understand.

For all that was innovative and new in the plan, there were parts of it that were neither. Despite what journalists may have written in 1997, scientists involved in the prediction project had not given up on the search for precursors or on efforts to put short-term earthquake prediction into practice. Their perspectives and their demands that support for monitoring for precursors continue may not have attracted much attention, but they had been part of the post-Kobe debates about what direction to take prediction-related research. Thirty-some years of work on short-term prediction wasn't about to be set aside that easily and, as the outline of a new direction for the field began to take shape after 1997, a number of seismologists and other experts lobbied to make sure support for it survived.

One sign of their success was the disappearance of the language characterizing short-term prediction as problematic and of recommendations that research associated with it be set aside in favor of other priorities, both of which had been highlighted in much of the post-1995 planning. In its place was language that, instead, made minimal reference to scientists' concerns about the viability of short-term prediction in general and downplayed concerns about the predictability of the Tōkai earthquake in particular. "Prediction is seen as an important means of minimizing damage caused by earthquakes," the new plan's authors insisted in 1999. They acknowledged that, while "to a certain extent it is possible to predict the location and maximum magnitude of a future earthquake," under most circumstances, "time is difficult to predict accurately enough to be able to issue a warning." The Tōkai earthquake, on the other hand, is described as an exception, "which can be predicted in the case of phenomena such as unusual changes in the earth's crust."[76] And, finally, there is this: "[I]f it is possible to predict an earthquake just prior to its occurrence, so that a warning can be issued, if appropriate preventive measures are taken, human casualties, fires and the occurrence of secondary disasters can be greatly reduced. For this reason it is desirable for research on earthquake prediction to continue at a steady pace."[77] While less than a ringing endorsement of its prospects, the Headquarters' proposals of 1999 affirmed short-term prediction's ongoing importance to the nation and committed the government to supporting the monitoring infrastructure the project relied on for "observation and research for the elucidation of movements in the earth's crust up to the occurrence of the earthquake."[78]

Given all the renaming of institutes and headquarters in 1995 and 1996, it would have been easy to miss some of the other continuities between the older, prediction-centric approach and the newer, different one. What happened after the Kobe disaster arguably wasn't so much a radical rearrangement of agendas and personnel as it was a rebranding effort.[79] There was not a lot of difference between the personnel leading the new, forecasting-friendly initiative and the ones who had been in charge of the old, short-term prediction project. Eleven of the twelve scientists in the Headquarters group assigned to oversee the new surveying and risk studies were Yochiren members, for example.[80] The agencies that had previously been involved in earthquake monitoring efforts under the auspices of the prediction program were the same ones conducting the new forecasting surveys. None of the research institutions or agencies that had been active in the prediction program surrendered control of their personnel or budgets as a result of Tanaka's push to "consolidate" the overall research infrastructure. Tanaka left her post as Science and Technology Agency director in a Cabinet reshuffle in August 1995, one month after the Diet approved the Special Measures Law on Earthquake Disaster Countermeasures.[81] That brought an end to the post-disaster flurry of political interest in earthquake science.

When the dust settled, the short-term prediction project was no worse for wear, on balance. Its advocates had kept their priorities intact, and beaten back the momentum toward reform that had been building since the early 1990s. The government allocated more money after 1995 than it ever had before to studying seismicity in Japan. Before the catastrophe in January, the 1995 budget for earthquake prediction-related spending was 10.8 billion yen; supplemental budgets passed soon afterward dedicated another 63.5 billion yen toward earthquake monitoring and prediction efforts.[82] Over the next several years, the JMA, the Science and Technology Agency, and the Geographical Survey Institute all significantly expanded their monitoring capabilities for everything from micro-seismometry to strong-motion observation systems.[83]

Grey Areas

Even insiders had a hard time convincing their colleagues and officials to make modest, incremental changes to the provisions they had agreed to in 1978, as Mogi Kiyō discovered during his abbreviated term as chairman of the Hanteikai. Mogi was a member of the Hanteikai from the very start; he knew as well as anyone when he became its chairman in 1991 that the committee's options for responding to reports of unusual measurements from the instruments monitoring the Tōkai district were the same then as they had been back in 1978. Those options were limited and binary. Committee members could either confirm that an earthquake appeared to be imminent or that it did not. If the latter, then everyone would go about their business as before. The first result, however, would presumably trigger an earthquake warning from the prime minister and be followed by all the interventions allowed by the Daishinhō. By the time Mogi took over the committee's leadership, he was anxious to give himself and his colleagues a little more flexibility in how they interpreted the information available to them. This was an issue he had been talking and writing about since the mid-1980s.[84]

What Mogi began asking for in 1992 was a third, intermediate option that would allow the Hanteikai to report something other than a "yes" or a "no" if the changes they were observing in the Tōkai district—whether in seismic activity, surface deformation, or other metrics—were significant but not definitive precursors of an imminent event. He wanted, in other words, to acknowledge that there were grey areas when it came to deciding whether something was a precursor to a major earthquake and that it made sense for the public to be able to respond accordingly.[85] As Mogi pointed out, they had absolutely no experience with a Tōkai earthquake in the form they now assumed it would take; they had records of what had happened in 1944 and 1946, but the next earthquake would not, presumably, be merely a repeat of those earlier events.[86] Under the circumstances, it made sense to Mogi that they be able to acknowledge their uncertainty. He recommended that an intermediate warning might lead to reduced speeds on highways and rail lines and enable other disaster-reduction measures that would fall somewhere between the "full stop" that would follow an emergency warning and business as usual.[87]

Mogi's proposal won support within the Hantekai and, initially, the different agencies involved were also behind the idea, but the JMA soon signaled that it would not allow changes to the original process to move forward. There were legal obstacles to consider—introducing another option might require a revision to the Daishinhō—as well as practical ones: What steps were people supposed to take if they were warned that an earthquake might occur soon but not immediately?[88] Lurking in the background, too, were concerns in the JMA and Land Agency about the downsides of admitting to a lack of certainty about what was and what was not a precursor to the Tōkai earthquake. To permit Mogi's modifications would be to publicly acknowledge that scientists' attitudes toward short-term prediction were changing, and this was a step that the JMA seemed reluctant to take. In addition, Mogi had also expressed concern about the potential costs associated with triggering an earthquake warning, especially if he and his colleagues were not certain that what they were seeing was a clear sign that a quake was coming. The JMA's stated policy was to err on the side of caution and insisted that a warning would go out if the data merited it. But Hanteikai members couldn't help but consider the consequences of making the wrong decision. According to one think-tank's estimates, by the mid-1990s, the costs to the economy of shutting down the Tōkai district would run to 700 billion yen a day.[89]

Mogi kept up his campaign for almost four years before finally resigning from the Hanteikai in frustration in March of 1996. His resignation and the reasons he gave for it were widely covered by the press.[90] In his public statements, Mogi did not imply that he had lost confidence in the possibility of recognizing signs that the Tōkai quake was about to occur; he continued to think that the short-term prediction project had merit. But Mogi did try to make it clear that the committee's task was made that much more difficult by the lack of options available to it and that the apparatus around preparations for the Tōkai earthquake needed to be changed.[91]

Several years later, policy makers appeared to finally heed Mogi's advice. In 2003, the Central Disaster Prevention Council released new policy guidelines that were supposed to cover everything from the procedures to be followed before the Tōkai earthquake to emergency response and recovery efforts afterward.[92] The General Principles Regarding Countermeasures for the Tōkai Earthquake recommended creating a three-tier system of warning announcements to replace the binary model followed by the Hanteikai up to that point, much as Mogi had proposed earlier. The new system worked like this: if sensors monitoring the Tōkai district detected changes at one location that might qualify as an earthquake precursor, the JMA would release a "Tōkai Earthquake Report" (Tōkai jishin kansoku jōhō). If sensors in more than one location detected anomalies, the JMA would summon the Hanteikai and issue a "Tōkai Earthquake Alert" (Tōkai jishin chūi jōhō). The committee would then review the data and, if warranted, issue its "Tōkai Earthquake Warning" (Tōkai jishin yochi johō), after which the prime minister could convene an emergency cabinet meeting and presumably follow that with his own emergency announcement.[93] Local officials were given guidelines about which emergency measures were appropriate for each of the three stages of warning. A Tōkai Earthquake Alert announcement, for example, would begin the mobilization process for first responders, send all children home from school, and ask residents to refrain from unnecessary travel. The more draconian (and costly) interventions allowed by the Daishinhō, would, if necessary, come later.[94] The new "General Principles" defined the responsibilities of the Hanteikai in ways that did not explicitly contradict the Countermeasures Act. Its members technically had the same options available to them as before, since the new steps the guidelines added to the assessment preceded any involvement by the Hanteikai.

The General Principles introduced another important break from previous approaches to preparing for the Tōkai earthquake, which was its designation of the detection of a specific precursory phenomenon as the new baseline for deciding whether or not an earthquake was imminent. When the Daishinhō first went into effect, scientists had deliberately cast a wide net in their search for potential earthquake precursors and kept watch for changes in seismicity, surface deformation, radon gas levels, and the like as possible indicators of a major rupture to come. The General Principles took a deliberate step away from those indicators in favor of a new one, "preslip," a term that describes a phenomenon that some earth scientists in the late 1990s and early 2000s were exploring as a potential earthquake precursor. "Preslip" refers to the very slow aseismic movement across a fault patch preparatory to its final, break-away rupture.[95] At the time, preslip was especially interesting to scientists because it seemed to scale with the size of the eventual earthquake; in other words, the bigger the preslip, the more powerful the event that followed. For scientists looking at the Tōkai earthquake, preslip arrived at an opportune moment; here was a phenomenon that was clearly part of the faulting process itself and not a byproduct of it. It could be observed and measured, under the right conditions, and, in the early 2000s, at least, the field was still in the early stages of understanding what its implications might be. Its usefulness as a precursory signal had not been ruled out. Better yet, the types of changes that preslip could be expected to produce at the earth's surface seemed to correlate well with observations of deformation at ground level that had preceded the 1944 Tōnankai earthquake.[96]

In 2003, the Central Disaster Prevention Council made it clear that "preslip" had become the key to predicting the Tōkai earthquake. The authors of the new report stressed that the instruments currently monitoring the district were there to watch for indications of preslip; it would be "difficult" (*konnan*), they said, to arrive at a warning on the basis of phenomena of any other kind.[97] This is a differently qualified statement than ones from the 1990s, in that it seems to imply that experts were confident in the connections they'd made between preslip and the Tōkai earthquake and that any difficulty would lie with expectations that they would be able to derive equally reliable conclusions from other phenomena, as well. The earlier discussion had highlighted how many things had to go exactly right for the prediction to take place; in 2003, the Geodesy Council projected a more confident assessment. And unlike Mogi's earlier plea to allow the Hanteikai to acknowledge its own uncertainty about how to interpret anomalous data from monitoring instruments, the General Principles implied that scientists had actually gotten much better at it. The new three tiers of warning suggested to the public that experts were more certain about what constituted a precursor than ever before and, thus, more certain that they would know when the Tōkai earthquake was about to occur.[98]

The share of people in Shizuoka who trusted short-term prediction to be able to warn them in time had been on a slow but steady decline since the mid-1990s, and the Geodesy Council's expression of faith in prediction doesn't seem to have reversed that trend. In 2009, there were only a handful of people in the prefecture—less than 1 percent—who trusted in prediction without question.[99] One out of ten believed that prediction was impossible under any circumstances; 59 percent had concluded that it might not be completely impossible but was, at best, extremely difficult. Only 20 percent of respondents believed that prediction would be correct even half the time.[100]

The diminishing trust in prediction's capabilities did not mean that people were any less worried about a Tōkai earthquake, however. As the thirtieth anniversaries of Ishibashi's announcement (in 2006) and of the passage of the Countermeasures Act (in 2008) came and went, the public was, if anything, even more concerned about the threat of a Tōkai earthquake than they had been previously. Half of all those who participated in Shizuoka's biennial public opinion survey in 2009 described themselves as "extremely concerned"—the most anxious the survey allowed one to be—about a possible Tōkai earthquake. That was the highest percentage of respondents expressing that degree of concern since the survey had started asking people how they felt about the earthquake.[101] Fifty-eight percent of the poll's respondents reported that they were more concerned about the earthquake in 2009 than they had been two or three years prior. In that same poll, when asked to describe what they thought might happen to themselves if the earthquake were to strike suddenly, one out of four people said they were worried that they might die. The only time a higher percentage of respondents in Shizuoka had reported that they feared the earthquake would kill them had been soon after the Hanshin-Awaji earthquake in 1995.[102] According to these metrics, the residents of Shizuoka had not grown complacent about the possibility of a Tōkai earthquake, nor did they seem to expect that better technology and thirty years of preparations would keep everyone safe.

Faith in Probabilities

Scientists in Japan continued the pursuit of short-term prediction after the Hanshin-Awaji disaster but the government also encouraged the development of other methods for reducing the uncertainty about future earthquakes. This was one of the priorities for the new Headquarters for Earthquake Research Promotion after 1995. With its support, earth scientists started working on the first-ever survey of all of the active faults in Japan, and on assembling the data they would need to calculate the probabilities of different seismic scenarios occurring in the decades ahead. HERP's approach differed from what had been happening up to that point in several ways. Monitoring for earthquake precursors had focused, from the start, on a relatively small number of sites, for example, with most of the attention on the Tōkai district. The post-1995 efforts to estimate medium- and long-term risk encompassed the entire country, not just parts of it. Once the surveying and analysis processes were complete, scientists would, in theory, be able to tell citizens what the earthquake risks were for just about any location in Japan. The move toward forecasting also implied a shift in research priorities for earthquake scientists. As University of Tokyo seismologist Mitsuhiro Matsuura explained, "The purpose of earthquake research is not to make a guess when an earthquake will strike, but to understand the scientific mechanism behind it."[103]

The time-frame and vocabularies for expressing these new risk calculations were also different from the ones associated with short-term prediction. A successful example of the latter would specify when in the future the earthquake would strike (usually on a scale delineated in hours, days, or, at most, weeks), how powerful it would be, and where it would occur. The new framework expressed risk in terms of probabilities: What was the probability that a given fault or plate boundary—a site that had produced earthquakes in the past—would produce a potentially destructive earthquake again over the next thirty years? What was the likelihood that a given location in Japan would experience shaking powerful enough to do significant damage over the next thirty years?

And, finally, the methods that scientists used to communicate their findings to the public changed, as well. The government released Japan's first nationwide probabilistic seismic hazard map in 2005 as a product of its commitment a decade earlier to put science to better use protecting the country from earthquakes. While there wasn't anything fundamentally novel about the thinking behind maps like these, the ones shared with the public in the early 2000s drew on the extensive surveys of Japan's known faults and many other studies of the country's seismicity that had been launched only after 1995.[104] The availability of that data, the complexity of the calculations that went into the maps' production, and the variety of formats the experts used to convey this information reinforced the idea that the maps represented new frameworks for reducing uncertainty and responding to earthquake risk.[105]

In broad strokes, the maps offered readers information about those risks from three different perspectives. One took the list of all the known inland active faults in Japan (a catalog that was itself a product of the new post-1995 research agenda), as well as all the offshore source faults, estimated how large the next major earthquake at that fault would be, and calculated the probability of its occurrence over the next thirty years.[106] At first, scientists experimented with describing the probabilities simply as low, medium, or high, but after complaints that those categories were not specific enough, they added numerical ranges as well.[107]

The second perspective started with a map of Japan divided into a grid, each unit of which represented a single square kilometer of territory. For every one of those units, scientists turned to historical records of past earthquakes and recent studies of the geological makeup and seismicity of the area to calculate the probability of that square kilometer of land experiencing shaking of a given intensity over the next thirty years. Level 6 on the JMA's 10-level scale was the most common standard for the maps but ones showing the probabilities for both less and more intense shaking were also available. At Shindo 6, the collapse of less-earthquake-resistant wooden homes was to be expected, unsecured furniture inside a home or apartment was likely to topple, and people inside would have a hard time remaining upright for very long.[108] Color-coded maps made it easy to see which parts of Japan were thought to be at greatest risk. Accompanying text listed the different earthquakes that scientists believed were likely to occur in a given area and explained how the likelihood and anticipated power of each had been factored into the overall rating conveyed by the map.[109]

A third set of maps was intended to provide more specific information to disaster planners and earthquake engineers than the aggregate results of the first two. "Seismic Hazard Maps for Specified Seismic Source Faults" used the predicted strong ground motion for a dozen or so potential earthquakes to show their effects (in terms of seismic intensity) on those areas in proximity to the fault zone.[110] The calculations that went into producing the maps were supposed to take the composition of the subsurface structure (and many other factors) into account on their way to determining how powerful the shaking would be at a given location, at a resolution of one square kilometer. Information in this format could help disaster planners and earthquake engineers focus their efforts on those parts of a city or district that the maps suggested would be hardest hit, if and when the anticipated earthquake occurred. Studies along these same lines had been part of disaster planning for Tokyo, the Tōkai district, and a handful of other sites since the 1960s, but this was the first time that parts of Japan not normally in the spotlight were afforded a similar degree of concern.

The utility of short-term prediction on the one hand and probabilistic forecasting on the other also differed. Short-term prediction was supposed to give communities time to get people out of harm's way once a warning went out; a secondary effect was to remind residents that they lived in a place where the risk of a destructive earthquake was high enough that such preparations were necessary. Long-term forecasting put a different perspective on how communities might think about and respond to earthquake risk, once it was expressed in probabilistic terms. Communities that previously had not been aware that they were at risk could take steps to educate themselves about disaster preparedness, perhaps harden vulnerable infrastructure, and so on. A clearer picture of the long-term risks could also influence plans for economic development or decisions about where (or how) to build new factories, power plants, schools, and the like. Because the calculations of probabilities were taking place across Japan, forecasting also invited comparisons and judgements about the relative safety of one part of the country over another.[111]

Short-term prediction and these newer approaches to mitigating earthquake risk did have qualities in common. Both were presented to the public as the products of scientific analysis by experts embedded in state-sanctioned research projects. Both also made an argument about science's capacity to describe the timing and potential effects of seismic events that hadn't happened yet, even if they did so using different vocabularies and temporal frames. The new forecasting effort was obviously not prediction in the way that it had been presented to the Japanese public up to that point—the new studies were not intended to raise alarms about any imminent event—but it was still making a claim that future earthquakes would occur at intervals that scientists could calculate in the present.

EPILOGUE

The Future of Disasters

For a while, it had looked to prediction's critics in Japan as if the combination of their arguments and the events of 1995 might be enough to alter the project's trajectory and open up new spaces for deliberations over how to think about and prepare for future earthquake disasters. The realization that it wasn't, and that the changes they had hoped for weren't coming, was frustrating for Geller, Ishibashi, and their like-minded colleagues. There really hadn't been any meaningful opportunities for debate, in which the arguments for keeping the search for precursors going could be weighed alongside those opposed to doing so, and after which the scientists with the most convincing evidence and arguments would have a say in shaping research agendas going forward. The most important deliberations after 1995 took place largely behind the scenes, away from the scrutiny of potential critics, and continued to privilege many of the priorities scientists had arrived at in the 1970s. Had the process been more transparent, the rejection of the reformists' efforts might not have been so frustrating. Geller, Ishibashi, and a small group of fellow scientists devoted even more energy to critiques of the prediction project after 1995 than they had before.

Ishibashi, for example, left the International Institute of Seismology and Earthquake Engineering and moved to Kobe in 1996, where, until his retirement in 2008, he was a professor at the Research Center for Urban Safety and Security at Kobe University. In many ways, the second half of his career was more closely aligned with what it had been like at its very beginning, as Ishibashi again took his concerns about Japan's vulnerabilities to earthquakes to the public and his colleagues on a regular basis, in articles, books, op-eds, and the like, urging them to see the dangers ahead and to take steps to prevent the worst harms from happening.[1] After the move to Kobe, he turned his attention, in particular, to Japan's incongruous reliance on nuclear power plants and to the alarming proximity of a good many of them to locations with long histories of producing large earthquakes.[2]

It was Ishibashi who introduced the phrase "nuclear power plant earthquake disaster" (*genpatsu shinsai*) into the public lexicon in 1997.[3] There were fifty-one active reactors in Japan at that point and, although the government insisted that all were designed to withstand the most powerful earthquakes that experts had said were possible, Ishibashi was quick to point out the problems with the assumptions behind those assurances. There were many, he said—for instance, the existence of intra-plate "directly-below" earthquakes that could happen (and had happened) in locations where there was nothing on the surface to indicate the presence of fault zones below. Similarly, Ishibashi pointed out that the government's experts were relying on records that went back a thousand or so years, at best, to come up with estimates of the most powerful earthquakes a given part of the lithosphere could produce, when the seismic history of those sites went back far longer than that.[4] There was simply no way to be sure that their calculations were correct.

Ishibashi also drew attention to what might happen if the government's experts were wrong and an earthquake did significant damage to a reactor complex. Using the Hamaoka nuclear power plant in Shizuoka as an example, he listed the many risk factors associated with its location (its proximity to Suruga Bay and vulnerability to a Tōkai earthquake among them) and then described a scenario in which the facility's reactor number three suffered a meltdown after a catastrophic earthquake.[5] With the containment vessel breached, Ishibashi posited that the subsequent radioactive plume would spread inland, creating a Chernobyl-like exclusion zone that would eventually displace millions.[6]

He kept these warnings and his recommendations in the public eye for years and was relentless in his insistence that Japan was at greater risk than most people thought. "I think the situation right now is very scary," Ishibashi told the *Japan Times* in 2004. "It's like a kamikaze terrorist wrapped in bombs just waiting to explode."[7] The government and the nuclear power industry (to the extent that there was a difference between the two) continued to insist that the plants were perfectly safe. Both were still taking that position when 3.11 happened. Much of what took place in Fukushima that day and those that followed made clear that Ishibashi's fears had been well founded. He certainly hadn't been wrong about the assumptions that had left the Fukushima Daiichi plant vulnerable, or about the long-term damage that would follow.

3.11

The March 11, 2011, Great East Japan Earthquake was unexpected. Where the Pacific plate meets and incrementally moves beneath the Okhotsk plate and northeastern Japan, earthquakes are not uncommon. The Japan Trench subduction zone is seismically active and both contemporary monitoring and historical sources have documented large numbers of events there, many of them unremarkable in terms of their effects on the communities along the coast or inland. At the other end of the range of documented earthquakes originating along the northeast coast, however, is a shorter list of exceptionally powerful and destructive events. Massive tsunami-generating earthquakes devastated communities along the Sanriku coast in 1611, 1896, and 1933. Further south along the fault zone, very powerful but less destructive events occurred in 1897, 1936, 1978, and again in 2005. The relative frequency of earthquakes along that segment of the subduction zone and careful study of the geology at the plate boundaries had led scientists to describe the likelihood of another event in the same location as very high.[8] They estimated that the next major earthquake in the area could be as powerful as the strongest on record, an M7.5 event. This was an environment that earthquake scientists believed that they understood reasonably well.

They were mistaken.

Earthquake scientists knew that they had been wrong about how the slow collision of plates along the Japan Trench would unfold almost as soon as the first shocks reached the shoreline on March 11. A dense network of instruments blanketing Japan allowed scientists to determine the earthquake's location and to estimate the earthquake's magnitude even before the fault had finished rupturing, and to continue their analyses in real-time from then on. The JMA issued tsunami warnings within minutes of the initial event, as early estimates of the earthquake's magnitude coupled with what the instruments were reporting about the locations of the faulting processes meant that one was almost certain to occur.[9] Two minutes after the initial shock, the JMA reported that its early earthquake warning system had estimated the earthquake's magnitude as 8.1 and that communities on the coastline should therefore anticipate tsunami heights that could reach six meters in Miyagi prefecture, and three meters in Fukushima.[10] The JMA's report significantly underestimated the earthquake's magnitude, in part because an M8.1 event was the most powerful event its early warning system could interpret. The system's instrumentation went off-scale once that limit was exceeded.[11] Within an hour of the first tremor, however, researchers using data from monitoring stations outside Japan and different methods of calculating magnitude were producing much higher estimates than the JMA's, quite close to the M9.0 figure that was ultimately assigned to the earthquake. By then, tsunami with run-up heights much, much higher than those predicted by the JMA were sweeping inland all along the coast.[12]

In the months that followed, Japanese earthquake scientists were consumed with explaining what had happened—how and where the fault had ruptured—and with questioning why they, as a community, had failed to imagine that such a massive earthquake could happen in that location. Initial answers to the first set of inquiries were not long in coming, in part because "the best scientifically recorded great earthquake to date" had generated so much data across many different instrument platforms.[13] Explaining why the earthquake was unexpected took longer but, to their credit, researchers did not hesitate to question their assumptions and the practices that had produced them.

What scientists had expected would happen when the next earthquake struck off the Miyagi-Fukushima coast was that slippage at the plate boundaries would follow patterns similar to those seen in earlier events. The geology of those boundary areas, the complicated, disrupted line of the boundary zone itself, and recent history set fairly clear parameters for what was likely and for what was possible. Earthquakes that had originated in the boundary zone off Miyagi and Fukushima prefectures since the late nineteenth century had been the product of ruptures along discrete segments of the fault plane, occurring at fairly regular intervals. The "stuck" parts of the fault seemed to let go a piece at a time, suggesting to observers that there was a relatively uniform upper limit on how much stress the fault in that region could contain. This model fit nicely with what was known about the composition of the plate edge. It was thought to be relatively weak and, therefore, incapable of remaining locked in place long enough to produce exceptionally powerful earthquakes, when it did rupture. The historical pattern from 1897 to 2005 had been one of earthquakes that fell in the M7.1 to M7.5 range. These are powerful and potentially damaging events, worrisome especially because of their capacity for generating large tsunami. During the production of new seismic hazard maps in 2005, scientists used M7.5 as the upper limit of the range of earthquakes that were possible along that length of the subduction zone. They also estimated that the probability of such an earthquake occurring over the next thirty years was among the highest of any calculated in Japan.

The actual mechanisms of the earthquake on 3.11 overthrew several of the key assumptions that scientists had used to model future behavior in the subduction zone. Instead of coming "unstuck" a section at a time, on March 11, the fault "ruptured across almost the entire width of the plate boundary megathrust," many hundreds of kilometers in length.[14] Some descriptions of what happened refer to the fault as having "unzipped" in a cascading series of ruptures, as one previously locked section after another gave way. The fault had both stored much more energy than earthquake scientists had thought possible and was releasing it in ways that they had not foreseen. A 300-by-200-kilometer section of the fault plane (an area roughly the same dimensions as the state of Massachusetts) slipped by at least a meter, with some parts of the plane shifting by as much as sixty, the largest single slip ever associated with a recorded earthquake.[15] Japan's eastern shoreline, as a whole, moved more than two meters closer to the United States, as large segments of the ocean floor in the fault zone rose by almost the same amount. On a map, the combined fault zones of all the earlier earthquakes that researchers had used to understand the dynamics of the region fit easily within the boundaries of the slippage defined by this single new event.[16]

One important consequence of the differences between the fault's behavior on March 11 and how experts had expected it to behave was that they had established an upper limit for the earthquake's magnitude that was much too low. The Great East Japan Earthquake was 150 times more powerful than communities along the coast had been told to expect. Put another way, it would take at least 150 of the M7.5 earthquakes the Japanese seismological establishment had forecast for the region to produce the same destructive force as the M9.0 earthquake that actually occurred.[17] On the ground, the ramifications of the underestimate were most visible in the differences between the heights of the tsunami that the anticipated M7.5 earthquakes would have produced and the much more massive wave-fronts that swept over the coastlines in Iwate, Miyagi, and Fukushima prefectures on March 11. The large majority of the people who died that day drowned; casualties from collapsed buildings and deaths otherwise attributable to the tremors themselves were comparatively few. Had experts understood that the subduction zone along that part of the Japan Trench was capable of producing such a large, tsunami-generative earthquake, and had they conveyed that information effectively, it is possible that national and local governments could have engineered more robust defenses, that more residents along the coast could have been quicker to flee when the alarm was raised, and that more lives could have been saved. That there is no way to know for certain how much difference a fuller understanding of the fault dynamics might have made has not been a comfort to earthquake scientists.

A month or so after the catastrophe, Kōketsu Kazuki, a University of Tokyo Earthquake Research Institute seismologist with a long-standing interest in earthquake forecasting and disaster mitigation, was speaking to journalists about the field's future in Japan. For weeks after the events of March 11, he told them, he had buried himself in his work, coming and going from his home with barely a word to his family, not bothering even to shave. At the same time, he was no longer sure that the values he had once thought so important meant anything anymore. "When the Great East Japan Earthquake struck," one of his interviewers asked, "what was your first reaction?" "That science had been defeated," was Kōketsu's reply.[18]

Prediction and Uncertainty after 3.11

Some of the blame for the multiple disasters of March 11, 2011, did eventually fall on the earth sciences community in Japan. Scientists misjudged how powerful an event involving the plate boundaries off the northeast coast could be. The initial reports from the strong-motion monitoring system that was supposed to provide real-time estimates of an earthquake's strength (itself a product of the post-Kobe push to develop a national observation network) also underestimated the event's actual size. Sentiments like Kōketsu Kazuki's, that science had been defeated, were no doubt shared among his colleagues.

Compared to what happened after the Kobe Earthquake, however, the impact of the March 2011 disasters on the earthquake prediction infrastructure in Japan was relatively limited, at least in the short term. There were no equivalents to the quick, top-down interventions like the ones that Tanaka Makiko pushed through in 1995. Observers have speculated that whatever questions policy makers and the public might have had about the state of seismology in Japan were set aside as the enormity of the catastrophe at the Fukushima Daiichi power plant came into focus.[19] Even if an argument could be made that seismologists bore some of the blame for what had happened there, the nuclear power industry and the bureaucrats who regulated it had demonstrated far more egregious failures of judgment and governance. Daily reminders of the consequences of those failures were there for anyone to see, as workers struggled to contain the release of radioactivity from the ruined reactors, as tens of thousands of people fled communities near the plant for fear of contamination, and as protestors all over Japan rallied against the government's plans to restart temporarily shuttered nuclear power plants.[20]

Those who were critical of the earthquake prediction project found themselves making some of the same points they had back in 1995. The significant expansion of the nation's seismic monitoring network after what happened in Kobe was, in part, an effort to improve the odds of spotting a precursor and, ideally, warn people ahead of time that an earthquake was imminent. As Geller repeatedly emphasized in 2011 and often afterward, however, not only had there been no warning spotted prior to the earthquake on March 11, there were no precursory signs to be found after the fact either.[21] "What the earthquake and geodetic monitoring network built over fifty years and at great cost revealed," he wrote, "was that there were no remarkable precursory phenomenon prior to 3.11."[22]

To its credit, the field as a whole did not hesitate to reflect on its shortcomings and on the disaster's implications for earthquake science. The Seismological Society of Japan launched two symposia after 3.11 to take up some of these concerns. The first, "Questioning the State of Seismology Today," was held in October 2011 and the second, "The Fiftieth Anniversary of 'The Blueprint:' Seismology's Progress to Date and Its Future Course," a year later. Both featured blunt criticisms of the field's trajectory and of the earthquake prediction project especially, along with a wide range of recommendations about where Japan's earthquake scientists should focus their efforts next. In a sign that the earth sciences community might want to foster a more open dialog about prediction's future in Japan than it had in the past, Robert Geller was the special keynote speaker at the conference on "Questioning the State of Seismology Today." The article based on his talk was the first chapter in the volume that came out of the symposium.[23] In 2012, Geller chaired the committee that organized a second symposium on the Blueprint, spoke at the event, and contributed a chapter to that conference volume, too. Speaking to a reporter about having been asked to appear at the 2011 symposium, Geller described the experience using the same term that many scientists and politicians had used to refer to the multiple catastrophes of March 11: "sōteigai," or "beyond the imaginable."[24]

In the longer-term, how earthquake scientists talked about the seismic risks that Japan would face in the future did change after 3.11. While not every shift in focus or tone needs to be attributed to the effects of the disaster, there are at least two developments in how earthquake prediction and forecasting were put to use that were accelerated in the aftermath of 3.11. The first was a move to reevaluate the worst-case scenarios for some of the events that scientists had said could happen soon, like the Tōkai and the Tokyo-inland earthquakes. The other was a fairly definitive break with the presumption that a short-term prediction of a Tōkai earthquake was, in fact, possible.

The Nankai Trough Earthquake

As part of its review of preparations for the Tōkai earthquake, in 2003 the Central Disaster Prevention Council (CDPC) had also acknowledged that, so long as it kept not happening, it would at some point be necessary to consider other scenarios involving the Nankai Trough. The council had pointed out that the Nankai Trough plate boundary from Suruga Bay to Tosa Bay had historically produced massive earthquakes every one hundred to one hundred fifty years. Given that pattern, and given that neither the 1944 Tōnankai nor the 1946 Nankai events along the trough had released as much energy as the 1854 Ansei earthquakes had, the council's experts began looking at several different scenarios for how the accumulating stress along the plate boundary would be resolved.

One possibility was that the long-awaited Tōkai earthquake would take place first, in which case the likelihood of an event involving both the Tōkai segment and another in the near future would go down, considerably. In the absence of a distinct Tōkai event, however, the CDPC concluded that they ought to start preparing for the possibility of another Nankai-Tōnankai earthquake sequence at some point in the next fifty years or so. That pattern would be in keeping with the historical record. Unfortunately, there was also the possibility of a third outcome, in which the plate boundary along the trough would unzip as it had in 1707, when all four segments ruptured simultaneously.[25] The council estimated that a simultaneous Nankai-Tōnankai earthquake would produce powerful shaking from Shizuoka to Osaka to Shikoku, ten-meter-high tsunami along the coast, and damaging waves all the way to Osaka Bay. In that worst case scenario, experts' estimates at the time were that as many as 17,800 people could perish.[26] The council decided, in 2003, to start work on disaster-reduction plans and infrastructure for those parts of the country freshly designated as at-risk by one or more of these future scenarios. For the time being, those plans would take a back seat to the continued focus on preparations for the Tōkai earthquake. If, after ten years, that disaster had not yet occurred, then the council agreed it would be time for the country to shift its focus away from where it had been since 1976, and toward these new threats. There was agreement that, one way or another, the communities closest to the Nankai Trough were the ones most at risk for the foreseeable future, with Tokyo and its environs running a close second.[27]

Those plans accelerated after 3.11. The CDPC began taking another look at the possible impact of a Nankai Trough earthquake within a few months of the disaster in the northeast, prompted, in part, by the realization that their previous assumptions about the size of such an event probably needed to be revised upward. New calculations described a significantly larger faulting area than researchers had allowed for earlier and experts examining the new data determined that it was possible for the movement along the plate boundaries in the vicinity of the Nankai Trough to produce an M9.1 event (which would release roughly thirty times as much energy as they had earlier anticipated). If such an earthquake were to occur, researchers concluded this time around that as many as 323,000 people could die in the Tōkai district alone.[28] In early 2013, the council's experts recommended that preparations for a Nankai Trough earthquake begin immediately; it proposed that communities along the coast at risk of being inundated improve evacuation routes, build raised evacuation shelters, and make sure that schools and hospitals were equipped with their own tsunami-resistant facilities. Older homes and buildings inland, meanwhile, were to be targeted for retrofitting and repairs to make sure they could withstand the violent shaking anticipated in their communities.[29] In November 2013, the Diet passed legislation making it possible for the government to designate communities threatened by one or more of the earthquake's hazards and, in March 2014, 707 communities along and near the Pacific coast, from Ibaraki to Okinawa, became eligible for government subsidies, as an incentive to get them to invest in disaster-reduction projects.[30]

As part of the shift away from preparing solely for the Tōkai earthquake, the council also appointed a group of scientists to assess whether a Nankai earthquake would be predictable, in any sense of the term. Was it amenable to short-term prediction, as the Tōkai earthquake was still presumed to be, or perhaps to forecasting in the longer term? The answers to those questions had a bearing on what sorts of disaster reduction efforts might be possible and to what extent preparations for a Nankai earthquake would resemble those that had come before, when it was just the Tōkai district at risk.

In their report back to the CDPC, the members of the working group dismissed short-term prediction as a useful tool for preparing for a Nankai earthquake. There were too many different ways in which the earthquake could happen and so many locations from which it might propagate, they concluded, that there was simply no way to offer a forecast or a short-term prediction with any degree of confidence at all. The most they would allow was that if "slow-slip" on a large scale or highly unusual changes at plate boundaries were observed, then it would be reasonable to indicate that the chances of an earthquake were higher than usual.[31] Lest there be any doubt about where they stood on the broader question of prediction's potential, however, the report's authors also pointed out that "recognition is growing that neither the basis for an announcement about the Tōkai earthquake nor its content, namely the emergency earthquake disaster reduction measures associated with the announcement of an emergency warning as stipulated by the Daishinhō, are commensurate with science's present capabilities."[32] Geller and others had been making this argument or ones just like it since the early 1990s but this was one of the first times that challenges to the 1978 system had come from within the house, so to speak.

While language like that stopped short of calling for the Hanteikai and the Countermeasures Act to be dismantled, it certainly seemed to be moving the CDPC in that direction. Tellingly, though, there were also caveats in the council's analysis that left the door open for continued research into prediction-related pursuits. At times, the shift in tone in the report from one perspective to the next can be quite jarring, as in a section on the prospects for earthquake forecasting (*jishin yosoku*) related to a Nankai Trough event that came near the end of the council's final report. The section starts out with a long list of reasons why a successful forecast is likely to be "difficult" and a brief acknowledgement of the very narrow range of circumstances in which there might be useful information obtained as a result of forecasting efforts. The report then shifts to a much more upbeat description of forecasting that seems to have little if any relationship to the assessment that proceeded it. "Earthquake forecasting," the next paragraph reads, "which saves lives from earthquakes and tsunami, is an important technology and it is necessary for research in that area to continue moving ahead."[33]

At any rate, the CDPC did not move quickly to address the 2013 report's implications. Years passed and more studies were conducted. In June 2016, not long after two powerful earthquakes struck just days apart from each other in Kumamoto, killing 258 people, the council appointed yet another working group, this one under the ERI's Hirata Naoshi, and charged it with assessing what useful information could be gleaned from observation and evaluation of the Nankai Trough and how it might shape the country's efforts to mitigate the next earthquake's effects. When the working group finished, more than a year later, its attitude toward short-term prediction was just as critical as the one from three years prior. A spokesman for the group stated unequivocally that "there was, at present, no scientific method to forecast when, where, or at what magnitude an earthquake will occur."[34] Statements like that put the council on the record acknowledging that short-term earthquake prediction of the kind that the Daishinhō relied upon was likely impossible. The council also seemed to accept that there were no reliable methods of determining ahead of time when a Nankai earthquake would occur. The report's authors repeatedly pointed out that a Nankai earthquake could take many different forms, depending on which segments of the fault plane were involved, in which order they ruptured, and so on. They had looked closely at some of those scenarios but confessed that others were possible.[35]

As had happened after the Hanshin-Awaji earthquake, however, advocates for the prediction project kept elements of it intact as the focus shifted from preparing for the Tōkai earthquake to getting ready for a Nankai Trough event. The CDPC's report came out in September 2017. On November 1, 2017, the Earthquake Assessment Committee (Hanteikai), created more than forty years earlier to watch for signs of an imminent Tōkai earthquake, was quietly replaced by the "Evaluation and Study Committee for Nankai Trough Earthquakes" (Nankai torafuzoi no jishin ni kansuru hyōkai kentōkai). Their membership rosters were identical; the committee's name may have changed but the scientists involved did not. The new committee was charged with assessing reports of unusual activity along the Nankai fault zone, which was not at all unlike what the Hanteikai's duties had been before, only over a wider territory. Unlike its predecessor, however, the new committee was not part of a system designed to produce an emergency announcement by the prime minister or trigger interventions like those permitted under the Daishinhō. (There had been some discussion, early on, about simply broadening the reach of the act to include the communities at risk from a Nankai earthquake but the proposal didn't get far.) Instead, having shelved its elaborate three-tiered warning protocol for a Tōkai earthquake, the JMA declared that it would switch to issuing "special alerts" (*rinji jōhō*) to the public if the new Evaluation and Study Committee agreed that the circumstances warranted it.[36]

The communities potentially at risk from a Nankai Trough earthquake—which included those once at risk only from the Tōkai earthquake—scrambled to make sense of what these new protocols meant for them. The government had formed the new committee and decided that it could issue special alerts, apparently, without a plan to tell the affected communities what they were supposed to do if one was announced. The uncertainty built into the JMA's new protocols and the lack of guidelines about how best to respond to them were frustrating for the public and local governments alike.[37] Did a JMA alert mean that everyone should evacuate right away? If so, how long would it be before it was safe to go back home? If evacuation wasn't the right response, what was? It was months before the government issued recommendations about who should leave threatened communities, when, and where residents ought to go if they did have to flee in response to a warning, but those instructions were eventually provided.[38]

That these guidelines came to be at all, delayed or not, signaled to the public that there was still a chance that they'd have some warning before the Nankai Trough earthquake struck. Some forty years after the Japanese government had turned to short-term earthquake prediction as a defense against a Tōkai earthquake, there were scientists and officials who were willing to insist that it still had its uses as the nation waited for its latest, even more dangerous, earthquake disaster to arrive. While, admittedly, a far cry from the days when experts had made it sound as if predicting the Tōkai earthquake was all but guaranteed, there were nevertheless echoes of that earlier faith in science's capacity to make the future knowable in how Japan chose to prepare for this new threat.

Some scientists expressed dismay at the government's new plans. Ishibashi, for example, was still doing what he could to remind the public of all the ways the country was still vulnerable to earthquakes. The nuclear industry was still a problem, obviously, but not the only one. Given his track record, people did tend to pay attention to the threats that worried Ishibashi most and, after 2011, one of them was the Nankai Trough earthquake. His 2014 monograph on the topic provided a historically grounded analysis of the harm earlier Nankai earthquakes had done, explained why scientists were increasingly concerned about one at present, and speculated about what its effects might be.[39] With Ishibashi, however, things were never so simple that there was only one earthquake to worry about—he frequently emphasized that the whole country, from Hokkaido to Kyushu, was at risk, not just the capital, the Tōkai district, or the Nankai region. The message that there was no safe place was a hard one to convince the public (or policy makers) to get their heads around.[40]

Another of the ways Ishibashi's analysis of how to respond to the hazards ahead differed from those of the CDPC was in his sharply diminished willingness to support the pursuit of short-term, deterministic earthquake prediction as an important tool in disaster mitigation.[41] It is telling that Ishibashi, who, in 1978 and for many years thereafter, had expressed a fair degree of confidence in prediction's potential, by 2020 was insisting that even the JMA's "special alert" system and the slim chance of a warning that it held out was misleading.[42] Ishibashi argued that the JMA's plans should be abandoned, along with the Daishinhō, to make it clear that the era of faith in short-term prediction was over.

Other earthquake scientists were just as skeptical about the whole "special alert" project. In 2019, Kansai University's Hayashi Yoshinari surveyed a hundred or so of the Seismological Society of Japan office-holders to ask what they thought about the JMA's preparations for the Nankai Trough earthquake. Collectively, that cohort concluded that the probability of the JMA being able to issue a "special alert" before the earthquake struck was less than 6 percent. And, even if the JMA did issue an alert, Hayashi's respondents estimated that the odds of it being correct—meaning that the Nankai Trough earthquake would actually follow the alert—were only about one in a hundred.[43]

The general public's perception of short-term prediction's capabilities was a bit more trusting, but there, too, skepticism had become the dominant take. In its 2017 survey of residents' opinions about the Nankai Trough and Tōkai earthquakes, Shizuoka's government asked respondents for their thoughts on whether or not the latter could be predicted. Roughly one in five of the respondents said that they believed it was completely impossible to do so; another 59 percent said that, while they weren't sure that prediction was completely impossible, it was certainly "difficult" (*konnan*). Fewer than one in a hundred believed that prediction would work all the time and only 17 percent thought that scientists could predict a Tōkai earthquake successfully at least half of the time.[44] In a sign that short-term prediction was no longer seen as a key part of the prefecture's expectations for defending its residents against a Nankai (or Tōkai) earthquake, when Shizuoka polled its residents about the earthquake threat again, two years later, it made no inquiries at all about whether people believed that the Nankai Trough earthquake could be predicted or not, or whether they had much faith in the JMA's plans to warn residents ahead of time.[45] In 2019, it was much easier for the prefecture to avoid those questions entirely than to have to engage with the likely answers to them. Had they been asked, certainly, some people would have responded that they still trusted experts to raise the alarm—that premise had been in place for more than forty years at that point and was unlikely to be abandoned overnight—but, for the first time since 1978, it was the government's position that such faith was misplaced.

What Explains Prediction's Persistence?

The pursuit of short-term earthquake prediction held on in Japan much longer than in other settings where it had once been embraced and despite the efforts of scientists in Japan to bring the whole enterprise to a close. The project's critics have often attributed the divergence to self-interest, not science, arguing that its supporters kept it going in order to keep funds flowing from the government. The project's advocates, on the other hand, insisted that there was as yet no proof that short-term prediction was impossible under the right circumstances and, thus, no scientific justification to stop pursuing it. Arguments like these connect the prediction project's trajectory to choices scientists themselves might make, based on their own assessments of its legitimacy and purpose. There are also other factors worth considering when it comes to explaining prediction's persistence in Japan.

One is simply that, after the 1950s, no one institution or professional body was in a position to decide whether prediction was worth pursuing or not. There was a time when sidelining prediction in favor of other ways of putting science to work against earthquake disasters was as simple as preventing Imamura Akitsune from putting his own research plans into effect but, with the distribution of the Blueprint, the prospects of doing something like that again via administrative fiat were all but nil. Researchers and the budgets that supported their work (directly or otherwise) were spread widely across many different institutions and agencies, from the late 1960s on. Even if gatekeepers in one ministry or one legislative committee had decided to pull the plug on prediction, those in other parts of the government need not have.

Critics occasionally compared prediction's supporters to advocates of cold fusion as part of an argument about how untenable the science behind both ideas had turned out to be. The differences between those camps are also illuminating, however. The "cold fusionists" in the United States were few in number and arguably at the periphery of the field from the start. Neither of those qualities applied to prediction's supporters in modern Japan, a group that eventually counted among its members many of the country's best known and most accomplished scientists. Discrediting and defunding cold fusion would have been relatively simple compared to what doing the same for earthquake prediction would have entailed.[46] It is also worth pointing out that the barriers to entry to becoming a self-proclaimed "authority" on earthquake prediction are low; private, commercial experts have, in recent years, joined the many amateurs making public claims about future earthquakes. The Japanese media often accommodates both.

A second factor to consider as part of an explanation for prediction and forecasting's ongoing appeal is that the practices and imagery associated with each were not static. The public-facing elements of the prediction project changed over time, sometimes significantly. The protocols associated with the Daishinhō and preparations for a Tōkai earthquake, for example, differed from the work the Yochiren had been doing up to that point and the circulation of probabilistic hazard maps represented yet another perspective on how scientists explained uncertainty and earthquake risk to the public. There were innovations that the public could see and that prediction's supporters were quick to draw attention to that could be interpreted as evidence that the project was getting closer to meeting its goals. Japan's Earthquake Early Warning (Kinkyū jishin sokuhō) system is a case in point. Launched in 2007, this system was designed to detect major seismic events as they happened and issue targeted, automated emergency alerts to locations that were likely to experience severe shaking, in the very brief window of time before the S-waves that would do that damage arrived. (These warnings reach the public in real time via television, radio, cell phones and online messaging.) While not a prediction, it was still a potentially lifesaving intervention, one made possible because the government built a national strong motion observation network after the Kobe earthquake, as part of its plans to support new probabilistic forecasting and mapping efforts.[47]

The nature of the controversy among scientists over prediction's future is a third factor that may help put its ongoing role in Japan into context. A central tenet of the arguments against both the search for precursors and probabilistic forecasting was that earthquakes are fundamentally unpredictable, non-linear, non-repeating events. This element of the critique defined earthquakes as such both because the processes that produced them did not produce information that could reveal whether an earthquake was imminent or not (so no precursory signals) and because they do not occur on a "seismic cycle" or any other quasi-periodic basis (so probabilistic forecasting is pointless).[48] Given these conditions, researchers pursuing prediction were confronted not by epistemic uncertainty, which could be reduced, but by its aleatory counterpart, which could not. The prediction project's supporters, on the other hand, spent decades arguing that it was possible to transform at least some of what they viewed as the predominantly epistemic uncertainty around earthquakes into useful information and analyzable levels of risk.

There is a model for the resolution of controversies in science in which the parties involved come to recognize that one side's position is the unique, correct one and the field moves on accordingly, but that was not the trajectory that the debates over short-term earthquake prediction in Japan were on in the 2020s.[49] Neither prediction's critics nor its supporters showed signs of a willingness to abandon their arguments outright, at least not publicly, as one would expect to see in what Bruce Lewenstein refers to as "the idealized scientific outcome" of "closure through sound argument."[50] That said, as Lewenstein and other science studies scholars have pointed out, controversies in science move toward closure along a range of routes and timelines. No one model prescribes how those processes could or should unfold, and idealized outcomes are less common than one might think.

This chapter documented the slow, sporadic decoupling of references to short-term earthquake prediction from disaster preparedness policies developed for both the Tōkai and Nankai earthquakes. A logical next step in that same direction would be for the government to repeal or significantly modify the Large Scale Earthquake Countermeasures Act, thereby fully severing plans to protect the Tōkai district from any direct reliance on a successful short-term prediction.[51] While such a step would not, on its own, signal that the controversy over prediction had been resolved, separating the arguments over the science from the debates over prediction's value as a tool for mitigating risk would likely have several important effects. Dropping or revising the Daishinhō, for example, could put Japan's approach to earthquake disaster preparedness much more in line with what the Americans are doing, since both would then be relying primarily on probabilistic seismic hazard mapping to assess long-term risk. The divergence between Japan and everyone else that Geller and others have lamented for many years would be significantly diminished if the Daishinhō were to be set aside.

It is less clear what such a step would mean for researchers in Japan and further work on earthquake prediction. In the United States, neither the abandonment of plans for a major prediction project in the 1970s nor the end of the Parkfield experiment closed the door entirely on scientists' pursuit of prediction. Although it is true that, after Parkfield, attempts to put prediction into practice were greeted with considerable skepticism by many seismologists, discussions of potential pathways toward prediction continued to appear with some frequency in professional journals.[52] In Japan, revision or suspension of the Daishinhō would presumably leave intact other expressions of support for short-term prediction's potential, such as the Yochiren. Self-proclaimed forecasting and prediction experts operating outside officially sanctioned channels, meanwhile, would be free to continue to provide the media with their own dramatic warnings about the Nankai Trough and other potential earthquake disasters.

When Tsuboi Chūji, Wadati Kiyo, and Hagiwara Takahiro issued the "Blueprint" in 1962, their plan was to dedicate a decade to determining whether or not short-term earthquake prediction might someday be possible. Scientists in Japan are still pursuing answers to that question six decades later. The controversies over the prediction project's methods and goals, meanwhile, are just as resilient. There have been limited forms of "closure" in the debates over prediction in the aftermaths of disasters, as when scientists acknowledged after 3.11 that predicting the next Tōkai earthquake might be impossible. Even those admissions, however, implicated only specific applications of the project's methods, not the premises that kept it going for so long and seem likely to continue to do so.

Back to the Future in Tokyo

After the city announced its plans to construct disaster shelters in the Kōtō delta but before construction of the first of the six complexes was complete, businessman Yoshino Yatarō took matters into his own hands and built his own. In 1977, he converted one of the factories he owned in the area into a private shelter for himself and up to four hundred of his employees, complete with a generator and stocks of food and water. Yoshino had experienced the disaster in 1923 and the fire-bombings in 1945, he said, and worried that he and the others would never make it to their distant designated evacuation site in time, when the next earthquake struck. The advice on one of the bunker's exterior walls read, "There is no place that is absolutely safe. Everyone please deal with the disaster as best you can. This shelter is a sample [*sampuru*]."[53] Yoshino couldn't have known it at the time but his small "sample" and the much bigger site at Shirahige-East were the only expressions of that era's faith in shelters to see the light of day.

At first, the obstacles to moving forward with the city's plans to build the shelter complexes were limited to issues with the sites themselves. Construction at Shirahige-East was repeatedly delayed while officials negotiated with local residents over the terms of their relocation, among other issues.[54] The land the city acquired for the shelter at Ōjima-Komatsugawa, at the eastern edge of the delta, meanwhile, turned out to be heavily contaminated with hexavalent chromium, a carcinogenic compound that had been used in a number of the factories that once occupied the area.[55] After several failed attempts at cleaning up the polluted topsoil, so that construction at the site could finally get underway, the city ended up simply burying the hazardous areas under thick layers of clean dirt.[56]

It would be hard to identify when, precisely, it became clear that none of the shelters other than the one at Shirahige-East would ever be built. The nation's attention to the risks to the Kōtō delta and the commitment to protect the communities there owed a great deal to Kawasumi's advocacy, so his death in 1972 may help explain why both waned later in the decade. Governor Minobe Ryōkichi's successor as governor—Suzuki Shunichi held the office between 1979 and 1995—had no particular stake in the shelters' fates and grassroots support for the disaster shelter proposal had been limited from the start. It was always a top-down initiative, and the delta's residents themselves were not particularly invested in making sure that the remaining five sites all looked like Shirahige-East when the dust settled.

The project's momentum fell off slowly and, because the city didn't entirely abandon the disaster-preparedness elements of the work at the five other sites along the way, officials could always point to one or the other of the ongoing construction projects as evidence of the administration's ongoing concern for the wellbeing of the delta's residents. At Ōjima-Komatsugawa, planners landscaped the fields over the buried toxic waste and added enough amenities that they could reasonably be described as parks. The eighteen hectares that the city reclaimed in this way is enough room, experts say, to safely hold 200,000 people in the event of disaster, which is the same capacity called for in the original plan. The city encouraged and helped fund the construction of public housing, schools, and other community infrastructure near the evacuation parks, a reflection, at some level, of Takayama's belief in the need to create pockets of safety in the city that were also sites for normal everyday life. In addition, the city's ongoing redevelopment plans reflect the thoughtful use of deliberately widened roads, the exploitation of natural barriers to fire within the site, and the location of tallish buildings along stretches of the zone's periphery.

A similar process unfolded at most of the other locations, as the shelter concept was quietly dropped in favor of less complex redevelopment projects, each of which offered at least some measure of protection against the hazards residents feared most. “Less complex” in this context usually meant parks like the one at Komatsugawa. The delta’s Sarue-onshi Park (completed in 1983), Kiba Park (1992), the green spaces along the Arakawa River near Yotsugi station, and the expansive open spaces and facilities at Shirahige-West are all legacies to one extent or another of the original disaster shelter plans.[57] Some of the parks are equipped with restrooms and other facilities that, in an emergency, would make it easier for large numbers of evacuees to remain there for many days. That they exist at all is proof that planners continued to believe that the delta remained at risk and that residents would need somewhere to flee to when the time came.

The scaling back of the shelter projects took place as experts were being forced to rethink some of the assumptions behind their plans for the Kōtō delta. That reassessment was, in large part, the product of new studies of the city’s vulnerabilities to earthquakes. Tokyo’s 1971 Earthquake Disaster Prevention Ordinance had required that officials conduct detailed, city-wide “hazard assessments” (*kikendo sokutei*) every five years; the results of the first of those surveys were released in 1975. The Tokyo Metropolitan Fire Department, Hamada Minoru, and other experts had been conducting studies of their own for many years before the city launched the first of the *kikendo sokutei* surveys. The results of those earlier studies, as we have seen, were shared with the public and played important roles in shaping the discussion around earthquake disasters from the mid-1960s on.

One of the things that was new and different about the “hazard assessments” was that they dealt comprehensively with topics that, up until then, had almost always been handled separately. Prior to the 1975 study, fire hazards, the likelihood that local buildings would collapse during an earthquake, and residents’ ability to find shelter had all been studied, but seldom by the same institutions or individuals and almost never in ways that allowed for their findings to be helpful to anyone but other planners or disaster experts. Research into each issue proceeded along parallel tracks, which made perfect sense if your key concern was how best to solve the evacuation problem or the fire problem. Imagined disasters, like the ones portrayed in works like NHK’s *Prepare for an Earthquake: Your Disaster-Preparedness Countermeasures,* offered their own attempts at a more holistic portrayal of an earthquake’s effects and, while they shaped the public’s understanding of how dangerous

Tokyo might be and probably convinced some to avoid it altogether, narratives like those were still a long way away from providing information that could help make the city a safer place to live and work. The new assessments, instead of problematizing each hazard and making it the focus of the inquiry, ended up problematizing each one of the several thousand city blocks they set out to scrutinize. By taking the perspective not of the hazards and what they would do but of the community and what would be done to it, these studies helped make Tokyo the problem they needed to solve.[58] Also new in 1975 was the attempt to distill the study's findings to the point that they could be reflected in a single "score" or ranking. In the earliest versions, the assessments assigned a score of 0–4 to each "community"; later iterations dropped the zero and shifted the range in favor of a scale that ranged from 1–5. In each case, the higher the number, the more dangerous the location was judged to be.

When the first hazard assessment was released by the city in late 1975, its results were not what most experts had expected. On the one hand, the study's findings confirmed that Kawasumi and others had been right to raise the alarm about the situation in the Kōtō delta and the lowlands more generally. Many of the sites with the highest numbers (3s or 4s) were clustered in the low-lying wards along the Arakawa and Sumida rivers, the same locations that had proven themselves so vulnerable to destruction in 1923 and again in 1945.[59] What was surprising was that, in addition to the locations that the experts already knew they needed to worry about, the site-by-site survey also characterized as considerably or even extremely hazardous a wide swath of neighborhoods in parts of the city that had until then not generally been thought of as vulnerable. The consensus among most disaster experts had been that the lowlands to the east were at risk in ways that the "upper" city to the west generally was not, in part because the latter rested on firmer ground but also because it was believed that the combination of densely packed wood structures, poor access to safe evacuation zones, and any number of other negative factors were unique to communities on the Kōtō delta and those nearby. They were right to be concerned about the delta and the lowlands, of course, but most observers had misjudged how often those same factors could be found in many other sectors of the city.

Overlaid onto a map of Tokyo, the results of the 1975 assessment revealed not just clusters of 3s and 4s in the usual, chronically hazardous neighborhoods but, instead, depicted an arc of at-risk sites, nestled between Ring Road 6 and Ring Road 7, that swept through the city from Ota ward in the southwest, north to Nerima, and then across and down to Edogawa in the southeast. An *Asahi* reporter described the pattern on the map as "donut shaped," a label that reflects the surprising thickness of the band of hazardous locations that encircled the city center.[60] There were only a handful of flagged sites in the wards at the middle of the donut—if one were to imagine the Imperial Palace as a marker of sorts for that center—but then their number expands significantly as one approaches the 6th ring road and drops off again once past the 7th.

That the distribution of the city's most dangerous locations looked like a donut had some important implications for how the city planned to proceed with its disaster prevention preparations. According to the provisions of the Earthquake Disaster Prevention Ordinance, those parts of the city that were expected to be hardest hit by a future earthquake were supposed to be designated by the governor as "danger zones" (*kiken chiiki*). Some fraction of the areas categorized as dangerous but known to be even more vulnerable than the rest would be labelled as "special danger zones" (*tokubetsu kiken chiiki*). The point of this process—apart from the obvious one of figuring out, in some systematic way, which parts of the city were the most vulnerable and which the least—was that, once designated, "danger zones" and "special danger zones" were potentially subject to more intense regulation and intervention, which might include projects to decrease the flammability of structures or to protect open space. In the absence of such designation, there were still ways for the government to exert its influence at the local level, but a big part of the ordinance's intent was expressed in the mechanisms it used to identify, label, and then act on the locales in Tokyo that needed the most help. The ordinance also promised that under some circumstances, designation would come with assistance from the Tokyo government. Exactly how much help or precisely what those circumstances might be were not spelled out, but all these pieces, together, outlined an ambitious and arguably long-overdue disaster-aware approach to city planning.

Minobe's administration never went through with plans to publicly designate parts of the city as "dangerous zones" or "special danger zones." One theory as to why the city did not take that step is that administrators were concerned about how much it might cost the Tokyo government to achieve whatever it was that the new ordinance implied might have to be done to make a dangerous district safe, especially now that so much more of the city was implicated.[61] The new ordinance had provided a robust set of goals but left out some important details, including where the money to pay for all the improvements it implied would be necessary was going to come from.[62] To complicate matters even more, the ordinance stipulated that any planning involving "special danger zones" or the establishment of evacuation sites had to be done in consultation with local residents. While in keeping with the Minobe administration's collaborative approach to policy-making, the city's recent experiences trying to negotiate the terms of the Shirahige-East complex and other encounters with local politics must have made planners hesitate before committing the city to promises it might not be able to keep.[63] Although other approaches to differentiating the most at-risk districts from others did go forward—the Tokyo Metropolitan Fire Department, for example, created "counter-measures strengthening areas" to help focus its efforts—in 1975 the city itself backed away from using all the disaster-reducing tools the ordinance had put at its disposal.

It instead pursued a series of initiatives in which the city's control over the agenda was a little more clear. Beginning in the early 1980s, the Tokyo Metropolitan Government released a series of master plans and improvement projects that were supposed to reduce the city's vulnerabilities to earthquake hazards. As understanding of the hazards changed, the plans and projects followed suit. In general, however, Tokyo's goals over the years have been fairly consistent and would be recognizable to anyone who had been following the debates around what to do on the Kōtō delta in the 1960s. In 1981, for example, the city released a "Master Plan for Disaster Management Facilities" (Toshi bōsai setsubi kihon keikaku), the main elements of which combined improvements to the city's major thoroughfares with programs that would make some street-facing buildings fire-resistant. For those parts of the city that had been shown to be most at risk from a major earthquake, these interventions would come together to create what planners described as "disaster-proof living zones" (*bōsai seikatsuken*). As Nakabayashi Itsuki, one of the country's leading disaster-planning researchers, has pointed out, the new "living zones" differed from the old delta shelter designs in that they didn't rely on large-scale redevelopment and new construction and were, instead, limited in scale, made use of existing buildings, and focused on the community already in place.[64] That said, there are clearly common threads running between the disaster shelters of the 1960s and the "living zones" of the 1980s. One is the decision to protect selectively; rather than simply provide incentives across the board for individual property owners to make their homes or buildings more fire-resistant, the disaster-proof living zones were focused on some parts of the city and not others. In addition, and not unlike what was happening in the delta, they were designed to foster a sense of community and an identity as "unit areas sharing the same destiny in a disaster."[65] The emphasis on protection from fire as disaster reduction's primary imperative was another shared attribute between the two. Finally, both emphasized the importance of embedding disaster resistance into the practices and spaces of daily life and treating the threat of catastrophe as ever-present.

A second factor that helps explain how thinking about future disasters in Tokyo changed was that seismologists had come to the conclusion that the greatest contemporary seismic threat to the capital district was probably not a repeat of the 1923 earthquake but, rather, another of the "directly beneath" (chokkagata) events that had happened often in the city's past. Disaster planners had assumed that their worst-case scenario would involve an earthquake equal in magnitude to the one in 1923 and with an epicenter in a similar location; their estimates of that future earthquake's intensity at different locations in the city was based on what had happened at the site in 1923. From the late 1980s on, the consensus around which earthquake the city was most likely to encounter next shifted away from worries over a repeat of a 1923-like event and toward the prospect of another iteration of a near-field (chokkagata) event, like the Ansei-Edo earthquake of 1855. Chokkagata events tended to be much less powerful in terms of overall energy release but their relatively shallow epicenters and the proximity of those epicenters to the capital region meant that their effects at the surface could still be catastrophic.[66] The Tokyo Metropolitan Government was in the middle of producing its first estimates of how much damage such an event would do to the city when the 1994 Northridge and 1995 Hanshin-Awaji earthquakes occurred. Both were near-field earthquakes; officials in Tokyo halted their calculations long enough to study what had happened in Kobe and then applied those findings to produce their best estimates of what a similar event would do to the capital district.[67]

After the Hanshin-Awaji earthquake, Tokyo's disaster experts drew on data from that event to model the effects of a similar chokkagata event on the capital district. Their best estimate in 1997 was that an M7.2 earthquake directly under the city would lead to the total collapse of around forty-three thousand structures and significant damage to another one hundred thousand. Unlike Kobe's experience, however, in Tokyo, planners expected that it would be the fires after the earthquake that would do the most harm. They estimated that by the time the fires were finally out, 9,600 hectares of the city would have burned, along with 378,000 buildings and homes. Of the seven thousand or so fatalities a chokkagata event would cause, two-thirds would be due to fire.[68] Food, water, and shelter would be in short supply; if the quake happened while family members were at school or work or otherwise away from their homes, the number of those in need of shelter the night after the earthquake could exceed 2.3 million people.[69] As those who had homes to return to did so, experts anticipated having to provide longer-term shelter for as many as 1.5 million of the city's inhabitants.

One observation to emerge out of the city's estimates in 1997 is that a chokkagata earthquake would be almost as deadly and almost as destructive—by some measures—as the long-awaited repeat of the Great Kantō Earthquake. Many more buildings and much more of the city as a whole would burn in a repeat of the 1923 M7.9 earthquake, but the number of fatalities experts anticipated arising from that disaster was not that much larger than would result from a near-field event (9,363 versus 7,159); the total numbers of buildings destroyed or badly damaged were also roughly the same across the two scenarios.[70] Given how extensively the destruction caused by the Hanshin-Awaji earthquake had been covered by the media, no doubt many residents of Tokyo had already imagined what had happened in Kobe happening closer to home and would not have been surprised to learn that their city would not fare well under similar circumstances.

The disaster in Kobe triggered another set of adjustments to Tokyo's disaster reduction planning. In late 2000, the city replaced its Earthquake Disaster Prevention Ordinance with a new Ordinance for Earthquake Disaster Measures (Tokyo-to shinsai taisaku jōrei). The latter placed much more emphasis on the role of individuals and civic organizations in disaster reduction than had the older regulations; it was also much more specific about city residents' responsibilities to each other and to the safety of the capital in general.[71] It called on them to see to it that buildings were fire- and tremor-resistant, that household items were secured to prevent them from falling during an earthquake, that everyone kept supplies of food and water on hand, and that they stayed up-to-date on evacuation routes and nearby evacuation sites.[72] The new ordinance also insisted that post-disaster recovery planning would be an important part of the city's preparations for the next earthquake. Kobe's excruciatingly slow recovery from its disaster had driven home how important it would be to have plans in place, both because doing so would make it that much easier to restore electricity, water, and other necessary infrastructure and because it sent a clear signal that recovery was a given and that life would go on.

The Great East Japan Earthquake shattered many of the assumptions that had gone into preparations for Tokyo's next earthquake disaster. Perhaps the most important of these was the belief that it was enough to be ready for the earthquake that seemed most likely. After 3.11, researchers instead became intent on finding seismicity's upper limits along any given fault or plate boundary, all over Japan, and determining the worst-case scenarios that could result if those limits were reached. For Tokyo, that meant abandoning its models for estimating how much damage a chokkagata earthquake or another Sagami Trough event could do. The first had taken the size of the Hanshin-Awaji event as a baseline and the second had presumed that the next Great Kantō Earthquake would be no more powerful than the last one. Neither assumption was easily defensible after March 2011; both the Tokyo Metropolitan Government and the Central Disaster Prevention Council set out to determine, as best they could, how powerful an earthquake the complicated nexus of plate boundaries under the capital district and those along the Sagami subduction zone could possibly produce. They also sought to calculate how much damage and how many casualties would result from these new worst-case scenarios.

The council released its findings in late 2013. The report seemed thorough, especially with regard to the potential effects of a near-field earthquake. Its authors looked at nineteen distinct scenarios for that event, the differences between them arising from where along the complicated plate boundaries and other potential fault zones under the capital district the earthquake originated and from variations in magnitude (which ranged from M6.8 to M7.3). This was a more involved analysis than the authorities had conducted in their last round of damage estimates in 2004 but it wasn't that dissimilar from that earlier model in terms of the predicted intensity of the shaking at the surface. It therefore came as a surprise to many that the council's 2013 estimates of how much harm the city could suffer differed considerably from the ones it had calculated less than a decade earlier.

In some areas, the new numbers suggested that the city would fare better than previously thought; estimates of the number of buildings destroyed by fire or shaking, for example, fell from eight hundred fifty thousand in 2004 to six hundred fifty thousand in 2013, a reflection in part of retrofitting efforts and the introduction of more fire-resistant building materials.[73] The projected economic impact of such a disaster also declined. One of the differences that caught people's attention, however, was that the authorities were now predicting that as many as 23,000 city residents could die as a result of the next near-field earthquake. That was more than twice as many fatalities as the 2004 study had reported.[74] The most recent calculations did try to account for fatalities that happened while people were trying to flee, something the previous study had not done, which was one of the factors that resulted in the higher estimates.[75] Even so, the context for discussions about estimates like these and the uses to which they were put had changed after 3.11, which immediately cast these new numbers in a different light.

"No more 'beyond expectations'" had become something of a rallying cry after 3.11, as seismologists and other earth scientists worked to describe worst-case scenarios in as much detail as they could. In some instances, as with studies of the effects of a potential Nankai Trough earthquake, there were suggestions that the experts had taken things too far and that the casualty numbers had been inflated in an effort to avoid another situation like the one scientists had found themselves in a few years earlier, when it appeared that they had badly underestimated how much damage an earthquake like the one on 3.11 could do.[76] In Tokyo's case, however, criticisms directed at the council argued that it may still have been too conservative in its approach and that the lessons of 3.11 had not been taken to heart. The new study's worst-case scenarios were premised on an M7.3 near-field earthquake or on an M8 Sagami Trough event, as previous ones had been. Recent models had set the upper limit for a Sagami Trough earthquake much higher than that, at M8.6, but the council's experts chose not to provide an estimate of what the effects of an earthquake of that size would do to Tokyo.[77] Kawata Yoshiaki, who had led the team of experts responsible for generating the Nankai Trough report, remarked somewhat sarcastically that the report "assumed from the start that most of the central infrastructure would survive, which is damage that is 'entirely within expectations [*sōteinai*].'"[78] In other words, the report had stopped short of addressing what might happen if a larger event were to occur and, thus, painted an overly optimistic image of such an earthquake's effects.[79]

For Tokyo's residents and those responsible for protecting them from the next earthquake disaster, one thing that had been consistent all along in the different studies was that fire remained the primary threat to life and limb. As Nakabayashi Itsuki put it in 2011, "A near-field earthquake beneath the capital district will strike not with a tsunami from the sea, but in a tsunami of fire."[80] Even in 2013, the CDPC estimated that as much as 70 percent of the fatalities in a near-field earthquake would be due to fire.[81] The city had made progress over the years in its efforts to replace older wooden structures with newer buildings that were both less likely to collapse and much less likely to burn, but many pre-1981 structures still remained. Tokyo's administration pledged to do more to deal with this longstanding problem.

At the same time, city officials also insisted that the most important defenses against the next earthquake disasters were in the hands of the public. Tokyo-ites were encouraged to make sure that they always had enough food and water on hand to last at least three days and, ideally, a week. If fires broke out after an earthquake, it would fall to people in neighborhoods all over the city to be sure that they were extinguished immediately. Residents of many of those neighborhoods should know in advance where their designated evacuation site was ánd what route they should take to get there. As of 2018, there were 234 such sites scattered throughout the city; the Tokyo Metropolitan Government's website and many publications provided full-color maps and detailed instructions on how to find your family's destination.

Those maps also identified those sectors of the metropolitan area where residents were instead told not to evacuate, but to instead shelter in place. They include areas at the city center, like Chiyoda-ku, Ginza, and Nihonbashi, where the risk of both violent shaking and fire had long been expected to be low. In addition to these "naturally" safe spaces, the city had also identified many others that met its minimum standards for size (at least sixty-four hectares) and flame-resistant construction (at least 70 percent) and didn't exceed the upper limit on the number of wooden buildings in the zone (no more than fifty). In 2018, there were around three dozen such sites, which together accounted for around one hundred square kilometers of the city's surface.[82] Most of the reclaimed land and man-made islands at the southern end of the Kōtō delta received this shelter-in-place designation. The rest of the delta's residents, however, were expected to head to one of twenty or so designated evacuation sites. Shirahige-East and West are both on that list, as are others of the original six "disaster shelter" sites chosen in the late 1960s.

Conclusion

A few months after 3.11, producers from one of Tokyo's smaller television broadcasters asked urban planner Murakami Suminao to appear in a special series they were making on disaster preparedness in the capital district. More specifically, they wanted Murakami to be part of a segment devoted to the apartment and shelter complex at Shirahige-East he had helped design some four decades earlier.[83] Murakami agreed and appears on camera across the street from the complex's immediately recognizable kilometer-long façade, where he explained to the reporter accompanying him how the need to find a place for the delta's residents to flee, and then to protect them from fire, had defined their approach to the shelter's design. The sprinklers, the gates, the park on the other side of the towering apartment buildings—all of which the program went on to show to viewers—had made sense in the context of their fears at the time. Murakami paused to take a few photographs of the building for himself. He seemed delighted to be encountering it again.

The roles that relationships between expertise, knowledge production, disasters, and strategies to mitigate future harm have played in facilitating some futures and precluding others has been well documented in other settings, in circumstances not so different from the ones Japan found itself in after the 1960s. The Cold War civil defense campaigns in the United States that held out the promise of a survivable nuclear war, for example, wedded patriotic zeal with scientific certainty and gave frightened Americans at least the illusion of agency over their fates.[84] Sociologist Lee Clarke coined the term "fantasy documents" to describe the sort of planning that accompanied preparations like these not just for nuclear war but for other twentieth-century situations characterized by high degrees of uncertainty—the threat of catastrophic oil spills, nuclear power plant accidents, and the like.[85] One of Clarke's key observations was that such documents serve as "public declarations that planners, or the planning organization, have thought carefully about some problem and have developed the necessary wisdom and power to establish dominion over it."[86] Clarke's case studies were all drawn from the United States, but the uncertainty and the historical contexts that made the production of "fantasy documents" so attractive there were qualities just as easily found in Japan in equal if not greater measure.

This book has shown how experts' attempts to "establish dominion" over future earthquake disasters in Japan came to be and what they owed to the similar efforts that preceded them. Murakami's invitation to visit Shirahige-East in the aftermath of the 3.11 disasters is a reminder of both the impact such efforts have had and of their limitations. Kawasumi's warnings about the earthquakes in Tokyo's future transformed how citizens thought about the uncertainty ahead and the city's preparations for its next disaster. Shirahige-East and its counterparts on the delta are monuments to his persuasive powers and to the faith in technocratic expertise that led to their construction. Murakami's post-3.11 pride in what he had helped create speaks to the excitement that had surrounded the project in its early years, when it seemed an answer to the hazards ahead had finally been found. But Shirahige-East is also a relic, an artifact of a strategy for dealing with uncertainty that disaster experts had to abandon almost as soon as they had embraced it. The optimism that had surrounded the project was short-circuited by the realization that the threats to the capital were more complex and widely distributed than originally thought. That pattern repeated itself more than once and not just in Tokyo, as the rest of the country, too, settled into permanent states of anticipation armed with frequently changing sets of tools with which to prepare for the next disaster.

Japan's demographics being what they are, as you read this there are likely still plenty of Japanese old enough to remember a time when earthquake disasters and their country's future were not so explicitly entangled. Before Kawasumi Hiroshi's announcement in 1964 and Ishibashi Katsuhiko's in 1976, powerful earthquakes in either Tokyo or the Tōkai would hardly have come as a complete surprise, given each location's long history with such events, but that is a very different relationship to the future than one in which disaster's arrival is understood to be only a matter of time. Once Kawasumi and Ishibashi raised their respective alarms, the calculus of expectations about what the future would be like changed for those regions' inhabitants and for much of the rest of the nation, alike. Japanese who were children in the 1960s and 1970s and those born since grew up in a country where the risks to the capital and the Tōkai district were a given, where scientists could describe the slow creep of tectonic plates, the forces acting on them, and the earthquakes they caused with ever-greater precision. Laws, regulations, and urban planning changed to accommodate these new risks, while disaster preparedness drills and evacuation maps connected them to the practices of everyday life. Japanese plans to preempt the worst effects of a major earthquake close to Tokyo or offshore from the Tōkai district have not, as of this writing, been directly tested by the disasters they anticipated. This is, of course, a good thing—one hopes those tests never come—but it is precisely in the spaces between disasters that efforts to reduce uncertainty and plans for catastrophe have had their greatest impact.

The ways of thinking about disasters, earthquake prediction, and uncertainty that came to prominence in Japan in the 1960s and 1970s continued to shape how it dealt with risk well into the early decades of the next millennium. How scientists and government actors organized themselves to deal with the earthquake threat, for example, changed relatively little. The Yochiren was still meeting regularly in 2023. Its membership was no longer as dominated by the old Imperial universities as it once had been, but the committee's functions and interactions with the public in the early 2020s were not so different from what they had been in 1970. The Hanteikai, created in 1978 to watch for signs of an impending Tōkai earthquake, performed that function for forty years, until a name change in 2017 signaled the committee's shift in focus to a "new" threat, the Nankai Trough earthquake. The Large Scale Earthquake Countermeasures Act, meanwhile, was still the law of the land in 2023.

The Japanese government continued to update its estimates of the likelihoods of the earthquakes it was telling the public to fear the most, among them the shuto chokkagata, or "directly-beneath the capital district" earthquake (70 percent probability within thirty years), for example, or a Nankai Trough event (70–80 percent likelihood within thirty years), along with detailed descriptions of the damage each would do. Although it became more common than it once was for the media and government spokesmen to publicly acknowledge that nowhere in Japan was safe from the effects of a powerful earthquake, some locations were clearly more important to the state than others. The capital, the Tōkai, and the Nankai districts continued to be singled out for special measures in ways that the rest of the country generally was not.[87]

Policy makers in Japan are still a long way from imagining a replacement for the strategies for dealing with uncertainty that they have relied on since the 1960s. After 3.11, the Japanese government continued to support efforts to refine probabilistic models and hazard mapping. It also encouraged local governments to focus on making "resilience" their goal as they prepared their communities for future disasters. Resilient communities will have anticipated the worst effects of a potential disaster and have plans to restore community "lifelines"—utilities, access to food and shelter, communications, and so on—as quickly as possible in the aftermath of one. Resilience doesn't rely on the sort of warnings that short-term prediction was meant to provide but it does assume that experts have a reasonably clear picture of what risks and hazards a given community is likely to face, going forward. Japan's probabilistic hazard maps support that assumption. The government has also discouraged efforts by victims of the tragedies in 2011 (and others since) to make the full extent of their suffering visible to others.[88] It isn't hard to imagine reasons why it has done so but, certainly, a reluctance to question the efficacy of Japan's contemporary earthquake disaster mitigation strategies, in the absence of politically palatable alternatives, has to be one of them.

Modern Japan's earthquake prediction project took shape right alongside the policies that fostered high-speed economic growth, rapid urbanization, and an unprecedented concentration of people and wealth in a megalopolis that stretches from Tokyo to Kobe. The tools that supported one aided the other. Applied science funded by the state, the construction of widespread communications and data networks, and deference to certain forms of expertise were all tools of late capitalism as much as of earthquake prediction. From the perspective of the 2020s it makes sense to wonder if both pursuits have outlived their usefulness and, at the same time, to be uncertain about what might come next. Can ways of thinking about disasters and uncertainty be separated from policies that have prioritized economic growth and the structures that support it? The two have been so closely entangled for so long that imagining them apart will be difficult.

Japan's future will never not have powerful earthquakes in it. At present, the stakes of an earthquake disaster in Japan are probably higher than they are just about anywhere else on the planet where serious effort has gone into preparing for such an event. The history of earthquake prediction in modern Japan and of the science and scientists behind it is, in many ways, a testimony to the optimism that the twentieth century so frequently and perhaps indiscriminately inspired. That some of that optimism was likely misplaced—toward the reduction of uncertainty around earthquakes as toward other achievements—has become unavoidably clear. What that history also reveals, however, is that changing how we think about the future, and prepare for it, was and is possible.

After the Great Kantō Earthquake, observers often described Tokyo's transformation into "a sea of fire" (*hi no umi*) as one of the most horrific effects of that disaster.[89] The same phrase surfaced frequently in the 1960s and 1970s in accounts of what would happen in the Kōtō delta in the event of the next powerful earthquake to strike the capital; I came across it again as residents speculated about Tokyo's fate after 2011.[90] When scientists in Japan launched the short-term earthquake prediction project and Kawasumi his plans for Tokyo, many were of a generation that knew what "seas of fire" were, what they looked like, their sound and the taste they left in the air, and what remained in their wake. The urgency with which they tried to translate what they knew about past earthquakes into useable information about future ones drew on their skills as scientists and reflected their fears. It was a product, too, of an optimism about the field's trajectory that is perhaps harder for us to grasp in the contemporary moment, while their dread of the terrible disaster just over the horizon feels all too familiar. Perhaps even for Imamura, but especially for his successors in the 1960s and 1970s, there were so many reasons to believe that seismicity's mysteries were theirs to solve. The institutions and practices they helped build and that still define how Japan thinks about earthquakes to come—the monitoring networks, the designation of at-risk sites, and so on—are products of that era and of their expectations about how quickly and how far science would take them.

NOTES

Introduction

1. Italy experienced twenty-seven earthquakes with at least one fatality directly attributed to them over that same time span, the United States twenty-six; Japan suffered forty-three before 1950 and another thirty-three between 1950 and 1999.

2. According to the National Centers for Environmental Information Database, total earthquake deaths in Japan between 1900 and 1999 numbered 167,474; between 684 (the first available records date from then) and 1899 there were 169,022 total earthquake deaths in Japan. See the Global Significant Earthquake Database, 2150 B.C. to present, NOAA, National Centers for Environmental Information, https://www.ngdc.noaa.gov/hazard/earthqk.shtml.

3. Gregory K. Clancey, *Earthquake Nation: The Cultural Politics of Japanese Seismicity, 1868-1930* (Berkeley: University of California Press, 2006), 212–17.

4. Mochizuki Toshio and Nakabayashi Itsuki, eds., *Anzen anshin no toshizukuri: anzen de anshinshite sumeru iezukuri machizukuri no tameni.* (Tokyo: Tokyo Toritsu Daigaku Shuppankai, 2001); Winfried Flüchter, "Tokyo before the Next Earthquake: Agglomeration-Related Risks, Town Planning and Disaster Prevention," *The Town Planning Review* 74, no. 2 (2003): 213–38; Robert K. Reitherman, *Earthquakes and Engineers: An International History* (Reston, VA: American Society of Civil Engineers, 2012), 110–35,163–79,197–203; Koshizawa Akira, *Tokyo toshi keikaku no isan: bōsai, fukkō, Orinpikku* (Tokyo: Kabushiki Kaisha Chikuma Shobō, 2014), 229–45.

5. For an analysis of the changing meanings associated with Bōsai no hi, see Mizuide Kōki, *'Saigo' no kiokushi: Medeiya ni miru Kantō daishinsai—Isewan taifū* (Tokyo: Jinbunshoin, 2019).

6. See chap. 3 in Clancey, *Earthquake Nation.*

7. In Japanese: *Shinsai Yobō Chōsakai*, or "Committee for the investigation of earthquake disaster prevention." See Clancey, *Earthquake Nation*, 160–61; Kim Boumsoung, *Meiji Taishō no Nihon no jishingaku: "rōkaru saiensu" o koete* (Tokyo: Tōkyō Daigaku Shuppankai, 2007), 63–65.

8. Kim Boumsoung, *Meiji Taishō no Nihon no jishingaku*, 64.

9. The most recent versions of these maps may be found on the website of the Headquarters for Earthquake Research Promotion: https://www.jishin.go.jp/.

10. See, for example, Seth Stein, Robert J. Geller, and Mian Liu, "Why Earthquake Hazard Maps Often Fail and What to Do About It," *Tectonophysics* 562–563 (2012): 1–25; Gregory Smits, "Probabilistic Earthquake Hazard Maps," in *Cartographic Japan: a History in Maps*, ed. Kären Wigen, Fumiko Sugimoto, and Cary Karacas (Chicago: University of Chicago Press, 2016).

11. Yan Y. Kagan, David D. Jackson, and Robert J. Geller, "Characteristic Earthquake Model, 1884–2011, R.I.P.," *Seismological Research Letters* 83, no. 6 (2012): 951–53.

12. Nakabayashi Itsuki, "Shuto chokka jishin taisaku no kihon hōkō," *Jishin jyānaru* no. 57 (2014), 9.

13. Another way to provide some perspective on those numbers is to take a quick look at some worst-case scenarios for anticipated destructive earthquakes in the United States. This is of course a crude juxtaposition—the Tōkai district bears little resemblance to the West Coast, Tokyo has three and a half times as many residents as Los Angeles, and so on, but the expected effects of some of the earthquakes thought most likely to occur in the near term in North America are worth thinking about, nevertheless. In 2008, the USGS projected that an M7.8 earthquake on the southernmost section of the San Andreas Fault would lead to 1,780 deaths in the Los Angeles metropolitan area. See Lucile M. Jones et al., "The ShakeOut Scenario: U.S. Geological Survey Open-File Report 2008-1150 and California Geological Survey Preliminary Report 25," http://pubs.usgs.gov/of/2008/1150/. To the north, in 2018, the USGS estimated that an M7.0 earthquake on the Hayward Fault could kill as many as 800 people in the San Francisco Bay Area; see Kenneth W. Hudnut et al., "The HayWired Earthquake Scenario—We Can Outsmart Disaster," *Fact Sheet* 3016 (2018): 3. Finally, and still further north, a Federal Emergency Management Agency exercise in 2015 projected that an M9 Cascadia Subduction Zone earthquake would cause as many as 14,600 deaths in Washington and Oregon, many of them due to tsunami; see Rebekah Paci-Green et al., "Cascadia Rising: Cascadia Subduction Zone Catastrophic Earthquake and Tsunami Functional Exercise Scenario Document," *Environmental Studies Faculty and Staff Publications* 28 (2015): 34–46, https://cedar.wwu.edu/envs_facpubs/28.

14. Kathleen J. Tierney, "From the Margins to the Mainstream? Disaster Research at the Crossroads," *Annual Review of Sociology* 33 (2007): 504–6.

15. Examples include Mike Davis, *Ecology of Fear: Los Angeles and the Imagination of Disaster*, 1st ed. (New York: Metropolitan Books, 1998); Theodore Steinberg, *Acts of God: The Unnatural History of Natural Disaster in America* (New York: Oxford University Press, 2000).

16. An early English-language overview of Japanese policies in a disaster studies context is Kumagai Yoshio and Nojima Yoshiteru, "Urbanization and Disaster Mitigation in Tokyo," in *Crucibles of Hazard: Mega-cities and Disasters in Transition*, ed. James K. Mitchell (New York: United Nations University Press, 1999).

17. Clancey, *Earthquake Nation*; Gennifer S. Weisenfeld, *Imaging Disaster: Tokyo and the Visual Culture of Japan's Great Earthquake of 1923* (Berkeley: University of California Press, 2012); J. Charles Schencking, *The Great Kantō Earthquake and the Chimera*

of National Reconstruction in Japan (New York: Columbia University Press, 2013); Gregory Smits, *Seismic Japan: the Long History and Continuing Legacy of the Ansei Edo Earthquake* (Honolulu: University of Hawaii Press, 2013); Alex Bates, *The Culture of the Quake: The Great Kanto Earthquake and Taishō Japan* (Ann Arbor: University of Michigan Center for Japanese Studies, 2015).

18. Janet Borland, *Earthquake Children: Building Resilience from the Ruins of Tokyo* (Cambridge, MA: Harvard University Asia Center, 2020).

19. The Japanese-language historical scholarship on disasters is abundant and varied and has developed on its own timeline. In addition to an extensive catalog of narrative accounts and source collections tied to specific events and their impacts, researchers in Japan have also done a great deal of analytical work on disasters' cultural, economic, and social effects in the modern era. Some examples of the latter include Harada Katsumasa and Shiozaki Fumio, *Tokyo Kantō Daishinsai zengo* (Tokyo: Nihon Keizai Hyōronsha, 1997); Kitahara Itoko, *Jishin no shakaishi: Ansei daijishin to minshū*, Kōdansha gakujutsu bunko (Tokyo: Kōdansha, 2000); Takemura Masayuki, *Kantō Daishinsai: dai Tokyo-ken no yure o shiru* (Tokyo: Kajima Shuppankai, 2003); Kitahara Itoko, *Nihon saigaishi* (Tokyo: Yoshikawa Kōbunkan, 2006); Takemura Masayuki, *Mizō no daisaigai to jishingaku: Kantō Daishinsai* (Tokyo: Kokonshoin, 2009); Yamashita Fumio, *Kakusareta daishinsai: Taiheiyō Sensō shi hiroku* (Sendai-shi: Tōhoku Daigaku Shuppankai, 2009); Narita Ryūichi, "Kantō daishinsai no metahisutori no tame ni," *Shisō* 866 (1996): 61–90.

20. One important exception to that rule can be found in the scholarship on the several slow disasters responsible for the environmental pollution crises of Japan's era of high-speed economic growth. Work by Timothy George, Brett Walker, and others on Minamata disease (caused by methyl-mercury poisoning), itai itai byo ("It hurts—it hurts" sickness, caused by cadmium poisoning), and Yokkaichi asthma (caused by sulphur dioxide emissions), for example, brought to light the costs in human lives and suffering of the largely unchecked economic growth of the 1960s and early 1970s especially, and the struggles its victims went through to obtain medical treatment and redress. See Frank K. Upham, "Unplaced Persons and Movements for Place," in *Postwar Japan as History*, ed. Andrew Gordon (Berkeley: University of California Press, 1992); Timothy S. George, *Minamata: Pollution and the Struggle for Democracy in Postwar Japan*, Harvard East Asian Monographs (Cambridge , MA: Harvard University Asia Center, 2001); Brett L. Walker, *Toxic Archipelago: A History of Industrial Disease in Japan* (Seattle: University of Washington Press, 2010); Simon Avenell, "From Fearsome Pollution to Fukushima: Environmental Activism and the Nuclear Blind Spot in Contemporary Japan," *Environmental History* 17 (2012): 244–76. For a discussion of "slow disasters," see Scott Gabriel Knowles and Zachary Loeb, "The Voyage of the Paragon: Disaster as Method," in *Critical Disaster Studies*, ed. Jacob A. C. Remes and Andy Horowitz (Philadelphia: University of Pennsylvania Press, 2021), 27–30.

21. Yoshikuni Igarashi, *Bodies of Memory: Narratives of War in Postwar Japanese Culture, 1945–1970* (Princeton, NJ: Princeton University Press, 2000); William M. Tsutsui, *Godzilla on My Mind: Fifty Years of the King of Monsters* (New York: Palgrave

Macmillan, 2004); Yoshikuni Igarashi, "Mothra's Gigantic Egg: Consuming the South Pacific in 1960s Japan," in *In Godzilla's Footsteps: Japanese Pop Culture Icons on the Global Stage*, ed. William M. Tsutsui and Michiko Itō (New York: Palgrave Macmillan, 2006); Susan J. Napier, "When Godzilla Speaks," in *In Godzilla's Footsteps: Japanese Pop Culture Icons on the Global Stage*, ed. William M. Tsutsui and Michiko Itō (New York: Palgrave Macmillan, 2006); Susan J. Napier, "Panic Sites: The Japanese Imagination of Disaster from Godzilla to Akira," *Journal of Japanese Studies* 19, no. 2 (1993): 327–51.

22. The novel has inspired numerous films, television shows, and at least one satirical take on its premise. The first movie based on the book was released in 1973, the second in 2006, and an anime version in 2020. The first television series aired in 1974 and the most recent, "Japan Sinks: People of Hope," was first broadcast in 2021. Tsutsui Yasutaka's short-story parody of Komatsu's book, "Nihon igai zenbu chinbotsu" (Everything Other Than Japan Sinks), appeared in *All Yomimono* a few months after the novel's publication, and was the basis for a 2006 film of the same title. See Komatsu Sakyō, *Nihon chinbotsu* (Tokyo: Kobunsha, 1973); William O. Gardner, "Narratives of Collapse and Generation: Komatsu Sakyō's Disaster Novels and the Metabolist Movement," *Japan Forum* 26, no. 3 (2014): 306–24; Christopher Bolton et al., "An Interview with Komatsu Sakyō," *Science Fiction Studies* 29, no. 3 (2002): 323–39; Jessica Langer, "Three Versions of Komatsu Sakyō's *Nihon chinbotsu* (Japan Sinks)," *Science Fiction Film and Television* 2, no. 1 (2009): 45–57; Napier, "Panic Sites," 332–36.

23. Igarashi, *Bodies of Memory*; Tsutsui, *Godzilla on My Mind: Fifty Years of the King of Monsters*; William M. Tsutsui and Michiko Itō, eds., *In Godzilla's Footsteps: Japanese Pop Culture Icons on the Global Stage* (New York: Palgrave Macmillan, 2006); Igarashi, "Mothra's Gigantic Egg"; Napier, "When Godzilla Speaks"; William M. Tsutsui, "Oh No, There Goes Tokyo: Recreational Apocalypse and the City in Postwar Japanese Popular Culture," in *Noir Urbanisms: Dystopic Images of the Modern City*, ed. Gyan Prakash (Princeton, NJ: Princeton University Press, 2010); Napier, "Panic Sites."

24. The Disaster Basic Countermeasures Act (Shinsai taisaku kihon hō) became law in 1961; the Large-Scale Earthquake Countermeasures Act (Dai kibo jishin taisaku tokubetsu sochihō) was approved in 1978.

25. Andy Horowitz and Jacob A. C. Remes, "Introducing Critical Disaster Studies," in *Critical Disaster Studies*, ed. Andy Horowitz and Jacob A. C. Remes (Philadelphia: University of Pennsylvania Press, 2021), 5.

26. Examples of scholarship on science in Japan before 1945 include James R. Bartholomew, *The Formation of Science in Japan: Building a Research Tradition* (New Haven, CT: Yale University Press, 1989); Clancey, *Earthquake Nation*; Hiromi Mizuno, *Science for the Empire: Scientific Nationalism in Modern Japan* (Stanford, CA: Stanford University Press, 2009); David G. Wittner and Philip C. Brown, *Science, Technology, and Medicine in the Modern Japanese Empire* (Abingdon, VA: Routledge, 2016). For the period after that, see Morris Low, *Science and the Building of a New Japan* (New York: Palgrave Macmillan, 2005); Lisa Onaga, "Reconstructing the Linear No-Threshold Model

in Japan: A Historical Perspective on the Technics of Evaluating Radiation Exposure," *Technology and Culture* 58 (2017): 194–204.

27. Jacob Darwin Hamblin and Linda M. Richards, "Beyond the *Lucky Dragon*: Japanese Scientists and Fallout Discourse in the 1950s," *Historia scientiarum. Second series* 25, no. 1 (2015): 36–56; Lisa Onaga, "Measuring the Particular: The Meanings of Low-Dose Radiation Experiments in Post-1954 Japan," *positions: asia critique* 26, no. 2 (2018): 265–304; Akira Kurosaki, "Japanese Scientists' Critique of Nuclear Deterrence Theory and Its Influence on Pugwash, 1954–1964," *Journal of Cold War Studies* 20, no. 1 (2018): 101–39.

28. That said, for a fascinating study of the role Cold War ideological battles played in the debates among scientists over plate tectonics theory in Japan, see Tomari Jirō, *Purēto tekutonikusu no kyozetsu to juyō: sengo Nihon no chikyū kagakushi*, Shohan. ed. (Tokyo: Tokyo Daigaku Shuppankai, 2008).

29. And a small audio archive as well—lectures of his about the Great Kantō Earthquake recorded by Nihon Victor in 1930 are available for listening courtesy of the National Diet Library as "Kantō daishinsai 1" (https://rekion.dl.ndl.go.jp/info:ndljp/pid/1319623) and "Kantō daishinsai 2" (https://rekion.dl.ndl.go.jp/info:ndljp/pid/1319624).

30. Imamura Akitsune, *Jishingaku* (Tokyo: Dai Nihon Tosho Kabushiki Kaisha, 1905).

31. Physicist Terada Torahiko also conducted numerous statistical analyses of the relationships between earthquake frequency and other natural phenomena. See Terada Torahiko, "Luminous Phenomena Accompanying Destructive Sea-waves (Tunami)," *Proceedings of the Imperial Academy* 9, no. 8 (1933): 367–69; Kobayashi Tadashi, *Terada Torahiko to jishin yochi* (Tokyo: Tokyo Tosho, 2003); Kurosawa Tairiku, *'Jishin yochi' no gensō* (Tokyo: Shinchōsha, 2014), 172–73; Terada Torahiko, "On a Measure of Uncertainty Regarding the Prediction of Earthquake Based on Statistics," *Proceedings of the Imperial Academy* 9, no. 6 (1933): 255–57.

32. 46th Diet, Shūgiin, Saigai Taisaku Tokubetsu Iinkai, July 3, 1964, p. 13. https://kokkai.ndl.go.jp/txt/104604339X01319640703.

33. "Kantō daishinsai sono toki watakushi wa," *Mainichi shinbun*, February 26, 1993; "Shinkin kokki," *Asahi shinbun*, September 26, 1985.

34. Ishibashi Katsuhiko, "Genpatsu shinsai—hametsu o sakeru tame ni," *Kagaku* 67, no. 10 (1997): 720–24.

35. Ishibashi Katsuhiko, *Nankai torafu kyodai jishin—rekishi, kagaku, shakai* (Tokyo: Iwanami Shoten, 2014).

36. Robert J. Geller, *Nihonjin wa shiranai 'Jishin yochi' no shōtai* (Tokyo: Futabasha, 2011); Robert J. Geller, *Gerā-san Nippon ni mono mōsu* (Tokyo: Tōkyōdō Shuppan, 2018).

37. The Institute supported only six researchers and staff when it opened its doors in 1925. By 1950 it had eighty and, as of 1975, 166. See Tokyo Daigaku Hyakunenshi Henshū Iinkai, ed., *Tōkyō Daigaku hyakunenshi, bukyokushi (3)*, (Tokyo: Tokyo Daigaku Shuppankai, 1987), 988–90.

38. They are Asada Toshi (1981–91), Mogi Kiyō (1991–2001), Ōtake Masakazu (2001–9), and Shimazaki Kunihiko (2009–12).

39. Matsuzawa Tōru from Tohoku University held the position on a temporary basis in 2012. His successor as chair was Kyoto University's Hirahara Kazurō, who remained in the post until 2019.

40. Robert J. Geller, "Sakete torenai deguchi sakusen," in *"Burupurinto" 50 shūnen jishin kenkyū no ayumi to kongo,* ed. Nihon Jishin Gakkai, Nihon Jishin Gakkai monogurafu (Tokyo: Nihon Jishin Gakkai, 2013), 20. Geller recalls being skeptical of earthquake prediction while an undergraduate at Caltech. He also points to the cold fusion "scandal" of the 1980s and its subsequent debunking as having shaped his decision to eventually speak out against earthquake prediction in Japan. See David Zierler, "Robert J. Geller (BS '73, PhD '77), Seismologist," *Caltech Heritage Project*, https://heritageproject.caltech.edu/interviews-updates/robert-geller.

41. Naomi Oreskes, "Why Predict? Historical Perspectives on Prediction in Earth Science," in *Prediction: Science, Decision Making, and the Future of Nature,* ed. Daniel R. Sarewitz, Roger A. Pielke, and Radford Byerly (Washington, DC: Island Press, 2000), 34.

42. Oreskes, "Why Predict?" 34. Emphasis in original.

43. Oreskes, "Why Predict?" 23. Emphasis in original.

44. Robert J. Geller, "Earthquake Prediction: a Critical Review," *Geophysical Journal International* 131 (1997): 425–50; Susan Elizabeth Hough, *Predicting the Unpredictable: the Tumultuous Science of Earthquake Prediction* (Princeton: Princeton University Press, 2010).

45. Joanne M. Nigg, "Predicting Earthquakes: Science, Pseudoscience, and Public Policy Paradox," in *Prediction: Science, Decision Making, and the Future of Nature,* ed. Daniel R. Sarewitz, Roger A. Pielke, and Radford Byerly (Washington, DC: Island Press, 2000).

46. Fa-ti Fan, "The People's War against Earthquakes: Cultures of Mass Science in Mao's China," in *Cultures Without Culturalism: The Making of Scientific Knowledge,* ed. Karine Chemla and Evelyn Fox Keller (Durham: Duke University Press, 2017); Fa-ti Fan, "'Collective Monitoring, Collective Defense': Science, Earthquakes, and Politics in Communist China," *Science in Context* 25, no. 1 (2012): 127–54.

47. Jamie L. Pietruska, *Looking Forward: Prediction and Uncertainty in Modern America* (University of Chicago Press, 2017), covers almost all of these areas. For climate and weather specifically see Katharine Anderson, *Predicting the Weather: Victorians and the Science of Meteorology* (Chicago: University of Chicago Press, 2005); Paul N. Edwards, *A Vast Machine: Computer Models, Climate Data, and the Politics of Global Warming* (Cambridge, MA: MIT Press, 2010); Aitor Anduaga Egaña, *Politics, Statistics and Weather Forecasting, 1840–1910: Taming the Weather*, Routledge Studies in the History of Science, Technology and Medicine (London: Routledge, Taylor & Francis Group, 2020). The ability to transform uncertainty into forms of risk—quantifiable and describable in ways that uncertainty was not—was what made insurance and any number of other transformative innovations in financial markets and trade possible. For prediction in financial markets

and the relationships between uncertainty and capitalism more generally, see Jonathan Levy, *Freaks of Fortune: the Emerging World of Capitalism and Risk in America* (Cambridge, MA: Harvard University Press, 2012); Jens Beckert, *Imagined Futures: Fictional Expectations and Capitalist Dynamics* (Cambridge, MA: Harvard University Press, 2016); Jonathan Levy, "Radical Uncertainty," *Critical Quarterly* 62, no. 1 (2020): 15–28.

48. Jenny Andersson, "The Great Future Debate and the Struggle for the World," *American Historical Review* 117, no. 5 (2012): 1411–30; Jenny Andersson, "Midwives of the Future: Futurism, Future Studies and the Shaping of the Global Imagination," in *The Struggle for the Long-term in Transnational Science and Politics*, ed. Jenny Andersson and Eglė Rindzevičiūtė (New York: Routledge, Taylor & Francis Group, 2015); Arthur P. Molella and Scott Gabriel Knowles, *World's Fairs in the Cold War: Science, Technology, and the Culture of Progress* (Pittsburgh, PA: University of Pittsburgh Press, 2019).

49. Paul Sabin, *The Bet: Paul Ehrlich, Julian Simon, and Our Gamble over Earth's Future* (New Haven, CT: Yale University Press, 2013).

50. Avenell, "From Fearsome Pollution to Fukushima."

51. Avenell, "From Fearsome Pollution to Fukushima," 252–53, 256.

52. *Silent Spring* was published in translation in Japan in 1964; a translation of *The Population Bomb* appeared in 1974.

53. Helene Joffe et al., "Stigma in Science: the Case of Earthquake Prediction," *Disasters* 42, no. 1 (2018): 81–100.

54. Kenneth Hewitt, "The Idea of Calamity in a Technocratic Age," in *Interpretations of Calamity from the Viewpoint of Human Ecology*, ed. Kenneth Hewitt (Boston: Allen & Unwin, 1983), 19–21.

55. Bruce V. Lewenstein, "Science Controversies: Can the Science of Science Communication Provide Management Guidance or Only Analysis?" in *The Oxford Handbook of the Science of Science Communication*, ed. Kathleen Hall Jamieson, Dan M. Kahan, and Dietram A. Scheufele (Oxford: Oxford University Press, 2017), 77.

56. Lewenstein, "Science Controversies," 77.

57. Levy, *Freaks of Fortune*; Beckert, *Imagined Futures*, 43–45, 289–91; Pietruska, *Looking Forward*; Levy, "Radical Uncertainty." Ulrich Beck's and Anthony Giddens' influential work in the 1990s, meanwhile, drew attention to risk in a different context by arguing that industrialized societies were being transformed by the emergence of new forms of it. See Anthony Giddens, *Modernity and Self-Identity: Self and Society in the Late Modern Age* (Stanford, CA: Stanford University Press, 1991); Ulrich Beck, *Risk Society: Towards a New Modernity* (London: Sage Publications, 1992); Arwen P. Mohun, "Constructing the History of Risk: Foundations, Tools, and Reasons Why," *Historical Social Research / Historische Sozialforschung* 41, no. 1 (2016): 30–47; Soraya Boudia and Nathalie Jas, "Introduction: Risk and 'Risk Society' in Historical Perspective," *History and Technology* 23, no. 4 (December 2007): 317–31.

58. William Ascher, "Scientific Information and Uncertainty: Challenges for the Use of Science in Policymaking," *Science and Engineering Ethics* 10, no. 3 (2004), 440; Warner Marzocchi and J. Douglas Zechar, "Earthquake Forecasting and Earthquake

Prediction: Different Approaches for Obtaining the Best Model," *Seismological Research Letters* 82, no. 3 (2011), 446.

59. David Spiegelhalter uses the example of "a lottery ticket (where the outcome depends on chance) and a scratch card (where the outcome is already decided, but you don't know what it is)" to illustrate the difference between the two types of uncertainty. See D. J. Spiegelhalter, *The Art of Statistics: How to Learn from Data*, first US ed. (New York: Basic Books, 2019), 240.

60. Taken further, this argument not only ruled out the possibility of reliable earthquake prediction but also called into question the validity of the science behind the production of probabilistic seismic hazard maps. See Max Wyss, "Do Probabilistic Seismic Hazard Maps Address the Need of the Population?," in *Geoethics: Ethical Challenges and Case Studies in Earth Sciences,* ed. Max Wyss and Silvia Peppoloni (Oxford: Elsevier, 2015); Francesco Mulargia, Philip B. Stark, and Robert J. Geller, "Why is Probabilistic Seismic Hazard Analysis (PSHA) Still Used?," *Physics of the Earth and Planetary Interiors* 264 (2017): 63–75; Kagan, Jackson, and Geller, "Characteristic Earthquake Model, 1884–2011, R.I.P."

61. Nicola Nosengo, "Scientists Face Trial Over Earthquake Deaths," *Nature*, https://doi.org/10.1038/news.2011.325.

62. Nosengo, "Scientists Face Trial Over Earthquake Deaths."

63. The public official was less fortunate; his conviction was upheld. See Joffe et al., "Stigma in Science."

64. For summaries of Wadati's work, see Cliff Frohlich, "Kiyoo Wadati and Early Research on Deep Focus Earthquakes: Introduction to Special Section on Deep and Intermediate Focus Earthquakes," *Journal of Geophysical Research* 92, no. B13 (December 10, 1987): 13777–88; Yasumoto Suzuki, "Kiyoo Wadati and the Path to the Discovery of the Intermediate-deep Earthquake Zone," *Episodes* 24, no. 2 (2001): 118–24. For Richter's life and work, see Susan Elizabeth Hough, *Richter's Scale: Measure of an Earthquake, Measure of a Man* (Princeton, NJ: Princeton University Press, 2007).

65. For a more complete discussion of "seismic moment" and the shortcomings of other measurements of magnitude, see Hough, *Richter's Scale*, 127–31.

66. Clancey, *Earthquake Nation*, 64–65, 154–56.

67. Kitahara Itoko, Matsuura Ritsuko, and Kimura Reo, eds., *Nihon rekishi saigai jiten* (Tokyo: Yoshikawa Kōbunkan, 2012), 776.

68. A list of the tiers and the effects associated with each is available at the JMA's website: http://www.jma.go.jp/jma/en/Activities/inttable.html.

Chapter 1

1. "Nankaidō" describes an area that includes all of Shikoku Island and the southern end of the Kii Peninsula.

2. The article is Imamura Akitsune, "Shigaichi ni okeru jishin no seimei oyobi zaisan ni taisuru songai o keigen suru kanhō," *Taiyō* 11, no. 12 (1905): 162–71. The book is Imamura Akitsune, *Jishingaku*.

3. By 1904, researchers sifting through historical accounts for references to earthquakes had constructed a catalog that included some two thousand earthquakes, the earliest from the fifth century CE. See Shinsai Yobō Chōsakai, *Dainihon jishin shiryō* (Tokyo: Maruzen, 1904); Clancey, *Earthquake Nation*, 104. An updated four-volume set published between 1941 and 1950 listed more than six thousand earthquakes.

4. Imamura Akitsune, "Shigaichi ni okeru jishin no seimei oyobi zaisan ni taisuru songai o keigen suru kanhō," 168.

5. Imamura Akitsune, "Shigaichi ni okeru jishin no seimei oyobi zaisan ni taisuru songai o keigen suru kanhō," 169.

6. Yamashita Fumio, *Jishin yochi no senkusha Imamura Akitsune no shōgai* (Tokyo: Seijisha, 1989), 98.

7. Imamura appealed to the *Niroku shinbun*, *Taiyō*, and the *Yorozu chōhō*, another widely read city paper, pleading with each not to misrepresent his work. See Yamashita Fumio, *Jishin yochi no senkusha*, 100.

8. Yamashita Fumio, *Jishin yochi no senkusha*, 100.

9. Yamashita Fumio, *Jishin yochi no senkusha*, 102–3.

10. Ōmori Fusakichi, "Tokyo ni okurubeki shōrai no jishin ni tsukite," *Shinsai Yobō Chōsakai hōkoku* 57 (1907): 2–17; Ōmori Fusakichi, "Tokyo to daijishin no fusetsu," *Taiyō* 12, no. 4 (1906): 173–76. Yamashita Fumio, *Jishin yochi no senkusha*, 102. Ōmori's use of the "fusetsu" label to describe Imamura's analysis was especially damaging. The phrase "mōgen fusetsu" (reckless remarks and groundless rumors) was often used by officials in the Meiji government to label seditious or otherwise politically unacceptable ideas or statements. See Hagiwara Takahiro, *Jishingaku hyakunen* (Tokyo: Tokyo Daigaku Shuppankai, 1982), 58.

11. Clancey, *Earthquake Nation*, 217–18; Ōmori Fusakichi, "Tokyo to daijishin no fusetsu."

12. Andrew Gordon, *Labor and Imperial Democracy in Prewar Japan* (Berkeley: University of California Press, 1991), 26–52.

13. Yamashita Fumio, *Jishin yochi no senkusha*, 110–12.

14. Clancey, *Earthquake Nation*, 217–19. For examples of Ōmori's reports describing his use of statistical analysis to find correlations between the variations in the frequency of earthquakes (primarily in Japan but in some instances elsewhere, as well) and times of day, months and seasons, barometric pressure, and other climatic phenomena, see Ōmori Fusakichi, "Jishin to tenki to no kankei ni tsukite," *Tōyō gakugei zasshi* 303 (1906): 320–25. Ōmori Fusakichi, "Jishin no fukuin ni tsukite," *Shinsai Yobō Chōsakai hōkoku* 68 (1910): 21–29; Ōmori Fusakichi, "Jishin no toki bunpu," *Shinsai Yobō Chōsakai hōkoku* 57 (February 1907): 38–42. An English-language version brings together the results of several of the Japanese articles: see Ōmori Fusakichi, "Notes on the Secondary Causes of Earthquakes," *Bulletin of the Imperial Earthquake Investigation Committee* 2, no. 2 (1908): 101–35.

15. See Clancey, *Earthquake Nation*, 221.

16. For Ōmori in Australia and after his return, see Tsuboi Chūji, "Ōmori Fusakichi to Nihon no jishingaku," *Chūō kōron* 80, no. 7 (1965): 383–88.

17. Imamura Akitsune, *Jishin no seifuku* (Tokyo: Nankōsha, 1926), 51.

18. Edward Thomas Mack, *Manufacturing Modern Japanese Literature: Publishing, Prizes, and the Ascription of Literary Value* (Durham, NC: Duke University Press, 2010); Weisenfeld, *Imaging Disaster*; Bates, *The Culture of the Quake*; Amy Bliss Marshall, *Magazines and the Making of Mass Culture in Japan* (Toronto: University of Toronto Press, 2019). The first set of official proposals to bring radio broadcasting to Japan was completed in August 1923; the first broadcasts began in March 1925. For discussions of the media's growth in the 1920s, see Satō Takumi, *"Kingu" no jidai: kokumin taishū zasshi no kōkyōsei* (Tokyo: Iwanami Shoten, 2002), 21; Gregory J. Kasza, "Democracy and the Founding of Japanese Public Radio," *Journal of Asian Studies* 45, no. 4 (1986): 745–67; Iwasaki Akira, "Atarashii media no tenkai," *Shisō* 624 (1976): 240–55.

19. Sari Kawana, "Mad Scientists and Their Prey: Bioethics, Murder, and Fiction in Interwar Japan," *Journal of Japanese Studies* 31, no. 1 (2005): 89–120; Mizuno, *Science for the Empire*. The success of Albert Einstein's six-week visit to Japan in late 1922 is another indicator of a growing interest in scientists and their perspectives. The "Relativity Boom" Einstein's visit provoked included its share of silliness and cleverly packaged Einstein-themed products but, at the same time, evidence of the public's appetite for engagement with him and his ideas is hard to overestimate. Some 14,000 people were said to have attended his public lectures. More than twenty books on relativity and Einstein were published in 1921 and 1922 alone. See Kaneko Tsutomu, *Ainshutain shokku* (Tokyo: Kawade Shobō Shinsha, 1981), 287.

20. Kaneko Tsutomu, *Ainshutain shokku*; Mizuno, *Science for the Empire*, 245.

21. Harada Mitsuo, *Omoide no shichijūnen* (Tokyo: Seibundō Shinkōsha, 1966), 236–37; Mizuno, *Science for the Empire*, 145.

22. Weisenfeld, *Imaging Disaster*; Bates, *The Culture of the Quake*.

23. Altogether, he published some forty magazine pieces in the quake's wake, including not one but two articles in *Kaizō*'s post-disaster collection. For a brief example of the enthusiastic press coverage Imamura continued to receive, see Yoyogi Shintarō, "(Gakkai zuiichi ninki gakusha) Imamura Akitsune hakushi," *Taiyō* 30, no. 1 (January 1924): 12–14.

24. Narita Ryūichi, "Kantō daishinsai no metahisutori no tame ni," 64.

25. Imamura Akitsune, "Daijishin chōsa nikki," *Kagaku chishiki* (October 1923): 12–38. See also Imamura Akitsune, "Konkai no daishinsai no keika o nobete shōrai ni oyobu," *Kaizō* 5, no. 10 (1923): 132.

26. Clancey, *Earthquake Nation*, 220. Ōmori had offered reassurances to the public after a series of quakes rocked Tokyo in December 1921, April 1922, and again in January 1923. In June 1923, not long before he left for Australia, Ōmori was quoted in the *Yorozu chōhō* to the effect that the city was safe from any major quakes in the foreseeable future.

27. Imamura Akitsune, "Daijishin chōsa nikki," *Kagaku chishiki* (October 1923), 36-37; Imamura Akitsune, "Daijishin chōsa nikki," (November 1923), 82. See also Tsuchida Hiroshige, *Kindai Nihon no "kokumin bōkū" taisei*, Shohan. ed. (Chiba-shi: Kanda Gaigo Daigaku Shuppankyoku, 2010), 73.

28. Yamashita Fumio, *Jishin yochi no senkusha*, 150.

29. Mizuno Hiroo, "Jishin yochi wa shakai no mukete dono yo ni hanashikakarareta kita ka—Kantō jishin kara Tōkai jishin made," *Kagakushi kenkyū* 43 (2004): 241.

30. The proposal's title in Japanese is *Jishin kenkyū no tokushu kikai setsuritsu ni kansuru kengian*.

31. Imamura's summary of the proposal appears in Imamura Akitsune, "Sono ato no daijishin chōsa," *Kagaku chishiki* (September 1924), 3–13.

32. Hagiwara Takahiro, *Jishingaku hyakunen*, 89; Imamura Akitsune, "Sono ato no daijishin chōsa," 5.

33. For an English-language description of Imamura's findings, see Imamura Akitsune, "On the Tiltings of the Earth Preceding the Kwanto Earthquake of 1923," *Proceedings of the Imperial Academy* 4, no. 4 (1928): 148–50.

34. Hagiwara Takahiro, *Jishingaku hyakunen*, 90. Kim Boumsoung describes the growing divergence between the field as Ōmori and Imamura understood it and the future that physicists like Nagaoka Hantarō others preferred in Kim Boumsoung, *Meiji Taishō no Nihon no jishingaku*, 109–11.

35. Suehiro's expertise was in shipbuilding engineering; Terada was a well-regarded physicist, with a parallel career as a widely published essayist.

36. Imamura Akitsune, *Jishin no seifuku*, 230; Hagiwara Takahiro, *Jishingaku hyakunen*, 91–94.

37. Terada Torahiko, "Jishin zakkan," in *Taishō Daishin-kasai shi* (Tokyo: Kaizōsha, 1924), 53–54.

38. Kim Boumsoung, *Meiji Taishō no Nihon no jishingaku*, 113–19, 129–34.

39. Imamura Akitsune, *Jishin kōwa* (Tokyo: Iwanami Shoten, 1924); Imamura Akitsune, *Jishin no seifuku*; Imamura Akitsune, *Namazu no zaregoto* (Tokyo: Sanseidō, 1941). The latter work was republished under a new title in 1949, as Imamura Akitsune, *Jishin no kuni* (Tokyo: Bungei Shunjū Shinsha, 1949).

40. Yamamoto Issei and Imamura Akitsune, *Hoshi to sora, kazan to jishin*, Nihon jidō bunko (Tokyo: Arusu, 1930), 49.

41. Publication resumed in 1948.

42. Imamura Akitsune, *Jishin no seifuku*, 74.

43. Yamamoto Issei and Imamura Akitsune, *Hoshi to sora, kazan to jishin*, 129.

44. Yamamoto Issei and Imamura Akitsune, *Hoshi to sora, kazan to jishin*, 174–75.

45. He helped convince the Ministry of Education to add content dealing with earthquake disasters to the elementary school curriculum, for example, which it finally did in the late 1930s. See Imamura Akitsune, *Namazu no zaregoto*, 54–55. It did so by adding the tale of "Inamura no hi" (The Fire of Rice Sheaves) to one of the textbooks elementary school children used to study Japanese. The story was based in part on the actions of one Hamaguchi Goryo in the aftermath of the 1854 Nankai earthquake. Lafcadio Hearn related a version of what had happened in Hamaguchi's coastal village that day in "A Living God," published in English in 1897. See Lafcadio Hearn, *Gleanings in Buddha Fields* (Boston: Houghton, Mifflin and Company, 1897). Nakai Tsunezō,

an elementary school teacher who had grown up near Hamaguchi's village of Hiro in Wakayama Prefecture, wrote the version of the tale that ended up in the textbook as "Inamura no hi." Lest instructors be unsure exactly what elements of "Inamura no hi" to emphasize in their classrooms, Imamura released his own guide to teaching the text in 1940. See Imamura Akitsune, "'Inamura no hi' no oshiekata ni tsuite," *Jishin* 12, no. 8 (1940): 360–74; Yamashita Fumio, *Jishin yochi no senkusha*, 229.

46. Imamura Akitsune, "On the Tiltings of the Earth Preceding the Kwanto Earthquake of 1923."

47. Tomari Jirō, *Nihon no jishin yochi kenkyū 130-nenshi: Meijiki kara Higashi Nihon Daishinsai made*, Shohan. ed. (Tokyo: Tokyo Daigaku Shuppankai, 2015), 167.

48. Imamura Akitsune, "Supplementary Note on the Result of Precise Levelling in Etigo and Sinano," *Proceedings of the Imperial Academy* 4, no. 3 (1928): 111.

49. Tomari Jirō, *Nihon no jishin yochi kenkyū 130-nenshi*, 167–68.

50. Tomari Jirō, *Nihon no jishin yochi kenkyū 130-nenshi*, 170–71.

51. Yamashita Fumio, *Kakusareta daishinsai*, 117–18.

52. Yamashita Fumio, *Kakusareta daishinsai*, 119; Tomari Jirō, *Nihon no jishin yochi kenkyū 130-nenshi.*

53. Yamashita Fumio, *Kakusareta daishinsai*, 118–19.

54. Tomari Jirō, *Nihon no jishin yochi kenkyū 130-nenshi*, 171–73; Imamura Akitsune, "Nankaidōoki daijishin no nazo," *Jishin* 5, no. 10 (1933), 624–26.

Chapter 2

1. Yamashita Fumio, *Kakusareta daishinsai.*

2. Parts of the shoreline closest to the fault zone had gained as much as half a meter in elevation above sea level because of the earthquake, while sites further inland lost as much or more. In Kōchi City and elsewhere, the new topography along the coast meant that floodwaters carried inland by the tsunami could not flow back out to sea. See Kitahara, Matsuura, and Kimura, *Nihon rekishi saigai jiten*, 522. Newsreel footage of the earthquake's effects may be viewed here: https://www.pref.kochi.lg.jp/sonaetegood/research/video.html.

3. Thomas J. Fitch and Christopher H. Scholz, "Mechanism of Underthrusting in Southwest Japan: A Model of Convergent Plate Interactions," *Journal of Geophysical Research (1896–1977)* 76, no. 29 (1971): 7260–92.

4. Yamashita Fumio, *Kakusareta daishinsai*, 92–94.

5. Yamashita Fumio, *Kakusareta daishinsai*, 101.

6. A January 1957 letter from Admiral G. L. Schuyler to Beno Gutenberg reports that he has been asked by AEC Chairman Lewis L. Strauss "if I could give him anything on a World War II proposal to give the Japanese an earthquake by a large submarine explosion near some undersea fault." In his reply a few days later, Gutenberg wrote, "During the years 1943–1944 I have received repeatedly letters suggesting to start an earthquake or a volcanic eruption in or near Japan by dropping a large bomb to act as a trigger force for release of strain in the earth's crust in the form of an earthquake. Most of

these suggestions came directly to me from a variety of people, but one or two through official channels marked 'secret.' In all of these instances, I have pointed out that the energy released in this way normally must be considered completely insufficient to start an earthquake." Gutenberg discusses his encounters with these proposals in Folder 3.12, Papers of Beno Gutenberg, California Institute of Technology Archives.

7. US strategists reportedly based their assessments of Tokyo's flammability on studies of the fires that followed the Great Kantō Earthquake. See David A. Fedman and Cary Karacas, "A Cartographic Fade to Black: Mapping the Destruction of Urban Japan during World War II," *Journal of Historical Geography* (2012): 12–13.

8. Casualty records for both disasters are incomplete; the government prohibited efforts to document what had happened. See Yamashita Fumio, *Kakusareta daishinsai*; Kitahara, Matsuura, and Kimura, *Nihon rekishi saigai jiten*, 510–11.

9. Yamashita covers this topic at length in Yamashita Fumio, *Kakusareta daishinsai*.

10. Observers outside of Japan, on the other hand, speculated openly about how much damage the earthquakes might have caused. See, for example, "Disastrous Quake in Central Japan," *New York Times*, December 8, 1944.

11. See John W. Dower, *Embracing Defeat: Japan in the Wake of World War II*, 1st ed. (New York: W. W. Norton, New Press, 1999).

12. John W. Dower, "'Ni' and 'F': Japan's Wartime Atomic Bomb Research," in *Japan in War and Peace: Selected Essays* (New York: New Press, 1993); Bowen C. Dees, *The Allied Occupation and Japan's Economic Miracle: Building the Foundations of Japanese Science and Technology 1945–52* (Surrey, UK: Japan Library, 1997); Nakayama Shigeru, "The American Occupation and the Science Council of Japan," in *The Orientation of Science and Technology: A Japanese View*, ed. Nakayama Shigeru, Collected Papers of Twentieth-century Japanese Writers on Japan (Folkestone: Brill, 2009).

13. Yamashita Fumio, *Kakusareta daishinsai*, 132–33.

14. Yamashita Fumio, *Jishin yochi no senkusha*; Yamashita Fumio, *Kakusareta daishinsai*, 130–31.

15. Usami Tatsuo, "Great Earthquakes of the Past," in *Earthquake Prediction Techniques: Their Application in Japan*, ed. Asada Toshi (Tokyo: University of Tokyo Press, 1982), 21. The Imperial Academy was renamed the Japan Academy (Nihon Gakushiin) the following year.

16. Yamashita Fumio, *Kakusareta daishinsai*, 133–34.

17. Musha Kinkichi, *Jishin namazu* (Tokyo: Akashi Shoten, 1995), 180; Yamashita Fumio, *Kakusareta daishinsai*, 133.

18. Tomari Jirō, *Nihon no jishin yochi kenkyū 130-nenshi*, 187–89.

19. Jishin Yochi Renrakukai, ed., *Jishin Yochi Renrakukai 10-nen no ayumi* (Tokyo: Kankōsha Kokudo Chiriin Nihon Sokuryō Kyōkai, 1979), 13.

20. Sassa, the oldest, had been born in 1900, Tsuboi and Wadati in 1902.

21. Wadati and Tsuboi worked with Terada Torahiko, and Hagiwara with Ishimoto Mishio.

22. Jishin Yochi Renrakukai, *Jishin Yochi Renrakukai 10-nen no ayumi*, 14.

23. Meeting of the Research and Liaison Committee for Earthquake Forecasting on 29 Aug 47, 18 September 1947, Topical File, Earthquake, 1945–1952, Records of Allied Operational and Occupation Headquarters, World War II, 1907–1966, Record Group 331.49, National Archives at College Park, College Park, MD.

24. Meeting of the Research and Liaison Committee for Earthquake Forecasting on 29 Aug 47.

25. I discuss these issues in more detail here: Kerry Smith, "Earthquake Prediction in Occupied Japan," *Historical Social Research* 40, no. 2 (2015): 105–33.

26. Hagiwara Takahiro, *Jishin yochi to saigai* (Tokyo: Maruzen, 1997), 67–69.

27. Hagiwara Takahiro, *Jishin yochi to saigai.*

28. "Memorandum and Suggestions Concerning Geophysical and Especially Seismological Research in Japan, 1947," Box 14, Folder 22, Papers of Beno Gutenberg, California Institute of Technology Archives. This was a topic Gutenberg had addressed before, although not in the Japanese context. In late 1935, he and Harry O. Wood coauthored a short article in *Science* in which they responded to press reports in the U.S. speculating about earthquake prediction's potential. "Any moderately successful method of prediction for scientific testing will be welcomed by all seismologists," they wrote, "but public prediction in the present state of knowledge is nothing short of a menace. See Harry O. Wood and B. Gutenberg, "Earthquake Prediction," *Science* 82, no. 2123 (1935): 220.

29. "Memorandum and Suggestions Concerning Geophysical and Especially Seismological Research in Japan, 1947."

30. Report on Development of Research and Liaison Committee for Earthquake Forecasting, 18 September 1947, Topical File, Earthquake, 1945–1952, Records of Allied Operational and Occupation Headquarters, World War II, 1907–1966, Record Group 331.49, National Archives at College Park, College Park, MD.

31. Preliminary Meeting on 11 August 1947, of the Research and Liaison Committee for Earthquake Forecasting at the Central Meteorological Observatory, August 12, 1947, Topical File, Earthquake, 1945–1952, Records of Allied Operational and Occupation Headquarters, World War II, 1907–1966, Record Group 331.49, National Archives at College Park, College Park, MD.

32. "Kantō ni daijishin setsu," *Mainichi shinbun*, September 11, 1947.

33. Jishin Yochi Renrakukai, *Jishin Yochi Renrakukai 10-nen no ayumi*, 22; Meeting of the Research and Liaison Committee for Earthquake Forecasting, 9 June 1948, June 19, 1948, Topical File, Earthquake, 1945-1952, Records of Allied Operational and Occupation Headquarters, World War II, 1907-1966, Record Group 331.49, National Archives at College Park, College Park, MD.

34. Jishin Yochi Renrakukai, *Jishin Yochi Renrakukai 10-nen no ayumi*, 24.

35. Jishin Yochi Renrakukai, *Jishin Yochi Renrakukai 10-nen no ayumi.*

36. Jishin Yochi Renrakukai, *Jishin Yochi Renrakukai 10-nen no ayumi*, 150–55; Meeting of the Research and Liaison Committee for Earthquake Forecasting on 29 August 1947.

37. Earthquake Forecasting Project: Engineer Section Reaction to Portion of Program Concerned with Geographic Survey Board, 18 September 1947, Topical File, Earthquake, 1945–1952, Records of Allied Operational and Occupation Headquarters, World War II, 1907–1966, Record Group 331.49, National Archives at College Park, College Park, MD.

38. Shigeru Nakayama, Kunio Gotō, and Hitoshi Yoshioka, *A Social History of Science and Technology in Contemporary Japan*, Japanese society series (Melbourne, Australia: Trans Pacific Press, 2001), 278.

39. Jishin Yochi Renrakukai, *Jishin Yochi Renrakukai 10-nen no ayumi*, 15–17.

40. In August 1947, the widely read *Nihon keizai shinbun* (*Japan Economic News*) featured side-by-side articles by Imamura Akitsune and Sagisaka Kiyōnobu, head of the Seismology Section at the Central Meteorological Observatory, each addressing this question: "Can Earthquakes be Foretold?" Imamura, not surprisingly, insisted that they could, while Sagisaka argued the opposite. Both were members of the Liaison Committee. An English translation appeared in the *Nippon Times* on August 7, 1947; a different translation circulated within SCAP.

41. "Yōjin wa kekkō da ga, sawagu no ataranu," *Yomiuri shinbun*, December 10, 1947; "Giant Quake Predicted for Kobe, Osaka, Kyoto," *Stars and Stripes*, December 11, 1947; "Kansai ni chikaku jishin," *Asahi shinbun*, December 9, 1947.

42. "Japanese Scientists Give Quake Safeguards," *Pacific Stars and Stripes*, December 16, 1947; "Kansai 'Quake Prediction Panics Area's Inhabitants," *Nippon Times*, December 23, 1947.

43. "Giant Quake Predicted for Kobe, Osaka, Kyoto."

44. Jishin Yochi Renrakukai, *Jishin Yochi Renrakukai 10-nen no ayumi*, 22; Meeting of the Research and Liaison Committee for Earthquake Forecasting, 9 June 1948.

45. Jishin Yochi Renrakukai, *Jishin Yochi Renrakukai 10-nen no ayumi*, 23; "Jishin wa yochi dekiru ka?" *Asahi shinbun*, July 5, 1948.

46. For a brief biographical sketch, see "Wadati Kiyō," *Asahi gurafu* 1423 (1951): 18.

47. "Yōjin wa kekkō da ga, sawagu no ataranu."

48. Kiyoo Wadati, "Daijishin ga kuru to iu hanashi," *Yomiuri shinbun*, December 15, 1947.

49. Kiyoo Wadati, "Daijishin ga kuru to iu hanashi."

50. It is possible to overstate the committee's role in calming the waters as the "deadline" for the Chichibu event approached, not least because the most detailed accounts of how the press and the public were eventually reassured that Inoue had been mistaken are themselves written by committee members.

51. Jishin Yochi Renrakukai, *Jishin Yochi Renrakukai 10-nen no ayumi*, 23.

52. Jishin Yochi Renrakukai, *Jishin Yochi Renrakukai 10-nen no ayumi*, 23–24.

53. The first significant postwar earthquake in the Kantō region occurred in 1987 and was centered in Chiba prefecture, east of Tokyo.

54. Hagiwara Takahiro, *Jishin yochi to saigai*, 90; Tomari Jirō, *Nihon no jishin yochi kenkyū 130-nenshi*, 209, 218–24.

55. Tomari Jirō, *Nihon no jishin yochi kenkyū 130-nenshi*, 209, 218–24.

56. Yamashita Fumio, *Jishin yochi no senkusha*, 270–71.

57. Yamashita Fumio, *Jishin yochi no senkusha*, 271.

58. Yamashita Fumio, *Kakusareta daishinsai*, 124.

59. Yamashita Fumio, *Jishin yochi no senkusha*, 273.

60. Yamashita Fumio, *Jishin yochi no senkusha*, 271.

61. Kawasumi Hiroshi, "Megumarenu kagakusha," *Yomiuri shinbun*, March 18, 1948.

Chapter 3

1. "'Daijishin' ka no shōbō enshū," *Yomiuri shinbun*, August 31, 1960; *Kagaku gijutsuchō nenpō*, 5th ed., Kagaku gijutsuchō nenpō (Tokyo: Kagaku gijutsuchō, 1961), 173.

2. "Daijishin—issei shōbō enshū," *Yomiuri shinbun*, September 1, 1959.

3. "Kugatsu ichi nichi o bōsai dā," *Yomiuri shinbun*, June 7, 1960.

4. The Cabinet approval came on June 17, 1960. See *Kagaku gijutsuchō nenpō*, 172–73. For a useful discussion of the origins of Bosai no hi, see Mizuide Kōki, "'Bōsai no hi' o meguru saigai no kioku," *Masu komyunikeshun kenkyū* 88 (2016): 157–75.

5. Only one Atlantic basin storm—Hurricane Allen in 1980—is known to have produced winds of such speed.

6. Kitahara, Matsuura, and Kimura, *Nihon rekishi saigai jiten*, 578.

7. Kitahara, Matsuura, and Kimura, *Nihon rekishi saigai jiten*, 573.

8. Kazama Norio, "Saigai taisaku kihonhō no seitei: bōsai seisaku netowaku no keisei," *Kinki daigaku hōgaku* 50, no. 1 (2002): 20.

9. Kitahara, Matsuura, and Kimura, *Nihon rekishi saigai jiten*, 575.

10. Kitahara, Matsuura, and Kimura, *Nihon rekishi saigai jiten*, 576. By comparison, the disasters of March 11, 2011, are thought to have costs equal to around 6 percent of the nation's GNP.

11. Chūō Bōsai Kaigi, Saigai Kyōkun no Keishō ni Kansuru Senmon Chōsakai Chūō Bōsai Kaigi, "1959 Isewan Taifu hōkokusho," 2008, https://www.bousai.go.jp/kyoiku/kyokun/kyoukunnokeishou/rep/1959_isewan_typhoon/pdf/10_chap4.pdf, 141.

12. Kitahara, Matsuura, and Kimura, *Nihon rekishi saigai jiten*, 582.

13. Kitahara, Matsuura, and Kimura, *Nihon rekishi saigai jiten*, 582.

14. Kitahara, Matsuura, and Kimura, *Nihon rekishi saigai jiten*, 579–80.

15. Kitahara, Matsuura, and Kimura, *Nihon rekishi saigai jiten*, 580.

16. Hikita Keiichirō, "Kuroi tsunami no ato o aruite," *Asahi shinbun*, October 9, 1959.

17. Among the reports issued between November 1959 and March 1960 are ones by the Science Council of Japan, the Science and Technology Agency, the Ministry of Education, and the Administrative Management Agency (Gyōseikanrichō). See Imai Makoto, "Saigai taisaku kihonhō ni tsuite (1)," *Jichi kenkyū* 37, no. 12 (1961): 90; Kazama Norio, "Saigai taisaku kihonhō no seitei," 29–30.

18. As measured by the number of lives lost and property damaged, the 1940s and 1950s were modern Japan's deadliest decades for typhoons. The first of the postwar

storms struck in mid-September 1945, sweeping across Kyushu before striking western Honshu; of the more 3,700 deaths attributed to the Makurazaki Typhoon, more than half occurred in Hiroshima prefecture, where the first atomic bomb had fallen a little over a month earlier. In 1947, Typhoon Katherine took more than 1,000 lives in the Kantō district. Almost every fall, for the next ten years, at least one and sometimes more typhoons lingered long enough to cause significant damage somewhere in Japan. The Tōya Maru Typhoon that struck central Japan in September 1954 killed 1,761; the Kanogawa Typhoon in 1958 left 1,269 people dead in the Kantō district. See Kitahara, Matsuura, and Kimura, *Nihon rekishi saigai jiten*, 513–20, 528–35, 544–46.

19. Kitahara, Matsuura, and Kimura, *Nihon rekishi saigai jiten*, 521; Tsukui Susumu, *Daisaigai to hō* (Tokyo: Iwanami Shoten, 2012), 5, 10.

20. Tsukui Susumu, *Daisaigai to hō*, 5. The 1899 Victims Relief Fund Law (Risai kyūjo kikinhō), itself a response to the 1891 Nōbi earthquake and 1896 Sanriku tsunami, required prefectures to establish and maintain emergency aid funds from which they would draw in order to temporarily provide shelter, food, and other necessities to their citizens. The law put prefectural governors in charge of managing relief efforts; the central government's role was confined to offering limited financial support to the neediest prefectures.

21. Kitahara, Matsuura, and Kimura, *Nihon rekishi saigai jiten*, 521; Tsukui Susumu, *Daisaigai to hō*, 5, 10.

22. Disaster Relief Act (Japan), Provisional Translation, 1947, http://www.ifrc.org/Docs/idrl/502EN.pdf.

23. Provisional Translation Disaster Relief Act (Japan).

24. Kazama Norio, "Saigai taisaku kihonhō no seitei," 22. Funding levels improved but legislation passed in the early 1950s meant that relief spending targeted frequently occurring small-scale localized events, not ones that devastated multiple communities or that would require exceptional amounts of money to repair. What happened in practice throughout the late 1940s and 1950s was that after almost every fire, flood, or earthquake, legislators rushed to introduce bills that increased the central government's share of the cost of reconstruction for that specific event in their home prefectures and districts. These so-called "special bills" were both opposed by the Finance Ministry and taken for granted by legislators' constituents. In the first few months after the Ise Bay Typhoon, legislators passed twenty-seven "special laws" to bring more funding into their districts. See Kazama, "Saigai taisaku kihonhō no seitei," 23.

25. Kawata Yoshiaki, "Saigai taisaku kihonhō to bōsai kihon keikaku," *Shizen saigai kagaku* 15 (1996): 84.

26. The strengthening of building codes is a notable exception to this pattern. Architects and engineers in Japan had long focused on the importance of designing and ultimately building structures that could survive earthquakes and resist fires; in that sense they were among the most proactive of all the professionals of this era who had to deal in some way with the risk of disaster. The extensive structural damage caused by the 1948 Fukui earthquake triggered a thorough reassessment of existing guidelines for

construction, for example. That review led to the passage of the Building Standards Act (Kenchiku kijunhō) in 1950, which, among other provisions, established stricter seismic design and fire-resistance requirements for new construction, replacing those spelled out in the 1919 Urban Building Law. Revisions to the act after the 1968 Tokachi-oki and the 1995 Hanshin-Awaji earthquakes incorporated changing understandings of best practices in seismic engineering.

27. 33rd Diet, Honkaigi, 3, October 28, 1959, https://kokkai.ndl.go.jp/txt/103305254X00319591028/3.

28. Kazama Norio, "Saigai taisaku kihonhō no seitei: bōsai seisaku netowaku no keisei," 15–16. In making these choices, Kishi and his cabinet were likely aware of proposals like those put forward by the National Governors Committee (Zenkoku chijikai) as early as 1952. The governors' plan for a comprehensive Emergency Disaster Measures Act (Hijō saigai taisakuhō) also sought to expand the focus of state intervention beyond relief to include the possibility of pre-disaster mitigation efforts and preparedness.

29. Kazama Norio, "Saigai taisaku kihonhō no seitei," 8.

30. Igarashi, *Bodies of Memory*, 137–38.

31. Nick Kapur, *Japan At the Crossroads: Conflict and Compromise After Anpo* (Cambridge, MA: Harvard University Press, 2018).

32. Igarashi, *Bodies of Memory*; Kapur, *Japan at the Crossroads.*

33. Scott O'Bryan, *The Growth Idea: Purpose and Prosperity in Postwar Japan* (Honolulu: University of Hawaii Press, 2009).

34. See Kazama Norio, "Saigai taisaku kihonhō no seitei," 47, for a discussion of opposition arguments.

35. For an overview of disaster countermeasures prior to the Basic Act, see Tsuchida Hiroshige, "Nijū seiki Nihon no bōsai," *Shigaku zasshi* 127, no. 6 (2018): 6–19.

36. Watanabe Ichirō, Aoki Sukehisa, Yagi Tsuruhei, and Ōtani Keiichi, eds., *Bōsai kankei hōrei (Eibun)*, Bōsai Kagaku Gijutsu Kenkyūjo kenkyu shiryō (Tsukuba: Kokuritsu Bōsai Kagaku Gijutsu Senta, 1978), 17.

37. Watanabe Ichiro et al, *Bōsai kankei hōrei (Eibun)*, 38, 41.

38. Watanabe Ichiro et al, *Bōsai kankei hōrei (Eibun)*, 28.

39. "Jishin taisaku wa jūten," *Asahi shinbun*, May 25, 1971. It added items relating to earthquake hazards, especially those that threatened large metropolitan areas, in 1971.

40. Jishin Yochi Renrakukai, ed., *Jishin Yochi Renrakukai 10-nen no ayumi* (Tokyo: Kankōsha Kokudo Chiriin Nihon Sokuryō Kyōkai, 1979), 167. The earliest mention I've been able to identify is by Suzuki Jiro (Tohoku University) in the minutes of the April 11, 1961, meeting of the Jishin yochi keikaku kenkyū gurupu.

41. Earthquake Prediction Research Group, *Prediction of Earthquakes: Progress to Date and Plans for Further Development*, ed. Kiyoo Wadati, Chuji Tsuboi, and Takahiro Hagiwara (Tokyo: Kokusai bunen insatsusha, 1962), 1.

42. Jishin Yochi Renrakukai, *Jishin Yochi Renrakukai 10-nen no ayumi*, 26.

43. Tomari Jirō, *Nihon no jishin yochi kenkyū 130-nenshi*, 221.

44. Jishin Yochi Renrakukai, *Jishin Yochi Renrakukai 10-nen no ayumi*, 26.

45. See, for example, O'Bryan, *The Growth Idea*.

46. Tomari Jirō, *Nihon no jishin yochi kenkyū 130-nenshi*, 223–24.

47. The Research Group replaced the original ad hoc committee.

48. This is according to lists of attendees at the four meetings of the Research Group in 1961. See Jishin Yochi Renrakukai, *Jishin Yochi Renrakukai 10-nen no ayumi*, 167, 169, 172. Nakamura had left Tohoku University in 1951 to take a position at Kumamoto University.

49. The eight included four different approaches to measuring crustal deformation, the monitoring of small and very small earthquakes, the use of controlled detonations to measure seismic wave velocity, the investigation of active faults, and geomagnetic and earth current studies.

50. Earthquake Prediction Research Group, *Prediction of Earthquakes*, 1; Hagiwara Takahiro, *Jishin yochi to saigai*, 115.

51. Earthquake Prediction Research Group, *Prediction of Earthquakes*, 4.

52. Earthquake Prediction Research Group, *Prediction of Earthquakes*, 4–5.

53. Earthquake Prediction Research Group, *Prediction of Earthquakes*, 8–9.

54. Earthquake Prediction Research Group, *Prediction of Earthquakes*, 17.

55. Earthquake Prediction Research Group, *Prediction of Earthquakes*, 1. Emphasis in original.

56. Earthquake Prediction Research Group, *Prediction of Earthquakes*, 5.

57. Earthquake Prediction Research Group, *Prediction of Earthquakes*, 6.

58. Earthquake Prediction Research Group, *Prediction of Earthquakes*, 2–3. Emphasis in original.

59. Earthquake Prediction Research Group, *Prediction of Earthquakes*, 6.

60. Earthquake Prediction Research Group, *Prediction of Earthquakes*, 7. This worked out to seventy stations distributed nationally and an additional thirty designed for the more carefully watched districts. In 1947, by comparison, Wadati and the Liaison Committee had proposed equipping just twelve new stations with tilt meters and pendulums.

61. Earthquake Prediction Research Group, *Prediction of Earthquakes*, 2–3.

62. Earthquake Prediction Research Group, *Prediction of Earthquakes*, 4–5.

63. Earthquake Prediction Research Group, *Prediction of Earthquakes*, 3. Emphasis in original.

64. Earthquake Prediction Research Group, *Prediction of Earthquakes*, 4–5.

65. Earthquake Prediction Research Group, *Prediction of Earthquakes*, 9.

66. Earthquake Prediction Research Group, *Prediction of Earthquakes*, 21.

67. Earthquake Prediction Research Group, *Prediction of Earthquakes*, 21.

68. Earthquake Prediction Research Group, *Prediction of Earthquakes*, 1. Emphasis in original.

69. Earthquake Prediction Research Group, *Prediction of Earthquakes*, 2, 20. The Blueprint floated the idea of creating a "Ministry of Territorial Welfare" (Kokudoshō), one of the tasks of which would be coordinating the prediction project. References to a "Ministry of Territorial Welfare" surfaced with some regularity in the Diet in 1960 and 1961 during discussion and debates having to do with the Disaster Countermeasures Basic Act.

70. Earthquake Prediction Research Group, *Prediction of Earthquakes*; Hagiwara Takahiro, *Jishin yochi to saigai*, 115.

71. National Science Foundation and Nihon Gakujutsu Shinkōkai, eds., *The United States–Japan Conference on Research Related to Earthquake Prediction Problems* (Tokyo: Earthquake Research Institute, March 1964), 1.

72. Columbia University hosted the next one in 1966, with meetings alternating thereafter between Japan and the United States. A seventh conference was held in 1988 at Morro Bay, California. See Carl Kisslinger, Takeshi Mikumo and Hiroo Kanamori, "U.S.-Japan Quake Prediction Research," *Eos, Transactions, American Geophysical Union* 69, no. 52 (1988): 1672–74.

73. National Science Foundation and Nihon Gakujutsu Shinkōkai, *The United States–Japan Conference*, 2–3. The four scientists who raised alarms about potential earthquakes between 1947 and 1949—Inoue, Nakamura, Sassa, and Yamaguchi—were among that number.

74. National Science Foundation and Nihon Gakujutsu Shinkōkai, *The United States–Japan Conference*, 13. This was a theme Oliver returned to later that year in *Science*. See Jack Oliver, "Earthquake Prediction," *Science* 144, no. 3624 (1964/06/12 1964): 1364–65.

75. National Science Foundation and Nihon Gakujutsu Shinkōkai, *The United States–Japan Conference*, 14.

76. Oliver, "Earthquake Prediction," 1364.

77. Hough, *Predicting the Unpredictable*, 62.

78. Robert A. Stallings, *Promoting Risk: Constructing the Earthquake Threat*, vol. Social problems and social issues. (New York: De Gruyter, 1995), 41–42, and Hough, *Predicting the Unpredictable*, 102, 105–6.

79. Frank Press, "The 10-Year Program for Earthquake Prediction Research in the U.S.," in *Proceedings of the Second United States–Japan Conference on Research Related to Earthquake Prediction*, ed. Robert Page (Palisades, NY: Lamont Geological Observatory, 1966), 9.

80. Press, "The 10-Year Program for Earthquake Prediction Research in the U.S.," 9.

81. The Ministries of Construction, Education, and Transport each had control over one or more of the units affected by the Blueprint's proposals.

82. Jishin Yochi Renrakukai, *Jishin Yochi Renrakukai 10-nen no ayumi*, 29.

83. Hagiwara Takahiro, *Jishin yochi to saigai*, 118.

84. Hagiwara Takahiro, *Jishin yochi to saigai*, 118–19.

85. Hagiwara later said that the budget issues had been worked out before the earthquake, but that the event was useful nevertheless. See Hagiwara Takahiro, Rikitake Tsuneji, "Jishin yochi keikaku hossoku no koro," *Jishin jyānaru* 1 (1986): 20.

Chapter 4

1. Hiroi Osamu, "Niigata jishin to saigai jōhō (1)," *Gekkan shōbō* 9, no. 6 (1987): 27.

2. Hiroi Osamu, "Niigata jishin to saigai jōhō (1)," *Gekkan shōbō* 9, no. 6 (1987): 26.

3. Kitahara, Matsuura, and Kimura, *Nihon rekishi saigai jiten*, 602.

4. For a thoughtful analysis of the bullet train's prompting of a "reimagination of identity on the levels of individual, city, and nation in a changing Japan," see Jessamyn Abel, *Dream Super-Express: A Cultural History of the World's First Bullet Train*, Studies of the Weatherhead East Asian Institute, Columbia University (Stanford, CA: Stanford University Press, 2022), 4.

5. Hiroi Osamu, "Niigata jishin to saigai jōhō (2)," *Gekkan shōbō* 9, no. 7 (1987): 51. BSN set up a mobile transmission unit at the prefectural disaster response headquarters in Niigata City, which allowed officials to speak more or less directly to listeners in the stricken area. Home Affairs Minister Akazawa Masamichi did so in a live broadcast at 6:00 p.m. on the 16th.

6. Simon Partner, *Assembled in Japan: Electrical Goods and the Making of the Japanese Consumer* (Berkeley: University of California Press, 1999), 247. Partner cites a 23.6 percent penetration rate in 1959, and an 88.7 percent penetration rate in 1963.

7. "Niigata—Yamagata—Akita ni kyōshin," *Asahi shinbun*, June 16, 1964; Kitahara, Matsuura, and Kimura, *Nihon rekishi saigai jiten*, 601.

8. See, for example, "Hisai no Niigata," *Asahi shinbun*, June 17, 1964; "Mi ni shimita Niigata jishin no kyōfu," *Asahi shinbun*, June 25, 1964.

9. Kojiro Kamei, "Fire Following Earthquake in Niigata City," in *General Report on the Niigata Earthquake of 1964*, ed. Kawasumi Hiroshi (Tokyo: Electrical Engineering College Press, 1968), 385–86, 392.

10. "Yukuri to taorete iku kenei apaato," Niigata Jishin Taisaku Renrakukai, http://www.hrr.mlit.go.jp/bosai/niigatajishin/contents/c24a.htm, accessed June 5, 2014. Residents returning to collect their belongings discovered that doors and windows worked as well as they had before the earthquake. If it weren't for the new, funhouse perspective that went with the skewed interior spaces, it might have been possible to imagine that nothing much had happened.

11. A good, if somewhat technical, discussion of these processes is in Yorihiko Ohsaki, "Building Damage and Soil Condition," in *General Report on the Niigata Earthquake of 1964*, ed. Kawasumi Hiroshi (Tokyo: Electrical Engineering College Press, 1968).

12. Normally, there were no events in the spring but, that year, planners had agreed to move what would have been the autumn session—always the most popular collection of events, by far—to early June in order to avoid conflict of any kind with the start of the Olympics in October.

13. Kenichiro Horii, "Highway Bridges," in *General Report on the Niigata Earthquake of 1964*, ed. Kawasumi Hiroshi (Tokyo: Electrical Engineering College Press, 1968), 442–47.

14. Akira Enami, "Shell Structures," in *General Report on the Niigata Earthquake of 1964*, ed. Kawasumi Hiroshi (Tokyo: Electrical Engineering College Press, 1968), 286–96.

15. For more on postwar urban reconstruction see André Sorensen, *The Making of Urban Japan: Cities and Planning From Edo to the Twenty-First Century*, Nissan Institute/Routledge Japanese Studies Series (London: Routledge, 2002); Carola Hein, Jeffry M. Diefendorf, and Yorifusa Ishida, *Rebuilding Urban Japan After 1945* (New York: Palgrave Macmillan, 2003).

16. See, for example, "Niigata jishin—Kore ga Tokyo dattara," *Asahi shinbun*, June 17, 1964.

17. See "Toki no hito," *Mainichi shinbun*, June 20, 1964, for a profile of Kawasumi and his take on the lessons of the Niigata quake.

18. 46th Diet, Shūgiin, Saigai Taisaku Tokubetsu Iinkai, 13, July 3, 1964, https://kokkai.ndl.go.jp/txt/104604339X01319640703.

19. Yanagita Kunio, "'69 nenmoku' o yogen sita otoko," *Bungei shunjū* 51, no. 13 (September 1973): 249–50.

20. Kawasumi contributed two papers to the joint US-Japan conference in March 1964; both focused on Imamura's contributions to earthquake prediction research. See Kawasumi Hiroshi, "A. Imamura's Note on Successful Earthquake Prediction," in *The United States–Japan Conference on Research Related to Earthquake Prediction Problems*, ed. National Science Foundation and Nihon Gakujutsu Shinkōkai (Tokyo, March 1964); Kawasumi Hiroshi, "Crustal Movements in the Historical Records," in *The United States–Japan Conference on Research Related to Earthquake Prediction Problems*, ed. National Science Foundation and Nihon Gakujutsu Shinkōkai (Tokyo, March 1964).

21. Kawasumi Hiroshi, "Shindo to shindokai," *Jishin* 15, no. 8 (1943): 1–7; Kawasumi Hiroshi, "Shindo to shindokai," *Jishin* 15, no. 1 (1943): 6–12. This was designated as M_k, or Kawasumi Magnitude.

22. Kawasumi Hiroshi, "Measures of Earthquake Danger and Expectancy of Maximum Intensity Throughout Japan as Inferred from the Seismic Activity in Historical Times," *Bulletin of the Earthquake Research Institute* 29 (1951): 469–82; Ishibashi Katsuhiko, "Status of Historical Seismology in Japan," *Annals of Geophysics* 47, no. 2/3 (2004): 349.

23. Yanagita Kunio, "'69 nenmoku' o yogen sita otoko," 251. Kawasumi's maps predate by more than a decade and a half C. Allin Cornell's 1968 paper on "Engineering Seismic Risk Analysis," which helped launch the development of probabilistic seismic hazard studies in the United States. See C. Allin Cornell, "Engineering Seismic Risk Analysis," *Bulletin of the Seismological Society of America* 58, no. 5 (1968): 1583–1606.

24. Kawasumi Hiroshi, "Arasuka jishin no omoide," *Yōbō jihō* 71 (October 1967): 11–16.

25. Yanagita Kunio, "'69 nenmoku' o yogen sita otoko," 251.

26. Yanagita Kunio, "'69 nenmoku' o yogen sita otoko," 252.

27. Sekino Masaru, ed., *Kōtokuin kokuhō dōzō Amida nyorai zazō shūri kōji hōkokusho* (Kamakura: Kōtokuin, 1961), 88. The one-page section on periodicity is titled "Kamakura kyōretsushin no shūkisei."

28. Usami Tatsuo, "Study of Historical Earthquakes in Japan," *Bulletin of the Earthquake Research Institute* 54 (1979): 399–439, references a 1963 article by Kawasumi in *Saigai taisaku kenkyukai*, published by the Songai Hokenryōritsu Santeikai Saigai Kagaku Kenkyūkai, which in its present form is the Non-life Insurance Rating Organization of Japan.

29. 65th Diet, Shūgiin, Shinsai Taisaku Tokubetsu Iinkai, Feb. 12, 1971, https://kokkai.ndl.go.jp/txt/106504339X00319710212.

30. Unlike Wadati Kiyō and other scientists who made relatively frequent appearances before Diet committees, Kawasumi had testified for legislators only once before his appearance in 1964. He spoke, on that occasion, about the effects of the 1946 Nankaidō earthquake. See 7th Diet, Sangiin, Kenchiku Iinkai, 8, March 3, 1950. https://kokkai.ndl.go.jp/txt/100714149X00819500303

31. Hayashi represented a district in Nagano and, as it happened, had been several years behind Kawasumi at the Suwa Middle School.

32. 46th Diet, Shūgiin, Saigai Taisaku Tokubetsu Iinkai, 13, July 3, 1964, https://kokkai.ndl.go.jp/txt/104604339X01319640703.

33. 46th Diet, Shūgiin, Saigai Taisaku Tokubetsu Iinkai, 13, July 3, 1964.

34. 46th Diet, Shūgiin, Saigai Taisaku Tokubetsu Iinkai, 13, July 3, 1964.

35. 46th Diet, Shūgiin, Saigai Taisaku Tokubetsu Iinkai, 13, July 3, 1964.

36. 46th Diet, Shūgiin, Saigai Taisaku Tokubetsu Iinaki, 13, July 3, 1964.

37. 46th Diet, Shūgiin, Saigai Taisaku Tokubetsu Iinaki, 13, July 3, 1964.

38. See "Daijishin wa okiru ka," *Yomiuri shinbun*, March 14, 1968, for tables and maps of the sifty-nine-year hypothesis. See also "Magunichūdo 7.5 (2)," *Yomiuri shinbun*, September 2, 1970.

39. Kawasumi Hiroshi, "Kantō nanbu jishin 69 nen shūki no shōmei to sono hassei no kinpakudo narabi ni taisaku no kinkyūsei to mondai ten," *Chigaku zasshi* 79, no. 3 (1970): 115–17.

40. Kawasumi Hiroshi, "Kantō nanbu jishin 69 nen shūki," 124. Each danger period thus accounted for 40 percent (twenty-seven years) of each sixty-nine-year period.

41. The gaps were in 956, 1025, 1163, 1370, and 1577.

42. Kawasumi Hiroshi, "Kantō nanbu jishin 69 nen shūki," 119.

Chapter 5

1. Parts of the complex were opened to the public for the first time in the early 1990s; efforts to document and commemorate the role Korean workers played in its construction continue.

2. One of the ERI's observatories in Hiroshima Prefecture was Japan's second participant in the network; it joined in 1966. By the early 1970s, there were some one hundred twenty sites in more than sixty countries in the network.

3. Jishin Yochi Renrakukai, *Jishin Yochi Renrakukai 10-nen no ayumi*, 30–31.

4. Kitahara, Matsuura, and Kimura, *Nihon rekishi saigai jiten*, 604.

5. Kuwano Shigeru, director, *Jishin yochi e no michi: Matsushiro jishin o megutte* (Tokyo: Iwanami Productions, 1967).

6. Kimura Kōzō, *Sanriku e nigeru* (Tokyo: Futami Shobō, 1974), 175.

7. "Matsushiro (Nagano) ni jishin keikoku," *Yomiuri shinbun*, October 9, 1965; "Kishōchō hatsu no 'jishin yohō,'" *Asahi shinbun*, October 9, 1965.

8. "It doesn't look good when someone from the same agency as you, using the same data as you, offers a point of view that is exactly the opposite of the one you reached," remarked the director of the Matsushiro Seismological Observatory. "It's like committing an error against your own baseball team." See "Ochitsuki haratta Matsushiro machi," *Asahi shinbun*, October 10, 1965.

9. Early in 1966, the ERI's Tsuboi said that February's sharp decline in earthquake frequency suggested that the cluster had run its course and that things would soon be back to normal. He was off the mark, as were others of his colleagues who, like him, tended to provide overly optimistic assessments of the cluster's duration or downplayed the possibility that some earthquakes could do significant harm. See Kitahara, Matsuura, and Kimura, *Nihon rekishi saigai jiten*, 606.

10. Ide Ichitarō speaking, 51st Diet, Shūgiin, Kagaku Gijutsu Shinkō Taisaku Tokubetsu Iinkai, 4, February 23, 1966, https://kokkai.ndl.go.jp/txt/105103913X004 19660223.

11. See, for example, Hagiwara's testimony before a Lower House Special Committee on the Policies for the Promotion of Science and Technology (Kagaku Gijutsu Shinkō Taisaku Tokubetsu Iinkai) on February 23 (https://kokkai.ndl.go.jp/txt/105103913X00419660223) and March 24, 1966 (https://kokkai.ndl.go.jp/txt/105103913X01119660324), and before the Special Committee on Disaster Countermeasures (Saigai Taisaku Tokubetsu Iiinkai), 51st Diet, Sangiin, 5, April 14, 1966 (https://kokkai.ndl.go.jp/txt/105114339X00519660414).

12. Edward F. Newren, *Japanese Science Films: A Descriptive and Evaluative Catalog of 16 mm Motion Pictures, 8mm Cartridges, and Video Tapes* (Bethesda, MD: American Science Films Association, August 1973), 1–2, 14. That English title was likely the work of the Japan Society for the Promotion of Science, which, along with the National Science Museum, was managing Japan's responsibilities in the newly inaugurated US-Japan Science Film Exchange Project. Between 1967 and 1970, the Society and the Museum dispatched one hundred eighty Japanese educational films to the US and their counterparts in the US—the NSF and the American Science Film Association—shipped two hundred sixty-five American films to Tokyo. The only version of *Jishin e no michi* that I've been able to view is the one that ended up in American hands, which includes English-language narration and adds titles in English when Japanese text appears on screen. I'm grateful to the University of Delaware Library staff for their help in making it possible to watch the film.

13. "Jishin o tsukamaeru," *Mainichi shinbun*, February 3, 1967.

14. Or, as one researcher referred to the decision to include research funding in the 1965 budget, "The origins of earthquake prediction research as a national project are

right there." See Jishin Yochi Renrakukai, *Jishin Yochi Renrakukai yonjūnen no ayumi* (Tsukuba-shi, Ibaraki-ken: Kokudo Chiriin, 2009), 1.

15. Rikitake Tsuneji, "Nenji keikaku no shinten jōkyō," in *Jishin yochi kenkyū shinpojiumu: (1967-nen 12-gatsu 1-nichi)* (Tokyo: Nihon Gakujutsu Kaigi, 1968). Hagiwara did point out that even at those less-than-hoped for levels, the amount of money dedicated to this research was much more than it had ever been. See Hagiwara Takahiro, "Naigai ni okeru jishin yochi kenkyū no genjō," in *Jishin yochi kenkyū shinpojiumu: (1967-nen 12-gatsu 1-nichi)* (Tokyo: Nihon Gakujutsu Kaigi, 1968), 3.

16. See "Tokachi—Sanriku kuri kaesu daijishin," *Asahi shinbun*, May 16, 1968, for an explicit comparison of the sizes of recent earthquakes.

17. The 1971 revisions to the Building Standards Act were triggered by the widespread damage to recently constructed buildings by the Tokachi-oki earthquake. See Kitahara, Matsuura, and Kimura, *Nihon rekishi saigai jiten*, 613.

18. "'Ebino jishin' ni omou koto," *Asahi shinbun*, February 24, 1968.

19. "'Ebino jishin' ni omou koto," *Asahi shinbun*, February 24, 1968.

20. "Jishin no yochi, motto kakujitsu ni dekinai ka," *Yomiuri shinbun*, May 28, 1968.

21. "Jishin yochi no kenkyū shisetsu jūjitsu seyo," *Yomiuri shinbun*, May 21, 1968.

22. "Jishin no yosokuami zenkokuteki ni seibi," *Asahi shinbun*, April 6, 1968. For more on Nakasone's role in promoting science and technology policy, see Morris Low, "Japan: The Political Economy of Japanese Science: Nakasone, Physicists, and the State," in *Scientists and the State: Domestic Structures and the International Context*, ed. Etel Solingen (Ann Arbor: University of Michigan Press, 1994), 113.

23. 58th Diet, Sangiin, Kessan Iinkai, 9, April 5, 1968, https://kokkai.ndl.go.jp/txt/105814103X00919680405.

24. Nakasone read off this list of what was to be observed in the Upper House of the Diet: "geodetic surveys, geomagnetism, geo-electrical currents, tidal gauges, the analysis of faults, changes in the movement of geological folds, changes in the earth's crust, analysis of the behavior of earthquakes as they happen, and so on." See 58th Diet, Sangiin, Honkaigi, 22, April 17, 1968, https://kokkai.ndl.go.jp/txt/105815254X02219680517.

25. In an interview shortly after the Tokachi-oki earthquake, Hagiwara credited the spending and work they'd done under the auspices of the plan so far only with "the emergence of an outlook that earthquake prediction will sooner or later be possible." See "Yurareru Nihon retto," *Yomiuri shinbun*, May 17, 1968.

26. 58th Diet. Shūgiin, Honkaigi, 35, May 17, 1968. https://kokkai.ndl.go.jp/txt/105805254X03519680517.

27. Morimoto Ryōhei, "Medo tatanu jishin yochi," *Sankei shinbun*, August 31, 1994.

28. Or, as one scientist put it to an *Asahi* reporter, "We're not saying that a ten-year plan will end with us being able to predict earthquakes. It's more like we hope to have a rough idea of whether earthquake prediction is possible or not." See "Jishin kuni Nihon no jishin kansoku," *Asahi shinbun*, June 1, 1968.

29. Sokuchigaku Shingikai, "Dainiji jishin yochi keikaku (Showa 44-nen—48-nen)," *Bunjutsusoku* 9 (1968): 2.

30. Jishin Yochi Renrakukai, *Jishin Yochi Renrakukai 10-nen no ayumi*, 35. Rikitake Tsuneji (1921–2004) trained as a geophysicist at Tokyo Imperial University; his research focused on geomagnetism. He rose through the ranks at the ERI (including a brief stint as director) until 1974, when he left to take a position at the University of Colorado; he returned to Japan in 1975. He subsequently held faculty positions at the Tokyo Institute of Technology and Nihon University. Rikitake was a prolific author; in addition to his more than two hundred forty single-authored journal articles, from the mid-1970s on he published a number of works on geophysics and earthquake prediction intended for general audiences. See Honkura Yoshimoto, "Tsuitō—Rikitake Tsuneji sensei," *Nihon Jishin Gakkai nyūsureta* 16, no. 4 (2004): 19.

31. Kimura, *Sanriku e nigeru*, 201–2.

32. See also "Jishin yochi: Barabara kenkyū o chōsei seyo," *Asahi shinbun*, April 13, 1966, for the discussion in the Diet.

33. Jishin Yochi Renrakukai, *Jishin Yochi Renrakukai 10-nen no ayumi*, 193–95. Hokkaido and Nagoya were usually represented by one faculty member each, Tohoku by two, and Kyodai by three (including one from the university's Bōsai kenkyūjo). Shizuoka University was added in 1975.

34. Jishin Yochi Renrakukai, "Jishin Yochi Renrakukai 10-nen no ayumi," 193–95. Utsu Tokuji was Hokkaido University's representative on the committee at its outset, for example, but when he changed jobs in 1972 and moved to Nagoya University, he replaced the scientist who had been his new employer's representative up to that point. Similarly, when Rikitake Tsuneji left the ERC for a brief stint in the US and was later hired by Tokyo Institute of Technology, his absence from the Yochiren was short lived. He rejoined the committee the very next year and remained on it for many years thereafter.

35. Hagiwara retired from the university in 1969.

36. The *Jishin Yochi Renrakukai kaihō* (Coordinating Committee for Earthquake Prediction Bulletin) was a routine channel for sharing data with scientists outside the committee.

37. Sokuchigaku Shingikai, "Dai niji jishin yochi keikaku (Showa 44-nen—48-nen)," 8.

38. Sokuchigaku Shingikai, "Dai niji jishin yochi keikaku (Showa 44-nen—48-nen)," 5. The specific phrase used was "*jishin ni kanren suru mono de aru*," or, "associated with an earthquake."

39. "Tokyo o tokutei kansoku," *Yomiuri shinbun*, July 17, 1968.

40. Jishin Yochi Renrakukai Jimukyoku, "Tokutei kansoku chiiki nado no sentei ni itaru keika," *Jishin Yochi Renrakukai kaihō* 3 (July 1970), 89–90.

41. Jishin Yochi Renrakukai Jimukyoku, "Tokutei kansoku chiiki nado no sentei ni itaru keika," *Jishin Yochi Renrakukai kaihō* 3 (July 1970), 90

42. Jishin Yochi Renrakukai Jimukyoku, "Tokutei kansoku chiiki nado no sentei ni itaru keika," *Jishin Yochi Renrakukai kaihō* 3 (July 1970), 90–91.

43. Jishin Yochi Renrakukai, *Jishin Yochi Renrakukai yonjūnen no ayumi*, 21.

44. "Tokutei chiiki fuyasu," *Asahi shinbun*, August 22, 1978. How that number was arrived at is not clear. The zones lacked clearly defined boundaries, which would have made calculating how many square kilometers were in each difficult.

Chapter 6

1. Naikakufu Daijin Kanbō Seifu Kōhōshitsu, *Shōbō gyōsei (dai ichibu) ni kansuru yoron chōsa* (1952), https://survey.gov-online.go.jp/s26/S26-04-26-01_chosahyo.html.

2. Naikakufu Daijin Kanbō Seifu Kōhōshitsu, *Daitoshi jūmin no bōsai ishiki ni kansuru yoron chōsa* (1971), https://survey.gov-online.go.jp/s46/S46-06-46-01.html.

3. Naikakufu Daijin Kanbō Seifu Kōhōshitsu, *Daitoshi jūmin no bōsai ishiki ni kansuru yoron chōsa*. Two years later, when a Tokyo-based marketing firm asked that city's residents to list what frightened them most, earthquakes ranked first (78 percent), followed by fires (53 percent), traffic accidents (42 percent), and pollution (41 percent.) See "Tomin no fūzoku yoron chōsa," *Asahi shinbun*, October 31, 1973.

4. Naikakufu Daijin Kanbō Seifu Kōhōshitsu, *Daitoshi jūmin no bōsai ishiki ni kansuru yoron chōsa*.

5. Naikakufu Daijin Kanbō Seifu Kōhōshitsu, *Shōbō ni kansuru yoron chōsa*, https://survey.gov-online.go.jp/s51/S51-07-51-05.html.

6. Naikakufu Daijin Kanbō Seifu Kōhōshitsu, *Shōbō jishin ni kansuru yoron chōsa* (1978), https://survey.gov-online.go.jp/s53/S53-10-53-10.html. In the 1980 version of that survey, 61 percent of respondents reported that those dangers had been on their mind. See Naikakufu Daijin Kanbō Seifu Kōhōshitsu, *Shōbō jishin ni kansuru yoron chōsa* (1980), https://survey.gov-online.go.jp/s55/S55-09-55-12.html.

7. 46th Diet, Shūgiin, Saigai Taisaku Tokubetsu Iinkai, 13, July 3, 1964, https://kokkai.ndl.go.jp/txt/104604339X01319640703.

8. Tokyo Shōbōchō, *Tokyo-to no daishinkasai higai no kentō: taisaku ni taisuru shiryō*, dai ippō (Tokyo: Tokyo Shōbōchō, 1961), 35.

9. Tokyo Shōbōchō, *Tokyo-to no daishinkasai higai no kentō*, dai ippō, 58.

10. Tokyo Shōbōchō, *Tokyo-to no daishinkasai higai no kentō*, dai ippō, 59. Kōtō and Sumida together account for about 9 percent of the total surface area of the twenty-three special wards; adding the other three wards mentioned brings their share to around 25 percent of the total.

11. Tokyo Shōbōchō, *Tokyo-to no daishinkasai higai no kentō*, dai ippō, 59. The wards in question are Arakawa, Adachi, Edogawa, Katsushika, Kōtō, and Sumida.

12. For a comprehensive analysis of Japan's responses to the threat of aerial bombardment, see Gennifer Weisenfeld's *Gas Mask Nation: Visualizing Civil Defense in Wartime Japan* (Chicago: University of Chicago Press, 2023).

13. "Toshi no higai suikei jisshō," *Yomiuri shinbun*, July 4, 1964.

14. Shōbō Shingikai, *Tokyo chihō (Kantō chihō minamibu) ni okeru daishinsai taisaku ni kansuru tōshin*, 1970, p. 69, https://www.fdma.go.jp/singi_kento/singi/council/.

15. "According to the earthquake periodicity theory," the Shingikai report's authors explained, "large earthquakes happen in the Tokyo district with a periodicity of

sixty-nine years ± thirteen years and, since forty-six years have passed since the Great Kantō Earthquake of 1923, we will enter into the next period of danger ten years from now." See Shōbō Shingikai, *Tokyo chihō (Kantō chihō minamibu) ni okeru daishinsai taisaku ni kansuru tōshin*, 81.

16. Tokyo Shōbōchō, *Tokyo-to no daishinkasai higai no kentō*, dai ippō; Tokyo Shōbōchō, *Tokyo-to no daishinkasai higai no kentō: taisaku ni taisuru shiryō, dai nippō* (Tokyo: Tokyo Shōbōchō, 1967), 75.

17. Tokyo Shōbōchō, *Tokyo-to no daishin kasai higai no kentō, dai nippō*, 72, 75. Firefighters estimated in 1967 that almost 16 million square hectares would burn in the first five hours of the earthquake.

18. Shōbō Shingikai, *Tokyo chihō (Kantō chihō minamibu) ni okeru daishinsai taisaku ni kansuru tōshin*, 116–18.

19. Shōbō Shingikai, *Tokyo chihō (Kantō chihō minamibu) ni okeru daishinsai taisaku ni kansuru tōshin*, 109.

20. Shōbō Shingikai, *Tokyo chihō (Kantō chihō minamibu) ni okeru daishinsai taisaku ni kansuru tōshin*, 109. The tanker counts in Kanagawa prefecture for the same time period were 913 and 2,290—in Saitama: 74 and 412.

21. "Magunichūdo M7.5 (1)," *Yomiuri shinbun*, September 1, 1970. The lead author for the *Yomiuri* series was popular science writer Isa Kyōzō, who was interested enough in the topic to devote an entire book to it; *The Second Great Kantō Earthquake: What To Do When It Happens?* came out in 1975. See Isa Kyōzō, *Daini Kantō daijishin: sono toki dō suru* (Tokyo: Yomiuri shinbunsha, 1975). The *Asahi* series was in part a commemoration of the fiftieth anniversary of the Great Kantō Earthquake; its first article appeared on August 23, 1973.

22. NHK Shakaibu, *Jishin ni sonaeru: anata no bōsai taisaku* (Tokyo: Nihon Hōsō Shuppan Kyōkai, 1971). *Prepare for a Typhoon* appeared in 1972.

23. NHK's *Prepare* targeted a broad, national audience but the Tokyo Metropolitan Fire Department published a similarly titled (but shorter) work—*Your Earthquake Countermeasures* (*Anata no jishin taisaku*)—before that, in 1969. See Tokyo Shōbōchō Shinsai Taisaku Honbu, *Anata no jishin taisaku—Tokyo no shinsai* (Tokyo: Tokyo Shōbōchō Shinsai Taisaku Honbu, 1969).

24. NHK Shakaibu, *Jishin ni sonaeru*, 10.

25. NHK Shakaibu, *Jishin ni sonaeru*, 10, emphasis in original.

26. NHK Shakaibu, *Jishin ni sonaeru*, 9.

27. NHK Shakaibu, *Jishin ni sonaeru*, 18.

28. NHK Shakaibu, *Jishin ni sonaeru*, 12–13.

29. NHK Shakaibu, *Jishin ni sonaeru*, 20.

30. NHK Shakaibu, *Jishin ni sonaeru*, 22.

31. USGS, Historic Earthquakes, https://earthquake.usgs.gov/earthquakes/eventpage/ci3347678/impact.

32. See, for example, "Rosanzerusu daijishin," *Asahi shinbun*, February 10, 1971; "'Rosu jishin' no kyōkun dō ikasu," *Asahi shinbun*, February 11, 1971.

33. Chūō Bōsai Kaigi, *1971-nen 2-gatsu 9-nichi Rosu Anzerusu jishin ni kansuru seifu gōdō chōsadan hōkoku (chūkan hōkoku)* (Tokyo: Chūō Bōsai Kaigi Jimukyoku, March 1971).

34. Kawasumi Hiroshi, "San Furunando (Rosanzerusu) jishin to sono kunren," *Tsuchi to kiso* 19, no. 8 (August 25, 1971): 34–40.

35. Kawasumi Hiroshi, "San Furunando (Rosanzerusu) Jishin to sono kunren." Similar claims of superior Japanese construction practices followed the 1994 Northridge earthquake and were proven inaccurate the following year in Kobe.

36. "Jishin ni yowai kindai toshi," *Asahi shinbun*, February 10, 1971.

37. "Rosu jishin, Tokyo nara kyōfu," *Asahi shinbun*, March 6, 1971.

38. "'Rosu jishin' no kyōkun dō ikasu." The scenario the paper described was one in which the earthquake struck in the fall, around dinner time, with winds in the city quite mild. It is not clear whether this was the same scenario that was expected to result in 560,000 deaths, as discussed below. In Kawasumi's Diet testimony in March, he suggested that numbers that high would only come about if the winds were quite strong. See 65th Diet, Sangiin, Yosan Iinkai, 16, March 17, 1971, https://kokkai.ndl.go.jp/txt/106515261X01619710317.

39. 65th Diet, Shūgiin, Saigai Taisaku Tokubetsu Iinkai, 3, February 12, 1971, https://kokkai.ndl.go.jp/txt/106504339X00319710212.

40. See "Tokyo daijishin ga okitara," *Asahi shinbun*, July 30, 1971; "Daishinsai no higai yosoku," *Asahi shinbun*, June 8, 1974; "Nennen yowamaru toshi bōsai," *Asahi shinbun*, August 31, 1976.

41. See also Kumagai Yoshio, "Jishinkasai no tokuchō to gensai ni mukete," *Yobō jihō* 230 (2007): 36–41, for later estimates.

42. Azuma Hideki, "Sengo tōshikeikaku o saikō suru: Takayama Eika no shōgai," *42nd NSRI toshi—kankyō fōramu* 282 (2011): 4.

43. Azuma Hideki, "Sengo tōshikeikaku o saikō suru," 16.

44. Murakami Suminao, "Goaisatsu (Bōsai Toshi Keikaku Kenkyūjo)," http://www.lusp.co.jp/corp1_3.html.

45. Murakami Suminao, "Kōtō chiku bōsai saikaihatsu keikaku ni tsuite," *Toshi keikaku* 60 (November 1969): 145–52; Murakami Suminao, "Dai kibō kaihatsu no jitsurei to shisutemu," *Kenchiku zasshi* 85, no. 1028 (1970): 669–73.

46. See, for example, Murakami Suminao, "Toshi bōsai to keikaku," *Kenchiku zashi* 82, no. 987 (1967): 683.

47. Murakami Suminao, "Toshi bōsai to keikaku," 684.

48. Murakami Suminao, "Kōtō chiku bōsai saikaihatsu keikaku ni tsuite," 146.

49. An interim design in 1967 described the construction of two massive "Disaster Prevention Belts," one running the length of the delta from north to south, the other spanning it from east to west. The two belts—each some five hundred meters wide—would intersect close to the center of the delta. The belts would ultimately house the most important institutions of daily life for residents—schools, libraries, stores, parks, medical care, and the like. Fuel, dangerous chemicals, and other hazardous materials

would all be sequestered well outside the "belt parks" and tall apartment complexes and other infrastructure on their external borders would serve as fire and wind breaks. Concerns about the design's huge price tag and the all-but-insurmountable property rights issues involved with acquiring so much land led the city to pursue the third, less grandiose shelter plan. See "Saigai no tsuyoi toshi saikaihatsu," *Yomiuri shinbun*, June 25, 1968; Murakami Suminao, "Kōtō chiku bōsai saikaihatsu keikaku ni tsuite," 146; Murakami Suminao, "Toshi bōsai to keikaku," 686.

50. "Kōtō deruta ni jōheki toshi," *Mainichi shinbun*, November 16, 1968.

51. "At first the government was really just a bystander to all this," Kawasumi remarked in October 1971, "but after this year's quake in LA, they finally began to act." See "Jishin no ABC o owatte," *Asahi shinbun*, October 16, 1971.

52. Ogawa Yujiro, "Toshi bōsai no genjō to hyōka," *Sōgō toshi kenkyū* 47 (1992): 94; "Sōkyū ni jishin taisaku," *Asahi shinbun*, March 5, 1971.

53. "Tokyo-to shinsai yobō jōrei (zenbun)," *Zaisei keizai kōhō*, no. 1434 (November 8, 1971): 9–10.

54. It continues to surface from time to time in more recent analyses of risk to the southern Kantō district. See, for example, Yasuda Yasoi, "Dai sanji Kantō Daishinsai no risuku shisutema bunseki to seisaku shimyurāshun 'Daisanji Kantō daishinsai no shimyurāshun' kenkyū hōkokou (1)," *Kantō Gakuin Daigaku keizai keiei kenkyūjo nenpō* 34 (2012): 123–46.

55. Yanagita Kunio, "'69 nenmoku' o yogen sita otoko," 246. It could well have been Yanagita himself who posed this question. The *Asahi* coverage makes no mention of Kawasumi. See "Go nen keikaku de yochi ami," *Asahi shinbun*, June 30, 1973.

56. Yanagita Kunio, "'69 nenmoku' o yogen sita otoko," 247.

57. Sekino Masaru, ed., *Kōtokuin kokuhō dōzō Amida nyorai zazō shūri kōji hōkokusho* (Kamakura: Kōtokuin, 1961).

58. Kawasumi Hiroshi, "Kantō nanbu jishin 69 nen shūki"; Kawasumi Hiroshi, "Kantō ni daijishin ga yatte kuru," *Nihon oyobi Nihonjin* 1489 (1970): 94–101.

59. See, for example, Usami Tatsuo, "Tokyo ga shindo V ijō no jishin ni osowareru kakuritsu," *Bulletin of the Earthquake Research Institute* 48 (1970): 331; Usami Tatsuo, "Earthquake Studies and the Earthquake Prediction System in Japan," *Journal of Disaster Research* 1, no. 3 (2006): 416–33. The latter is an English-language version of an article that first appeared in 1974.

60. See Usami Tatsuo, "Tokyo ga shindo V ijō no jishin ni osowareru kakuritsu," 335–36, for a discussion of standard deviations in the different cases.

61. "Magunichūdo 7.5 (2)", *Yomiuri shinbun*, September 2, 1970.

62. "Magunichūdo 7.5 (2)," *Yomiuri shinbun*, September 2, 1970.

63. "Kawasumi Hiroshi shi," *Asahi shinbun*, December 13, 1972.

64. See "Magunichūdo M7.5 (12)," *Yomiuri shinbun*, September 17, 1970, number twelve in that newspaper's "M7.5" series featuring Kawasumi, or "Jishin no ABC o owatte" for a roundtable discussion at the end of that paper's "Jishin ABC" article series.

65. Usami Tatsuo, "Earthquake Studies and the Earthquake Prediction System in Japan," 417.

66. "Kawasumi Hiroshi shi."

67. "Jishin rettō (12)," *Asahi shinbun*, August 4, 1973.

68. "Jishin rettō (12)." The prominent seismologist Takeuchi Hitoshi makes similar observations about Kawasumi's methods and goals here in Takeuchi Hitoshi, "'Nihon Dajishin' ni sonaeru," *Keizai hyōron* 25, no. 2 (1976), 86–87.

69. 65th Diet, Sangiin, Saigai Taisaku Tokubetsu Iinkai, 6, May 14, 1971, https://kokkai.ndl.go.jp/txt/106514339X00619710514.

70. 65th Diet, Sangiin, Saigai Taisaku Tokubetsu Iinkai, 6, May 14, 1971.

Chapter 7

1. Mogi Kiyō, *Jishin yochi o kangaeru* (Tokyo: Iwanami Shoten, 1998), 115; "Yochi sareta Surugawan Enshūnada chōōgata jishin no zenbō," *Shūkan Gendai* 11, no. 50 (1969): 138.

2. "Yochi sareta Surugawan Enshūnada chōōgata jishin no zenbō," 138–42.

3. 71st Diet, Shūgiin, Kagaku Gijutsu Shinkō Tokubetsu Iinkai, April 26, 1973, https://kokkai.ndl.go.jp/txt/107103913X01019730426.

4. Utsu Tokuji, "Hokkaido shūhen ni okeru daijishin no katsudo to Nemuro nanpō-oki jishin ni tsuite," *Jishin Yochi Renrakukai kaihō* 7 (1972): 7, 10. Utsu presented some of his findings at the August 1971 Yochiren meetings and published them in the Yochiren's *Kaihō* the following March.

5. 71st Diet, Shūgiin, Kagaku Gijutsu Shinkō Tokubetsu Iinkai, April 26, 1973; Ishibashi Katsuhiko, "Surugawan jishinsetsu koshi," *Kagaku* 73, no. 9 (2003): 1059.

6. Rikitake Tsuneji, "Focal processes and the prediction of earthquakes: Probability of earthquake occurrence as estimated from crustal strain," *Tectonophysics* 23, no. 3 (1974): 299–312.

7. Rikitake Tsuneji, "Focal processes and the prediction of earthquakes," 311.

8. Rikitake Tsuneji, "Focal processes and the prediction of earthquakes," 311.

9. Hagiwara makes a similar point in "Jishin yochi mada nazo darake," *Yomiuri shinbun*, June 25, 1968.

10. "Keikoku pitari Nemuro oki," *Yomiuri shinbun*, June 18, 1973.

11. "Keikoku pitari Nemuro oki."

12. "Keikoku pitari Nemuro oki."

13. "Keikoku pitari Nemuro oki"; Ishibashi Katsuhiko, "Surugawan jishinsetsu koshi," 1059.

14. "Megalopolis," the term Jean Gottmann had used to describe the northeastern seaboard of the United States in the late 1950s, was very quickly adopted by Japanese scholars to do the same for an emerging "Tōkaido Megaroporisu." For a useful analysis of the emergence of the concept, see Jeffrey Hanes, "From Megalopolis to Megaroporisu," *Journal of Urban History* 19:2 (February 1993): 56–94. The "Tōkaidopolis," as

historian Louise Young describes it, "represents an immense concentration of human, financial, and productive resources," the cities of which "house the institutions of the central government, provide headquarters for most large industrial and financial institutions, are home to the nation's most prestigious universities, and bring together all of the major publishing and communications industries." See Louise Young, *Beyond the Metropolis: Second Cities and Modern Life in Interwar Japan* (Berkeley: University of California Press, 2013), 246–47.

15. Sorensen, *The Making of Urban Japan*, 170–71.

16. Andō Masataka, "Possibility of a Major Earthquake in the Tokai District, Japan and its Pre-estimated Seismotectonic Effects," *Tectonophysics* 25 (1975), 70.

17. Andō, "Possibility of a Major Earthquake in the Tokai District," 70; Rikitake Tsuneji, "Daijishin wa tsune ni yochi sarete iru," *Chūō kōron* 88, no. 9 (1973): 176–82. For earlier discussion of seismic gaps in the Japanese context, see Utsu Tokuji, "Hokkaido shūhen ni okeru daijishin no katsudo to Nemuro nanpō-oki jishin ni tsuite."

18. Andō Masataka, "Possibility of a Major Earthquake in the Tokai District," 81.

19. Andō Masataka, "Possibility of a Major Earthquake in the Tokai District," 81.

20. "Tōkai chihō jishin kansoku o kyōka," *Asahi shinbun*, March 1, 1974; "Jishin okosu osore," *Mainichi shinbun*, March 1, 1974. Jishin Yochi Renrakukai, *Jishin Yochi Renrakukai 10-nen no ayumi*, 223; Ishibashi Katsuhiko, "Surugawan jishinsetsu koshi," 1060.

21. Not everyone in the earth sciences community accepted Andō's conclusions unequivocally, but none of his critics were able to refute them. See Ishibashi Katsuhiko, "Surugawan jishinsetsu koshi," 1060.

22. Ishibashi Katsuhiko, "Surugawan jishinsetsu koshi," 1060.

23. Ishibashi Katsuhiko, "Surugawan jishinsetsu koshi," 1060.

24. Andō Masataka, "Possibility of a Major Earthquake in the Tokai District," 82–83.

25. Ishibashi Katsuhiko, "Surugawan jishinsetsu koshi," 1060. For an English-language summary of Ishibashi's theory and the reasoning behind it, see Ishibashi Katsuhiko, "Specification of a Soon-to-Occur Seismic Faulting in the Tokai District, Central Japan, Based Upon Seisometrics," in *Earthquake Prediction: An International Review*, ed. David W. Simpson and Paul G. Richards (Washington, DC: American Geophysical Union, 1981).

26. Ishibashi Katsuhiko, "Surugawan jishinsetsu koshi," 1060.

27. Ishibashi Katsuhiko, "Tōkai chihō ni yosō sareru daijishin no saikentō—Surugawan jishin no kanosei," *Jishin Yochi Renrakukai kaihō* 17, no. 4–13 (1977): 129.

28. Ishibashi Katsuhiko, "Tōkai chihō ni yosō sareru daijishin no saikentō," 129.

29. Ishibashi wrote a brief report on the theory, "The Focal Region of a Great Earthquake Expected in the Tōkai Region—Regarding a Suruga Bay Earthquake," and shared it with the Yochiren just before its meeting in late May. A summary of its key findings was published in the committee's in-house newsletter months later as Ishibashi Katsuhiko, "Tōkai chihō ni yosō sareru daijishin no saikentō."

30. Ishibashi Katsuhiko, "Surugawan jishinsetsu koshi," 1061.

31. Ishibashi Katsuhiko, "Surugawan jishinsetsu koshi," 1061.

32. Ishibashi Katsuhiko, "Surugawan jishinsetsu koshi," 1061.

33. Ishibashi Katsuhiko, "Tōkai chihō ni yosō sareru daijishin no saikentō," 127.

34. Ishibashi Katsuhiko, "Tōkai chihō ni yosō sareru daijishin no saikentō," 126–27.

35. Ishibashi Katsuhiko, "Tōkai chihō ni yosō sareru daijishin no saikentō," 128.

36. Kawabata Nobumasa, "Daikibo jishin taisaku tokubetsu sochihō seitei tōji no naibu jijyō," *Kagaku* 73, no. 9 (2003): 1045.

37. Kawabata Nobumasa, "Daikibo jishin taisaku tokubetsu sochihō seitei tōji no naibu jijyō," 1046.

38. As Tamura's colleague later put it, "Not even the Nikkei, the Sankei, Tokyo, or the other Kyōdō News Service affiliated newspapers, or television broadcasters like NHK, did anything at all with the story that first reported Ishibashi's theory. Few people in the three biggest cities—Tokyo, Osaka, Nagoya—saw it." See Koiwai Tadamichi, "Ryōiki no kakudai motomete kagaku kisha yonjūnen," *Shuzai nōto*, https://www.jnpc.or.jp/journal/interviews/29889/.

39. "Jishin ki ni naru sakkon desu ga yochi wa? Taisaku wa?" *Asahi shinbun*, August 24, 1976.

40. As quoted in Kawabata Nobumasa, "Daikibo jishin taisaku tokubetsu sochihō seitei tōji no naibu jijyō," 1044.

41. Kawabata Nobumasa, "Daikibo jishin taisaku tokubetsu sochihō seitei tōji no naibu jijyō," 1046–47.

42. "Jimoto yappari fuan," *Mainichi shinbun*, September 4, 1976.

43. Kawabata Nobumasa, "Daikibo jishin taisaku tokubetsu sochihō seitei tōji no naibu jijyō," 1047.

44. Ishibashi Katsuhiko, "Surugawan jishinsetsu koshi," 1062.

45. "Tōkai daijishin setsu," *Asahi shinbun*, November 6, 1976.

46. "Tōkai daijishin setsu."

47. "Tōkai daijishin setsu."

48. The Don Quixote comments appears in "Tōkai jishin," *Asahi shinbun*, October 6, 1976.

49. "Tōkai jishin."

50. Tsubokawa's explanation was technical, detailed, and hard to follow; Asada and the others generally used simpler language in their interactions with legislators.

51. 78th Diet, Shūgiin, Kagaku Gijutsu Shinkō Taisaku Tokubetsu Iinkai, 3, October 13, 1976, https://kokkai.ndl.go.jp/txt/107803913X00319761013.

52. 78th Diet, Shūgiin, Kagaku Gijutsu Shinkō Taisaku Tokubetsu Iinkai, 3, October 13, 1976.

53. Indeed, Hagiwara was in enough of a bad temper by November that at least one reporter referred to it in print. See "Tōkai daijishin kuru mae ni yureru," *Mainichi shinbun*, November 11, 1976.

54. 78th Diet, Shūgiin, Kagaku Gijutsu Shinkō Taisaku Tokubetsu Iinkai, 3, October 13, 1976.

55. 78th Diet, Shūgiin, Kagaku Gijutsu Shinkō Taisaku Tokubetsu Iinkai, 3, October 13, 1976.

56. 78th Diet, Shūgiin, Kagaku Gijutsu Shinkō Taisaku Tokubetsu Iinkai, 3, October 13, 1976.

57. Addendum 2 to the minutes of the November 29, 1976, meeting of the Yochiren. See Jishin Yochi Renrakukai, *Jishin Yochi Renrakukai 10-nen no ayumi*, 241–42.

58. Addendum 2 to the minutes of the November 29, 1976 meeting of the Yochiren. See Jishin Yochi Renrakukai, *Jishin Yochi Renrakukai 10-nen no ayumi*, 241–42.

59. Kawabata Nobumasa, "Tōkai jishin bōsai taisei no seiritsu hiwa," *Nihon Saigai Jōhō Gakkai dai 11 kai benkyōkai*, http://www.jasdis.gr.jp/_userdata/07benkyokai/011-2.pdf; "Yamamoto Keizaburō-shi shikyo," *Asahi shinbun*, May 9, 2006; Yamamoto Keizaburō, "Tōkai jishin to shutoken chokka jishin," *Keidanren geppō* 9 (1980): 67–68.

60. Records of all of Asada's appearances in the Diet are from the Diet Library's online Kokkai kaigiroku kensaku shisutemu.

61. Hatano had run against and lost, badly, to Minobe Ryōkichi in the 1971 Tokyo gubernatorial election. He was elected to the Upper House in 1974.

62. 78th Diet, Sangiin, Yosan Iinkai, 1, October 4, 1976, https://kokkai.ndl.go.jp/txt/107815261X00119761004.

63. 78th Diet, Sangiin, Yosan Iinkai, 1, October 4, 1976.

64. "Yochi muzukashii Tokyo chokkagata," *Asahi shinbun*, October 5, 1976.

65. For an example of coverage that did make that distinction, see "Tōkai jishin ari uru," *Mainichi shinbun*, October 5, 1976. For examples that were less nuanced, see "Yochi muzukashii Tokyo chokkagata" and "Tōkai jishin."

66. "Jishin yochi o kyōka sai hensei," *Yomiuri shinbun*, October 8, 1976.

67. "Tōkai jishin"; "Bukimi na 'Surugawan jishin' setsu," *Mainichi shinbun*, October 7, 1976; "Tōkai daijishin wa chokkagata," *Yomiuri shinbun*, October 8, 1976.

68. "Tōkai jishin"; "Bukimi na 'Surugawan jishin' setsu"; "Tōkai daijishin wa chokkagata."

69. "Magunichūdo 8 (1)," *Asahi shinbun*, October 19, 1976. Each article in the series included a different "Q and A."

Chapter 8

1. "Tensei jingo," *Asahi shinbun*, October 19, 1976; "Jimoto yappari fuan."

2. "Urete urete," *Mainichi shinbun*, October 21, 1976.

3. "Urete urete"; "Satto zukin kaburu no," *Yomiuri shinbun*, October 27, 1976.

4. "Itsu kuru ka ienu genjō," *Mainichi shinbun*, October 21, 1976. "Satto zukin kaburu no."

5. "Itsu kuru ka ienu genjō."

6. "Itsu kuru ka ienu genjō."

7. "Seifu, yatto taiōsaku," *Mainichi shinbun*, October 27, 1976.

8. Hough, *Predicting the Unpredictable*, 108–24; Kelin Wang, Qi-Fu Chen, Shihong Sun, and Andong Wang, "Predicting the 1975 Haicheng Earthquake," *Bulletin of the*

Seismological Society of America 96, no. 3 (2006): 757–95; Fan, "'Collective Monitoring, Collective Defense': Science, Earthquakes, and Politics in Communist China."

9. Wang, "Predicting the 1975 Haicheng Earthquake," 781, 786–87.

10. They apparently first learned of what the Chinese were trying to do after Bruce Bolt wrote about the PRC's efforts as part of his longer discussion of seismology in China following his trip there in 1973. See Bruce A. Bolt, "Earthquake Studies in the People's Republic of China," *Eos, Transactions, American Geophysical Union* 55, no. 3 (1974), 114; Oike Kazuo, Shichi Ryūichi, and Asada Toshi, "Chūgoku ni okeru jishin yochi (Tokubetsu hōkoku)," *Zisin: Journal of the Seismological Society of Japan* 28, no. 1 (1975), 75.

11. Asada Toshi, "Chūgoku ni okeru jishin yochi," *Nature* 30, no. 1 (1975): 36–46; Yoshida Minoru and Asada Toshi, "Jishin ni taisuru Chūgoku no katsudō," *Kagaku Asahi* 36, no. 10 (1976): 62–65. As Fa-ti Fan points out, the ongoing Cultural Revolution shaped every aspect of China's pursuit of earthquake prediction, from the important roles assigned to "citizen scientists" to how the story of what had happened in Haicheng was constructed. See Fan, "'Collective Monitoring, Collective Defense': Science, Earthquakes, and Politics in Communist China."

12. "Chūgoku no jishin yochi to Nihon," *Asahi shinbun*, December 5, 1975.

13. The best single account of the history of earthquake prediction in the United States, and one I draw on extensively here, is Hough, *Predicting the Unpredictable*.

14. Christopher H. Scholz, Lynn R. Sykes, and Yash P. Aggarwal, "Earthquake Prediction: A Physical Basis," *Science* 181, no. 4102 (1973): 803–10.

15. Hough, *Predicting the Unpredictable*, 69. The researcher most closely associated with the dilatancy model was Christopher Scholz; his ideas and later Scholz himself attracted a great deal of attention among earth scientists and the press in Japan. "Earthquake Prediction: A Physical Basis" appeared in *Science* in September 1973; *Chūō kōron* published a translation of it in October. See C. H. Scholz, "Kono fukeiki na zenchō," *Chūō kōron* 88, no. 9 (1973): 169–75. Over the next several years, dilatancy and Scholz came up often in Japanese scientific conferences, within the Yochiren, and in the Diet's discussions of earthquake prediction. Japanese seismologists may have been more interested in Scholz's ideas, and for longer, than was the case elsewhere. For an example of the type of coverage Scholz enjoyed in Japan, see the *Yomiuri*'s profile of "Dr. Catfish" on the occasion of his 1977 visit. See "'Namazu' hakase," *Yomiuri shinbun*, January 20, 1977; "Ningen tōjō," *Yomiuri shinbun*, January 21, 1977.

16. Hough, *Predicting the Unpredictable*, 96–97.

17. Hough, *Predicting the Unpredictable*, 102.

18. Hough, *Predicting the Unpredictable*, 97.

19. Panel on the Public Policy Implications of Earthquake Prediction of the Advisory Committee on Emergency Planning, Commission on Sociotechnical Systems, National Research Council, *Earthquake Prediction and Public Policy* (Washington, DC: National Academy of Sciences, 1975); Panel on Earthquake Prediction of the Committee on Seismology, Assembly of Mathematical and Physical Sciences,

National Research Council, *Predicting Earthquakes: A Scientific and Technical Evaluation—with Implications for Society* (Washington, DC: National Academy of Sciences, July 1976).

20. Hough, *Predicting the Unpredictable*, 103-104.

21. Earthquake Hazards Reduction Act, Pub. L. No. 95-124, S. 126, 95th Cong. (1977), https://www.govtrack.us/congress/bills/95/s126.

22. The *Asahi* framed the situation for readers this way: "Question: If there is a prediction, will an earthquake warning be issued? Answer: No, it won't. There is no mechanism to issue a warning." See "Magunichūdo 8 (6)," *Asahi shinbun*, October 19, 1976.

23. "Daijishin yochi dekite mo," *Yomiuri shinbun*, August 25, 1976. "Jishin yochi to bōsai no taisei o isoge," *Asahi shinbun*, October 10, 1976.

24. 78th Diet, Shūgiin, Kagaku Gijutsu Shinkō Taisaku Tokubetsu Iinkai, 3, October 13, 1976, https://kokkai.ndl.go.jp/txt/107803913X00319761013LH.

25. "Tōkai daijishin kuru mae ni yureru," *Mainichi shinbun*, November 11, 1976; "Tōkai jishin no yochi taisei kyōka," *Asahi shinbun*, December 18, 1976. "Tōkai jishin 'chokuzen yohō' no tame ni," *Yomiuri shinbun*, December 18, 1976.

26. Sokuchigaku Shingikai, "Dai sanji jishin yochi keikaku no ichibu minaoshi ni tsuite," *Gakujutsu geppō* 28, no. 7 (1975): 490–93.

27. Rikitake Tsuneji, *Earthquake Forecasting and Warning* (Tokyo: Center for Academic Publications Japan, 1982), 183.

28. "Tōkai jishin 'chokuzen yohō' no tame ni."

29. "Dēta shūchū hakaru," *Asahi shinbun*, April 5, 1977; "Jishin yochi ni kakeru roku nin," *Asahi shinbun*, April 6, 1977.

30. "Jishin yochi ni kakeru roku nin," *Asahi shinbun*, April 6, 1977.

31. "Dēta shūchū hakaru"; "Jishin yochi ni kakeru roku nin."

32. Rikitake Tsuneji, *Earthquake Forecasting and Warning*, 186; "'Ihen' ranku betsu ni happyō," *Asahi shinbun*, April 19, 1977.

33. "'Ihen' ranku betsu ni happyō."

34. "Tōkai jishin M8.4 de dai kunren," *Asahi shinbun*, June 8, 1977; "Itsu ka kono hi," *Asahi shinbun*, June 8, 1977; "Fuanami ni, hinan basho e," *Yomiuri shinbun*, June 8, 1977.

35. "Tōkai jishin M8.4 de dai kunren."

36. The prefecture's representatives in the Diet introduced Tōkai-earthquake-related legislation of their own around that same time; the LDP's Harada Shōzō, for example, introduced the "Large-scale Earthquake Prediction Policy Special Measures Draft Law" (*Daikibō jishin yochi taisaku tokubetsu sochi hōan*) in November. It was an unexciting piece of legislation and called on the government to create committees and councils and support the earthquake prediction effort in more or less the same ways that it did already. See "Daijishin tokubetsu rippō no mondaiten," *Asahi shinbun*, January 19, 1978; "Harada-an tatakidai ni," *Yomiuri shinbun*, November 2, 1977.

37. Kawabata's is the best description of Yamamoto's campaign. See Kawabata Nobumasa, "Daikibo jishin taisaku tokubetsu sochihō seitei tōji no naibu jijyō," 1048–49.

38. Shizuoka Shinbunsha, ed., *Chinmoku no Surugawan* (Shizuoka-shi: Shizuoka Shinbunsha, 2018), 83.

39. Shizuoka Shinbunsha, *Chinmoku no Surugawan*, 83.

40. Shizuoka Shinbunsha, *Chinmoku no Surugawan*, 83.

41. The draft law also stipulated that neither the national government nor any other public body could be held liable for losses arising from the prime minister's announcement. See "Shushō ni kinkyū meireiken," *Yomiuri shinbun*, December 8, 1977; "Yochi taisei no kyōka e," *Asahi shinbun*, December 8, 1977.

42. "Daijishin tokubetsu rippō no mondaiten."

43. Shizuoka Shinbunsha, *Chinmoku no Surugawan*, 88.

44. Rikitake Tsuneji, *Earthquake Forecasting and Warning*, 99.

45. Rikitake Tsuneji, *Earthquake Forecasting and Warning*, 273.

46. Kitahara, Matsuura, and Kimura, *Nihon rekishi saigai jiten*, 639.

47. "Sono shunkan—tatte orezu," *Asahi shinbun*, January 14, 1978; "Ōshima chūshin ni rettō daijishin," *Asahi shinbun*, January 14, 1978.

48. "Sono shunkan—tatte orezu"; "Ōshima chūshin ni rettō daijishin."

49. Rikitake Tsuneji, "Yochi, bōsai ni te tsukuse," *Mainichi shinbun*, January 16, 1978.

50. 84th Diet, Sangiin, Saigai Taisaku Tokubetsu Iinkai, 3, Feb. 9, 1978, https://kokkai.ndl.go.jp/txt/108414339X00319780209.

51. See 84th Diet, Sangiin, Saigai Taisaku Tokubetsu Iinkai, 3, Feb. 9, 1978; for similar sentiments, see 84th Diet, Shūgiin, Saigai Taisaku Tokubetsu Iinkai, 4, February 16, 1978, https://kokkai.ndl.go.jp/txt/108404339X00419780216.

52. Rikitake Tsuneji, *Earthquake Forecasting and Warning*, 275.

53. Rikitake Tsuneji, *Earthquake Forecasting and Warning*, 275–76.

54. The text of the chyron that ran at 2:34 on SBS read: "The Shizuoka Prefectural Office issued the 'aftershock information' to the southern and middle areas of Izu district this afternoon. Local citizens are requested to pay attention to information that follows and to behave calmly in case of a strong shock." See Rikitake Tsuneji, *Earthquake Forecasting and Warning*, 179.

55. The prefectural government reported receiving five hundred calls by early evening and the police received another four hundred, or so. Even the Shizuoka Railway office fielded more than a hundred. See "'Yoshin jōhō' de panikku," *Yomiuri shinbun*, January 19, 1978.

56. "Yoshin jōhō panikku," *Asahi shinbun*, January 21, 1978.

57. Rikitake Tsuneji, *Earthquake Forecasting and Warning*, 272.

58. 84th Diet, Sangiin, Saigai Taisaku Tokubetsu Iinkai, 3, Feb. 9, 1978, https://kokkai.ndl.go.jp/txt/108414339X00319780209.

59. For coverage of the report, see "Jishin panikku kōshite okita," *Asahi shinbun*, September 16, 1978. The report itself appeared as Tokyo Daigaku Shinbun Kenkyūjo, ed., *Jishin yochi to shakaiteki hannō* (Tokyo: Tokyo Daigaku Shuppankai, 1979).

60. 84th Diet, Sangiin, Saigai Taisaku Tokubetsu Iinkai, 3, Feb. 9, 1978.

61. "Daijishin taisaku de tokubetsu rippō," *Asahi shinbun*, January 17, 1978. Sakurauchi was a successful career politician with a history of leadership positions in the LDP and appointments as Minister of International Trade and Industry and as Minister of Agriculture and Forestry before joining the Fukuda Cabinet.

62. "Yochi taisei ni fuan," *Asahi shinbun*, May 1, 1978.

63. "Daikibō jishin taisaku tokubetsu sochi hōan," *Asahi shinbun*, March 22, 1978.

64. The first time was on February 16, when he was joined by Hagiwara, and the second on was on April 19, when he spoke alongside Asada. See 84th Diet, Shūgiin, Saigai Taisaku Tokubetsu Iinkai, 4, February 16, 1978, https://kokkai.ndl.go.jp/txt/108404339X00419780216.

65. 84th Diet, Shūgiin, Saigai Taisaku Tokubetsu Iinkai, 4, February 16, 1978. Shizuoka released its first official damage and casualty estimates for a Tōkai earthquake in November 1978; as Yamamoto had anticipated, the report concluded that in the absence of a successful prediction, the earthquake would take 10,927 lives. That figure would rise if deaths due to secondary hazards were taken into account. See "Yuragu shinario (5)," *Asahi shinbun*, January 12, 1994. Shizuoka continued to adjust its methods and categories in earthquake disaster damage studies that followed the first in 1993, 2001, and 2013, respectively.

66. 84th Diet, Shūgiin, Saigai Taisaku Tokubetsu Iinkai, 4, February 16, 1978.

67. 84th Diet, Shūgiin, Saigai Taisaku Tokubetsu Iinkai, 4, February 16, 1978.

68. Construction on the first of an eventual five nuclear reactors at the Hamaoka Nuclear Power Plant began in 1971; that unit began commercial operation in early 1976. Construction on the second reactor started in 1974; it was connected to the grid in May 1978.

69. "Meteorological Service Act," Article 11-2, Japanese Law Translation Database System, https://www.japaneselawtranslation.go.jp/en/laws/view/1968/en.

70. Rikitake Tsuneji, *Earthquake Forecasting and Warning.*

71. The new act's first article, "Purpose," borrowed its structure and some of its language outright from Article One of the Disasters Countermeasures Basic Law.

72. Rikitake Tsuneji, *Earthquake Forecasting and Warning*, 289.

73. Rikitake Tsuneji, *Earthquake Forecasting and Warning*, 348.

74. Suehiro was a former president of the Seismological Society of Japan; in 1978 he was deputy-director general in the JMA's Observation Department. He was also the son of the ERI's co-founder, Suehiro Kyoji. The Land Agency's Yotsuyanagi Osamu was an occasional participant in those conversations, as well.

75. As a member of the Upper House committee lamented in June, deliberations about the Daishinhō without Hagiwara present were like eating sashimi without soy sauce. See 84th Diet, Sangiin, Saigai Taisaku Tokubetsu Iinkai, 10, June 2, 1978, https://kokkai.ndl.go.jp/txt/108414339X01019780602.

76. 84th Diet, Shūgiin, Kagaku Gijutsu Shinkō Taisaku Tokubetsu Iinkai, 4, March 22, 1978.

77. 84th Diet, Shūgiin, Kagaku Gijutsu Shinkō Taisaku Tokubetsu Iinkai, 4, March 22, 1978.

78. Asada Toshi, "Jishin yochi wa kyōdai kagaku ka?," *Shizen* 33, no. 4 (1978): 48–55.

79. 84th Diet, Shūgiin, Kagaku Gijutsu Shinkō Taisaku Tokubetsu Iinkai, 4, March 22, 1978.

80. 84th Diet, Shūgiin, Kagaku Gijutsu Shinkō Taisaku Tokubetsu Iinkai, 4, March 22, 1978.

81. 84th Diet, Sangiin, Saigai Taisaku Tokubetsu Iinkai, 3, Feb. 9, 1978, https://kokkai.ndl.go.jp/txt/108414339X00319780209.

82. See 84th Diet, Sangiin, Saigai Taisaku Tokubetsu Iinkai, 3, Feb. 9, 1978.

83. See 84th Diet, Sangiin, Saigai Taisaku Tokubetsu Iinkai, 3, Feb. 9, 1978.

84. 84th Diet, Shūgiin, Kagaku Gijutsu Shinkō Taisaku Tokubetsu Iinkai, 4, March 22, 1978.

85. 84th Diet, Shūgiin, Kagaku Gijutsu Shinkō Taisaku Tokubetsu Iinkai, 4, March 22, 1978.

86. 84th Diet, Shūgiin, Kagaku Gijutsu Shinkō Taisaku Tokubetsu Iinkai, 4, March 22, 1978.

87. 84th Diet, Shūgiin, Kagaku Gijutsu Shinkō Taisaku Tokubetsu Iinkai, 4, March 22, 1978.

88. 84th Diet, Shūgiin, Saigai Taisaku Tokubetsu Iinkai, 10, April 19, 1978, https://kokkai.ndl.go.jp/txt/108404339X01019780419.

89. 84th Diet, Shūgiin, Saigai Taisaku Tokubetsu Iinkai, 10, April 19, 1978.

90. 84th Diet, Shūgiin, Kagaku Gijutsu Shinkō Taisaku Tokubetsu Iinkai, 6, March 29, 1978; https://kokkai.ndl.go.jp/txt/108403913X00619780329.

91. 84th Diet, Shūgiin, Naikaku Iinkai, 10, April 6, 1978, https://kokkai.ndl.go.jp/txt/108404889X01019780406

92. 84th Diet, Shūgiin, Naikaku Iinkai, 10, April 6, 1978.

93. 84th Diet, Sangiin, Kessan Iinkai, 9, April 7, 1978, https://kokkai.ndl.go.jp/txt/108414103X00919780407.

94. 84th Diet, Sangiin, Kessan Iinkai, 9, April 7, 1978.

95. 84th Diet, Shūgiin, Saigai Taisaku Tokubetsu Iinkai, 12, April 21, 1978, https://kokkai.ndl.go.jp/txt/108404339X01219780421.

96. Tokyo University seismologist Suzuki Jirō makes this point here: 84th Diet, Shūgiin, Saigai Taisaku Tokubetsu Iinkai, 10, April 19, 1978, https://kokkai.ndl.go.jp/txt/108404339X01019780419.

97. 84th Diet, Shūgiin, Saigai Taisaku Tokubetsu Iinkai, 8, April 13, 1978, https://kokkai.ndl.go.jp/txt/108404339X00819780413.

98. 84th Diet, Shūgiin, Saigai Taisaku Tokubetsu Iinkai, 8, April 13, 1978.

99. 84th Diet, Shūgiin, Saigai Taisaku Tokubetsu Iinkai, 8, April 13, 1978.

100. 84th Diet, Shūgiin, Saigai Taisaku Tokubetsu Iinkai, 8, April 13, 1978. He also remarked that day that, "It is now the case that the Hanteikai can definitely [*kakujitsu ni*] predict Magnitude 8 [earthquakes]."

101. 84th Diet, Shūgiin, Saigai Taisaku Tokubetsu Iinkai, 9, April 18, 1978, https://kokkai.ndl.go.jp/txt/108404339X00919780418.

102. Mogi also provides excerpts from Asada and Rikitake's statements. See Mogi Kiyō, *Jishin yochi o kangaeru*, 164–72.

103. Mogi Kiyō, *Jishin yochi o kangaeru*, 177.

104. Mogi's approach to highlighting these differences was one of the inspirations for my own; some of the excerpts that Mogi cites are ones I turned to as well. See Mogi Kiyō, *Jishin yochi o kangaeru*, 164–66.

105. When the Yochiren published a history of its first ten years, in 1979, Suehiro Shigeji co-wrote a number of its sections, including the two that dealt with the formation of the Hanteikai and the passage of the Daishinhō. Mogi reads these as downplaying the uncertainties surrounding prediction in favor of a more optimistic narrative. See Jishin Yochi Renrakukai, *Jishin Yochi Renrakukai 10-nen no ayumi*, 48–52.

106. See, for example, the Komeitō spokesmen's remarks in the Upper House on June 2, 1978: 84th Diet, Sangiin, Saigai Taisaku Tokubetsu Iinkai, 10, https://kokkai.ndl.go.jp/txt/108414339X01019780602.

107. 84th Diet, Shūgiin, Saigai Taisaku Tokubetsu Iinkai, 14, April 28, 1978, https://kokkai.ndl.go.jp/txt/108404339X01419780428; 84th Diet, Sangiin, Saigai Taisaku Tokubetsu Iinkai, 10, June 2, 1978.

108. 84th Diet, Shūgiin, Saigai Taisaku Tokubetsu Iinkai, 14, April 28, 1978.

109. 84th Diet, Shūgiin, Saigai Taisaku Tokubetsu Iinkai, 14, April 28, 1978.

110. 84th Diet, Shūgiin, Saigai Taisaku Tokubetsu Iinkai, 14, April 28, 1978.

Chapter 9

1. Nishimura Kiyoshi, dir. *Tokyo Daijishin: Magunichūdo 8.1.* Yomiuri Terebi, 1980; Ōmori Kenjirō, dir. *Jishin rettō*. Tōei Eiga, 1980. In the television drama, Sonny Chiba costars as private secretary to the finance minister, tasked with making sure the latter's daughter leaves Tokyo safely before the earthquake hits. See "Magunichūdo 8.1," *Asahi shinbun*, April 17, 1980. Takeuchi Hitoshi—discussed later in this chapter—was an advisor to *Jishin rettō*, as he had been for *Japan Sinks*. Takeuchi disdained the pursuit of short-term earthquake prediction as a waste of time; one can only speculate as to how much of the film's dim view of the council and its members might reflect Takeuchi's influence.

2. "Asada Toshi," *Asahi shinbun*, April 3, 1981; "Watakushi no ichibyō," *Yomiuri shinbun*, August 12, 1979; "Namazu ni makenai," *Yomiuri shinbun*, July 20, 1981.

3. "Namazu ni makenai."

4. "Jishin da, kaji da, sore hinan," *Mainichi shinbun*, September 1, 1979; "1200 mannin ga bōsai kunren," *Asahi shinbun*, September 1, 1981; "'Bōsai no hi,'" *Asahi shinbun*, September 1, 1984; "Jissen jūshi sita keredo," *Asahi shinbun*, September 1, 1984. The same schedule was used in 1980, 1981, and 1982, and perhaps in later years as well. See Kokudochō, ed., *Bōsai hakusho, Shōwa 58 nenban* (Tokyo: Ōkurashō Insatsukyoku, 1983), 130.

5. Kokudochō, *Bōsai hakusho, Shōwa 58 nenban*, 65; "Sandē—Bōsai no hi," *Yomiuri shinbun*, September 2, 1985.

6. "'Jishin yochi' 'keihō' ni oku no gokai," *Yomiuri shinbun*, September 11, 1980. However, only one in four said they actually recognized the law's full official title and only 16 percent said the same about the Earthquake Assessment Committee.

7. "Daijishin," *Yomiuri shinbun*, December 28, 1980; Naikakufu Daijin Kanbō Seifu Kōhōshitsu, "Shōbō—jishin ni kansu yoron chōsa," 1980, https://survey.gov-online.go.jp/s55/S55-09-55-12.html. The share of those expecting a major earthquake rose to around half of those polled in 1987. Shizuoka, Aiichi, and Mie constituted the Tōkai district for polling purposes; the Kantō area included Ibaragi, Tochigi, Gumma, Saitama, Kanagawa, and Tokyo. See Kokudochō, *Bōsai hakusho, Shōwa 58 nenban*, 8.

8. The other choices available to respondents were, "Prediction can't really be done well" (26.5 percent), "Prediction is absolutely impossible" (2.4 percent), and "Don't know" (12.5 percent). See Naikakufu Daijin Kanbō Seifu Kōhōshitsu, "Shōbō—jishin ni kansuru yoron chōsa," 1978, https://survey.gov-online.go.jp/s53/S53-10-53-10.html.

9. Sokuchigaku Shingikai and Jishin Kazan Bukai, *Jishin yochi keikaku no jisshi jōkyo nado no rebyū ni tsuite (Hōkoku)* (Tokyo: Sokuchigaku Shingikai, June 1997), 110.

10. Kokudochō, ed., *Bōsai hakusho, Shōwa 56 nenban*, (Tōkyō: Ōkurashō Insatsukyoku, 1981), 77. The number of instruments measuring changes in surface deformation, nationally, went from just twenty-six in 1970 to ninety-four a decade later; by 1990, 171 were in operation. Public universities and research institutes together managed just over fifty seismometers throughout Japan for the detection of very low levels of seismicity in 1970; by 1980 they controlled almost three times as many and, by 1990, 275. See Sokuchigaku Shingikai and Jishin Kazan Bukai, *Jishin yochi keikaku no jisshi jōkyo nado no rebyū ni tsuite (Hōkoku)*, 107–8.

11. "Tōkai daijishin kiken chiiki shitei," *Yomiuri shinbun*, August 7, 1979; "11 shichōson o tsuika shitei," *Asahi shinbun*, July 19, 1979.

12. Officials in Aiichi's Shinshiro City initially objected to their community having been included on the list but they, too, eventually went along with the designation. See "Tōkai daijishin kiken chiiki shitei," *Yomiuri shinbun*, August 7, 1979.

13. In 2002, after research suggested that the fault zone for a Tōkai earthquake was approximately twenty percent larger than originally thought and that it extended about 50 km further west than earlier estimates, the boundaries of the "Area under Intensified Measures against Earthquake Disaster" were adjusted accordingly. That revision added another ninety-six communities to the original list; one of them was the city of Nagoya. As of 2002, some twelve million people resided within the zone's newly drawn borders. See Tomari Jirō, *Nihon no jishin yochi kenkyū 130-nenshi*, 466.

14. Shizuoka Shinbunsha, *Chinmoku no Surugawan*, 95; Kawabata Nobumasa, "Daikibo jishin taisaku tokubetsu sochihō seitei tōji no naibu jijyō," 1050. The law's full and unwieldy title is: Jishin bōsai taisaku kyōka chiiki ni okeru jishin taisaku kinkyū setsubi jigyō ni kakaru kuni no zaisei ue no tokubetsu sochi ni kansuru hōritsu, or Special Fiscal Measures Act for Urgent Improvement Projects for Earthquake Countermeasures in Areas under Intensified Measures against Earthquake Disaster.

15. Yamamoto's constituents voted him back into office twice, once shortly after the passage of the Daishinhō and for the last time in 1982. He served as Shizuoka's governor until 1986.

16. "Shūin o tsūka," *Asahi shinbun*, May 14, 1980.

17. Shizuoka Shinbunsha, *Chinmoku no Surugawan*, 96–97.

18. He did so in a book he wrote for general readers: Takeuchi Hitoshi, *Jishin no kagaku* (Tokyo: Nippon Hōsō Shuppan Kyōkai, 1973).

19. "Jishin rettō," *Asahi shinbun*, August 3, 1973; "Jishin yochi wa taisetsu da ga," *Yomiuri shinbun*, June 20, 1974.

20. For his list, see Takeuchi Hitoshi, "'Nihon Dajishin' ni sonaeru"; Takeuchi Hitoshi, "Jishin yochi wa jishin taisaku ni arazu," *Keizai hyōron* 26, no. 2 (1977): 122–33; Takeuchi Hitoshi, "Jishin yochi yori ido no setsubi o," *Kagaku Asahi* 38, no. 9 (1978): 81–86.

21. Rikitake Tsuneji, "Jishin yochi wa yumemonogatari de wa nai," *Kagaku Asahi* 38, no. 11 (1978): 72–76.

22. Rikitake Tsuneji, "Jishin yochi wa yumemonogatari de wa nai," 76.

23. Rikitake Tsuneji, "Jishin yochi wa yumemonogatari de wa nai," 76. See also Tomari Jirō, *Nihon no jishin yochi kenkyū 130-nenshi*, 356–57.

24. Takeuchi retired from the University of Tokyo in 1981 and was immediately appointed as editor-in-chief of the just-launched popular science magazine *Newton*. He held that post until 2004 and remained a prolific author until his death that year.

25. Tomari Jirō, *Nihon no jishin yochi kenkyū 130-nenshi*, 356–57, 380.

26. Nihon Jishin Gakkai Nihon Gakujutsu Kaigi Jishingaku Kenkyū Renraku Iinkai, ed., *Jishin yochi kenkyū shinpojiumu (1994)* (Tokyo: Nihon Jishin Gakkai, 1994), ix. The 1994 symposium was cosponsored by the Science Council of Japan's National Committee for Seismology and the Seismological Society of Japan.

27. Ishibashi Katsuhiko, "'Jishin yochi keikaku' no kaitai-saihensei: Atarashi 'jishin yochi kenkyū keikaku' to 'jishin saigai keigen' no teian," in *Jishin yochi kenkyū shinpojiumu (1994)*, ed. Nihon Gakujutsu Kaigi Jishingaku Kenkyū Renraku Iinkai and Nihon Jishin Gakkai (Tokyo: Nihon Jishin Gakkai, 1994), 121.

28. Ishibashi Katsuhiko, "'Jishin yochi keikaku' no kaitai-saihensei: Atarashi 'jishin yochi kenkyū keikaku' to 'jishin saigai keigen' no teian," 121.

29. Ishibashi made more or less the same argument for a wider audience the following year. See Ishibashi Katsuhiko, "Kagakuteki 'jishin yochi' o mezashite," *Kagaku* 65, no. 9 (1995): 573–81.

30. Ōkubo Maki, "Tatakau jishin gakusha, mono iu kinzoku zairyō gakusha," *Ronza* (September 2012), https://webronza.asahi.com/national/articles/a.html. The International Institute of Seismology and Earthquake Engineering was a unit of the Construction Ministry's Building Research Institute (Kenchiku kenkyūjo) while Ishibashi worked there.

31. Ishibashi Katsuhiko, *Daichi dōran no jidai: jichin gakusha wa keikokusuru* (Tokyo: Iwanami Shoten, 1994).

32. He had made similar observations about the risk to Tokyo some years earlier. See Ishibashi Katsuhiko, "Daijishin ga 'seikai toshi Tokyo' o chokugeki suru," *Chūō kōron* 102, no. 11 (1987): 124–38.

33. Kabata Hisatoshi, "Ishibashi Katsuhiko-san, Jishingakusha," *Asahi shinbun*, March 4, 1995.

34. As opposed to a term-limited or renewable appointment.

35. For more on Geller's experiences at Caltech and as a scientist in Japan, see Zierler, "Robert J. Geller (BS '73, PhD '77), Seismologist."

36. Robert J. Geller, "Problems of Tenure in Japan," *Nature* 345, no. 6274 (1990): 380.

37. Robert J. Geller, "Shake-up for Earthquake Prediction," *Nature* 352, no. 6333 (1991): 275–76.

38. Robert J. Geller, "Shake-up for Earthquake Prediction," 275.

39. Geller, "Shake-up for Earthquake Prediction," 276.

40. Geller, "Shake-up for Earthquake Prediction," 275.

41. Kazuo Hamada and Robert J. Geller, "Unpredictable Earthquakes (Scientific Correspondence) (With Geller's reply)," *Nature* 353, no. 71 (1991): 611.

42. Tomari Jirō, *Nihon no jishin yochi kenkyū 130-nenshi*, 382; Kazuo Hamada and Robert J. Geller, "Unpredictable Earthquakes," 612.

43. Tomari Jirō, "Beijin gakusha no ronbun ni "yureru" Jishingakkai," *Aera* (December 3,1991): 66. Tomari was a journalist with the *Asahi shinbun* for many years before returning to the University of Tokyo to earn a doctorate in the History of Science. His history of earthquake prediction came out when I was already well launched on this book project; it is excellent. See Tomari Jirō, *Nihon no jishin yochi kenkyū 130-nenshi.*

44. Tomari Jirō, "Beijin gakusha no ronbun ni "yureru" Jishingakkai."

45. These are in addition to articles Geller himself published in Japanese; see, for example, Robert J. Geller, "Jishin gakusha no shakaiteki kōken," *Nikkei saiensu* (September 1993): 3.

46. David Swinbanks, "Trying to Shake Japan's Faith in Forecasts," *Nature* 356 (1992), 464.

47. Denis Normile, "Japan Holds Firm to Shaky Science," *Science* 264 (1994): 1656–58.

48. Normile, "Japan Holds Firm to Shaky Science."

49. Normile, "Japan Holds Firm to Shaky Science."

50. Robert J. Geller, "Jishin yochi saikō," in *Jishin yochi kenkyū shinpojiumu (1994)*, ed. Nihon Gakujutsu Kaigi Jishingaku Kenkyū Renraku Iinkai and Nihon Jishin Gakkai (Tokyo: Nihon Jishin Gakkai, 1994), 131.

51. Robert J. Geller, "Jishin yochi saikō," 131, 136.

52. Hirata Naoshi, "Sekkyokuteki ni 'Keiken o tsumu' kenkyū o suishin shiyō," in *Jishin yochi kenkyū shinpojiumu (1994)* (Tokyo: Nihon Jishin Gakkai, 1994), 155.

53. Tomari Jirō, *Nihon no jishin yochi kenkyū 130-nenshi*, 415.

54. Tomari Jirō, *Nihon no jishin yochi kenkyū 130-nenshi*, 396.

55. Tomari Jirō, *Nihon no jishin yochi kenkyū 130-nenshi*, 396–97.

56. Tomari Jirō, *Nihon no jishin yochi kenkyū 130-nenshi*, 401–2.

57. Tomari Jirō, *Nihon no jishin yochi kenkyū 130-nenshi*, 367–68.

58. Tomari Jirō, *Nihon no jishin yochi kenkyū 130-nenshi*, 365.

59. Tomari Jirō, *Nihon no jishin yochi kenkyū 130-nenshi*, 393–94.

60. "Ikasenakatta yochi kenkȳu," *Mainichi shinbun*, February 18, 1995; "Senmonka no mikata," *Mainichi shinbun*, January 18, 1995; "Chokuzen yochi wa kanō ka," *Mainichi shinbun*, January 31, 1995.

61. Tomari Jirō, *Nihon no jishin yochi kenkyū 130-nenshi*, 414–15.

62. Tomari Jirō, *Nihon no jishin yochi kenkyū 130-nenshi*.

63. David Swinbanks, "Japanese Science Agency Loses Iron Lady," *Nature* 376, no. 6541 (1995): 542; David Swinbanks, "Kobe Disaster Divides Earthquake Researchers," *Nature* 373 (1995): 373.

64. Tomari Jirō, *Nihon no jishin yochi kenkyū 130-nenshi*, 421.

65. Tomari Jirō, *Nihon no jishin yochi kenkyū 130-nenshi*, 424. See also "Jishin yochiren gurari," *Asahi shinbun*, June 4, 1995.

66. The National Research Institute for Earth Science and Disaster Prevention's Earthquake Prediction Research Center changed its name to the Earthquake Research Center (Jishin chōsa kenkȳu senta) in 1996, for example. See Tomari Jirō, *Nihon no jishin yochi kenkyū 130-nenshi*, 451, for additional examples.

67. Katō Naoyuki, "'Jishin yochi' ni taisuri Nihon gakkai no torikumi," in *"Burupurinto" 50 shūnen jishin kenkyū no ayumi to kongo*, ed. Nihon Jishin Gakkai, Nihon Jishin Gakkai monogurafu (Tokyo: Nihon Jishin Gakkai, 2013), 45–46; David Swinbanks, "Earthquake Prediction Likely to Grow' After New Review," *Nature* 374, no. 6519 (1995): 205; Hasegawa Akira, "Jishingaku kenkyūsha—jishingaku comyunitei no shakaiteki yakuwari," in *Jishingaku no ima o tō*, ed. Tōhoku Chihō Taiheiyō-oki Jishin Taio Rinji Iinkai (Tokyo: Nihon Jishin gakkai, May 2012), 19; Matsuzawa Toru, "M9 o sōtei suru tame ni nani ga kaketeita no ka? Kongo dō sureba yoi no ka?," in *Jishingaku no ima o tō*, ed. Tōhoku Chihō Taiheiyō-oki Jishin Taio Rinji Iinkai (Tokyo: Nihon Jisshin Gakkai, May 2012), 12.

68. Sokuchigaku Shingikai and Jishin Kazan Bukai, *Jishin yochi keikaku no jisshi jōkyo*; see also Tomari Jirō, *Nihon no jishin yochi kenkyū 130-nenshi*, 451–58.

69. David Swinbanks, "Quake Panel Admits Prediction is 'Difficult,'" *Nature* 388, no. 6637 (1997): 4; "Jishin yochi 'genjō de wa konnan,'" *Yomiuri shinbun*, March 23, 1997.

70. See "Sokuchishin tokubetsui ga 'jishin yochi konnan,'" *Yomiuri shinbun*, March 25, 1997; "Kiro ni tatsu jishin kenkyū," *Yomiuri shinbun*, September 18, 1997.

71. Tomari Jirō, *Nihon no jishin yochi kenkyū 130-nenshi*, 455.

72. These goals were spelled out in Headquarters for Earthquake Research Promotion, *The Promotion of Earthquake Research: Basic Comprehensive Policy for the Promotion of Earthquake Observation, Measurement, Surveys and Research*, 1999, https://www.jishin.go.jp/main/w_060_f-e.htm. See also Saegusa Asako, "Japan's Quake Strategy Urged to Switch to Long-term Forecasting," *Nature* 393, no. 6682 (1998): 202; Saegusa Asako, "Japan to Try to Understand Quakes, Not Predict Them," *Nature* 397 (1999): 284.

73. Headquarters for Earthquake Research Promotion, *The Promotion of Earthquake Research*, 14.

74. Headquarters for Earthquake Research Promotion, *The Promotion of Earthquake Research*, 6.

75. Headquarters for Earthquake Research Promotion, *The Promotion of Earthquake Research*, i.

76. Headquarters for Earthquake Research Promotion, *The Promotion of Earthquake Research*, 3.

77. Headquarters for Earthquake Research Promotion, *The Promotion of Earthquake Research*, 3.

78. Headquarters for Earthquake Research Promotion, *The Promotion of Earthquake Research*, 17.

79. David Swinbanks, "Critics Attack Hasty Review of Quake Prediction Efforts," *Nature* 374, no. 6525 (1995): 754; Shimamura Hideki, "Kuni no jishin kenkyū taisei o tō," *Asahi shinbun*, September 25, 1997.

80. As of early 1996. See "Jishin yochi wa hontō ni kanō ka?" *Aera* (January 29, 1996): 26.

81. Swinbanks, "Quake Panel Admits Prediction Is 'Difficult.'" Tanaka's stint as STA director helped launch her reputation as Japan's "Iron Lady." She later served as Foreign Minister in Junichiro Koizumi's Cabinet in 2001.

82. Tomari Jirō, *Nihon no jishin yochi kenkyū 130-nenshi*, 426.

83. Tomari Jirō, *Nihon no jishin yochi kenkyū 130-nenshi*, 426–27.

84. Mogi Kiyō, "Two Grave Issues Concerning the Expected Tokai Earthquake," *Earth Planets Space* 56 (2004): li-lxvi.

85. "Keisei sengen," *Asahi shinbun*, August 30, 1995.

86. "Keisei sengen," *Asahi shinbun*, August 30, 1995.

87. Mogi Kiyō, "Two Grave Issues Concerning the Expected Tokai Earthquake," lix; "'Chūihō' no tesishō," *Asahi shinbun*, March 10, 1993.

88. "Yuragu shinario (2)," *Asahi shinbun*, January 5, 1994.

89. Mogi Kiyō, *Jishin yochi o kangaeru*, 180–83; "Tōkai jishin keisei sengen hatsurei wa naze konnan ka," *Aera* (April 1, 1996): 18.

90. See, for example, Mogi Kiyō, "Fujūbun na taisei minaosu kikai ni," *Asahi shinbun*, March 29, 1996. Tomari Jirō discusses Mogi's concerns in Tomari Jirō, "Tōkai jishin e no taisaku," *Asahi shinbun*, July 19, 1995; and Tomari Jirō, "Mogi kaichō ni kiku," *Asahi shinbun*, March 16, 1993.

91. Mogi Kiyō, "Two Grave Issues Concerning the Expected Tokai Earthquake," lxi. After his resignation, he continued to call for changes in his books and magazine articles and, by his own count, raised the problem fourteen times between June 1995 and March 2003 in newspaper editorials in Shizuoka.

92. Chūō Bōsai Kaigi, *Tōkai jishin taisaku taikō* (2003), https://www.bousai.go.jp/jishin/tokai/pdf/taikou.pdf.

93. Tomari Jirō, *Nihon no jishin yochi kenkyū 130-nenshi*, 466–67.

94. Tomari Jirō, *Nihon no jishin yochi kenkyū 130-nenshi*, 466–67.

95. Gregory C. Beroza and William L. Ellsworth, "Properties of the Seismic Nucleation Phase," *Tectonophysics* 261, no. 1 (1996): 209–27; D. Kilb and J. Gomberg, "The Initial Subevent of the 1994 Northridge, California, Earthquake," *Journal of Seismology* 3, no. 4 (1999): 409–20.

96. Shimamura Hideki, *"Jishin yochi" wa usodarake* (Tokyo: Kōdansha, 2008), 185–91.

97. Chūō Bōsai Kaigi, *Tōkai jishin taisaku taikō*, 35.

98. For a powerful critique of the General Principles, see Shimamura Hideki, *Kōnin "jishin yochi" o utagau* (Tokyo: Kashiwa shobo, 2004); Shimamura Hideki, "Daishinhō no gensō," *Ekonomisuto* 82, no. 67 (2004): 33.

99. Shizuoka-ken Kikikanribu Kiki Jōhōshitsu, *Heisei 21-nendo Tōkai jishin ni tsuite no kenmin ishiki chōsa*, Jishin taisaku shiryō, vol. 242 (2010), 104, https://www.pref.shizuoka.jp/_res/projects/default_project/_page_/001/042/987/ishiki2009.pdf.

100. Shizuoka-ken Kikikanribu Kiki Jōhōshitsu, *Heisei 21-nendo Tōkai jishin ni tsuite no kenmin ishiki chōsa*, Jishin taisaku shiryō, vol. 242 (2010), 104.

101. Shizuoka-ken Kikikanribu Kiki Jōhōshitsu, *Heisei 21-nendo Tōkai jishin ni tsuite no kenmin ishiki chōsa*, 13. Another 46 percent indicated that they were "somewhat concerned."

102. At which time 29 percent chose that response. See Shizuoka-ken Kikikanribu Kiki Jōhōshitsu, *Heisei 21-nendo Tōkai jishin ni tsuite no kenmin ishiki chōsa*, 89.

103. Saegusa Asako, "Japan to Try to Understand Quakes, Not Predict Them."

104. The "Kawasumi Maps" of the 1950s had also conveyed estimates of relative risk based on estimates of how much time passed between earthquakes at a given location and estimates of how powerful the shaking would be if another one just like it were to occur again. In the United States, Allin Cornell pioneered methods for performing probabilistic seismic risk analysis in the late 1960s. See Cornell, "Engineering Seismic Risk Analysis."

105. For critical assessments of the maps, see Gregory Smits, *When the Earth Roars: Lessons from the History of Earthquakes in Japan* (Lanham, MD: Rowman & Littlefield, 2014), 153–54; Smits, "Probabilistic Earthquake Hazard Maps."

106. Headquarters for Earthquake Research Promotion Earthquake Research Committee, *National Seismic Hazard Maps for Japan (2005)*, https://www.jishin.go.jp/main/chousa/06mar_yosoku-e/NationalSeismicHazardMaps.pdf, 19. The thirty-year window was standard for these and other probabilistic maps because it was thought to be "the time period that common citizens will acknowledge."

107. Tomari Jirō, *Nihon no jishin yochi kenkyū 130-nenshi*, 434.

108. Headquarters for Earthquake Research Promotion Earthquake Research Committee, *National Seismic Hazard Maps for Japan (2005)*, 19. Because most buildings were expected to last fifty years before having to be replaced, mapmakers also calculated the probability of intense shaking over that time period.

109. Headquarters for Earthquake Research Promotion Earthquake Research Committee, *National Seismic Hazard Maps for Japan (2005)*, 19; Shinichi Kawai, Hiroyuki

Fujiwara, Shin Aoi, Nobuyuki Morikawa, Shigeki Senna, Kyoko Kobayashi, Toru Ishii, Toshihiko Okumura, and Yuzuru Hayakawa, "National Seismic Hazard Maps of Japan," *Bulletin of the Earthquake Research Institute* 81 (2006): 221–32.

110. Headquarters for Earthquake Research Promotion Earthquake Research Committee, *National Seismic Hazard Maps for Japan (2005)*, 59–102. In 2005, calculations were complete for a dozen of these scenarios.

111. The Yochiren finally did away with its system of designating some parts of the country for different levels of observation in 2008.

Epilogue

1. Ishibashi was a proponent of "de-concentrating" the capital district by dispersing public and private sector employers and institutions to other prefectures, among other proposals for restructuring the Japanese economic and social structures. He discusses some of his ideas in Uchihashi Katsuto and Ishibashi Katsuhiko, "Taidan: 'Mō hitotsu no Nihon' koso ga kongenteki na shinsai taisaku da," *Sekai* 721 (2003): 194–203.

2. "Itsu kuru ka: Tōkai jishin setsu kara 30 nen (4)," *Mainichi shinbun*, August 30, 2006.

3. Ishibashi Katsuhiko, "Genpatsu shinsai—hametsu o sakeru tame ni."

4. Ishibashi Katsuhiko, "Genpatsu shinsai—hametsu o sakeru tame ni," 721.

5. Ishibashi Katsuhiko, "Genpatsu shinsai—hametsu o sakeru tame ni," 723. Construction on the plant began in 1971, well after Mogi Kiyō's warning about a potential M8 earthquake in the area and the Yochiren's decision to designate the district for "intensive observation." The first reactor went online in March 1976, a few months before Ishibashi's announcement. See Mogi Kiyō, "Two Grave Issues Concerning the Expected Tokai Earthquake."

6. The only way to prevent an outcome like this from happening, Ishibashi argued, was for Japan to abandon nuclear power entirely in favor of safer alternatives. See Ishibashi Katsuhiko, "Genpatsu shinsai—hametsu o sakeru tame ni," 724.

7. Leuren Moret, "Japan's Deadly Game of Nuclear Roulette," *Japan Focus* 10, no. 54 (2012).

8. Ide Satoshi, "Asuperitii—undōgata—jishin yochi," in *Jishingaku no ima o tō*, ed. Tōhoku Chihō Taiheiyō-oki Jishin Taio Rinji Iinkai (Tokyo: Nihon Jishin Gakkai, May 2012), 15.

9. Jeroen Ritsema, Thorne Lay, and Hiroo Kanamori, "The 2011 Tohoku Earthquake," *Elements* 8 (2012): 183.

10. Ritsema, Lay, and Kanamori, "The 2011 Tohoku Earthquake," 185.

11. Ritsema, Lay, and Kanamori, "The 2011 Tohoku Earthquake," 184.

12. The tsunami that inundated the Fukushima-Daiichi nuclear complex had a reported height of fourteen meters; waves approaching forty meters were reported along the Miyagi prefecture coast.

13. Ritsema, Lay, and Kanamori, "The 2011 Tohoku Earthquake," 184.

14. Ritsema, Lay, and Kanamori, "The 2011 Tohoku Earthquake," 186.

15. Ritsema, Lay, and Kanamori, "The 2011 Tohoku Earthquake," 184.

16. Takeshi Sagiya et al., "Rebuilding Seismology," *Nature* 472 (2011): 146–48.

17. Robert J. Geller, "Bōsai taisaku to jishin kagaku kenkyū no arikata: risettto no jiki," in *Jishingaku no ima o tō*, ed. Tōhoku Chihō Taiheiyō-oki Jishin Taio Rinji Iinkai (Tokyo: Nihon Jishin Gakkai, 2012).

18. Kōketsu Kazuki, "Jishin no kagaku no yukue wa? Higashi Nihon Daishinsai kara ikagetsu, jishin gakusha wa nani o omou ka," *Tokushū: Kenkyūsha intabyū dai 8-kai* (April 2011): 2.

19. Tomari makes this point in Tomari Jirō, *Nihon no jishin yochi kenkyū 130-nenshi*, 516.

20. Richard J. Samuels, *3.11: Disaster and Change in Japan* (Ithaca, NY: Cornell University Press, 2013).

21. Robert J. Geller, "Bōsai taisaku to jishin kagaku kenkyū no arikata: risettto no jiki," in *Jishingaku no ima o tō*, ed. Tōhoku Chihō Taiheiyō-oki Jishin Taio Rinji Iinkai (Tokyo: Nihon Jishin Gakkai, 2012), 6.

22. Geller, "Sakete tōrenai deguchi sakusen." Geller made similar points in an August 2011 appearance on the Nippon BS Broadcasting Corporation's "INside-OUT" talk show, which is available to view on YouTube, here: https://youtu.be/zZFpt-NGY2M.

23. Geller, "Bōsai taisaku to jishin kagaku kenkyū no arikata: risettto no jiki."

24. "(Bōsai shin) Sōtegai no M9, habahiroi shiya," *Asahi shinbun*, October 27, 2011. For later examples of the inclusion of Geller's perspective within semi-official publications dealing with prediction and seismology in Japan, see Geller, "Sakete tōrenai deguchi sakusen"; Robert J. Geller, "Machigatta gakusetsu ni tayoru na," in *Jishin Chōsa Kenkyū Suishin Honbu 20-nen no shiryōshū*, ed. Jishin Yochi Sōgō Kenkyū Shinkōkai (Tokyo: Monbukagakusho, 2015); Robert J. Geller, "Hon shimpojiumu kaisei ni kanshite no hanseiten," in *Nankai torafu jishin rinji jōhō: kagakuteki de-ta ya chiken to katsuyō*, ed. Nihon Jishin Gakkai, Nihon Jishin Gakkai monogurafu (Tokyo: Nihon Jishin Gakkai, 2020).

25. Chūō Bōsai Kaigi, *Tōkai jishin taisaku taikō*, 1.

26. Tomari Jirō, *Nihon no jishin yochi kenkyū 130-nenshi*, 467–68.

27. "Jishin yochi zentei ippen," *Asahi shinbun*, August 27, 2008; "Kyōdai jishin ga kuru (5)," *Asahi shinbun*, August 8, 2006.

28. Most of the deaths in that scenario, around 230,000, were attributed to tsunami. These results were reported in Chūō Bōsai Kaigi Bōsai Taisaku Suishin Kentō Kaigi Nankai Torafu Kyōdai Jishin Taisaku Kentō Wākingu Gurūpu, *Nankai torafu kyodai jishin no higai sōtei ni tsuite (Dai ichiji hōkoku)*, 21, https://www.bousai.go.jp/jishin/nankai/taisaku_wg/pdf/20120829_higai.pdf.

29. Tomari Jirō, *Nihon no jishin yochi kenkyū 130-nenshi*, 534–35.

30. I'm referring here to the 2013 Act on Special Measures for Promotion of Nankai Trough Earthquake Disaster Management, or Nankai torafu jishin ni kakaru jishin bōsai taisaku no suishin ni kansuru tokubetsu sochihō.

31. Nankai Torafuzoi no Daikibō Jishin no Yosaku Kanōsei ni Kansuru Chōsabukai, *Nankai torafuzoi no daikibō jishin no yosaku kanōsei ni tsuite*, 7, https://www.bousai.go.jp/jishin/nankai/taisaku_wg/pdf/20130528_houkoku_s3.pdf.

32. Nankai Torafuzoi no Daikibō Jishin no Yosaku Kanōsei ni Kansuru Chōsabukai, *Nankai torafuzoi no daikibō jishin no yosaku kanōsei ni tsuite*, 2.

33. Chūō Bōsai Kaigi Bōsai Taisaku Suishin Kentō Kaigi Nankai Torafu Kyōdai Jishin Taisaku Kentō Wākingu Gurūpu, *Nankai torafu kyōdai jishin taisaku ni tsuite (saishū hōkoku)*, 56, https://www.bousai.go.jp/jishin/nankai/taisaku_wg/pdf/20130528_honbun.pdf.

34. "Jishin taisaku 3 kēsu gutaika," *Mainichi shinbun*, September 8, 2017.

35. "Jishin taisaku 3 kēsu gutaika."

36. "Nankai torafu 'rinji jōhō' to wa," *Asahi shinbun*, May 28, 2018; "Tōkai jishin: Yochi jisshitsu shūryō," *Mainichi shinbun*, October 31, 2017.

37. "Kyū wari, jūmin no hinan kentō," *Asahi shinbun*, May 28, 2018.

38. "Nankai torafu jishin," *Asahi shinbun*, April 22, 2019; "Nankai torafu jishin 'rinji jōhō,'" *Asahi shinbun*, April 23, 2019.

39. Ishibashi Katsuhiko, *Nankai torafu kyodai jishin*.

40. Ishibashi Katsuhiko, "Daishinhō haishi—jishin kanrenhō ipponka to, hassei yosoku no hojō shudanka o," in *Jishin hassei yosoku to daishinhō oyobi jishin bōsai kenkyū*, ed. Monogurafu 'Jishin Hassei Yosoku to Daishinhō Oyobi Jishin Bōsai Kenkyū' Henshū Iinkai, Nihon jishin gakkai monogurafu (Tokyo: Nihon Jishin Gakkai, December 2017), 30.

41. See, for example, Ishibashi Katsuhiko, "'Nankai torafu jishin rinji jōhō' taisei e no gimon," in *Nankai torafu jishin rinji jōhō: kagakuteki de-ta ya chiken to katsuyō*, ed. Nihon Jishin Gakkai, Nihon jishin gakkai monogurafu (Tokyo: Nihon Jishin Gakkai, March 2020), 25–28.

42. Ishibashi Katsuhiko, "'Nankai torafu jishin rinji jōhō' taisei e no gimon," 25.

43. Yoshida Masashi, "Nankai torafu jishin no sonaeru 'me,'" *Yomiuri shinbun*, September 1, 2019.

44. Shizuoka-ken Kikikanribu Kiki Jōhōka, *Nankai torafu jishin (Tōkai jishin) ni tsuite no kenmin ishiki chōsa*, Jishin taisaku shiryō, vol. 290 (2017): 105, https://www.pref.shizuoka.jp/_res/projects/default_project/_page_/001/042/987/h29kennminnisikityousahoukokusyo.pdf.

45. Shizuoka-ken Kikikanribu Kiki Jōhōka, *Reiwa gannen-do Nankai torafu jishin ni tsuite no kenmin ishiki chōsa*, Jishin taisaku shiryō, vol. 306 (2020): 40, https://www.pref.shizuoka.jp/_res/projects/default_project/_page_/001/042/987/nannkaitorahukennminnisikityousa.pdf

46. John R. Huizenga, *Cold Fusion: The Scientific Fiasco of the Century* (Rochester, NY: University of Rochester Press, 1992); Gary Taubes, *Bad Science: The Short Life and Weird Times of Cold Fusion*, 1st ed. (New York: Random House, 1993); Bruce V. Lewenstein, "From Fax to Facts: Communication in the Cold Fusion Saga," *Social Studies of Science* 25, no. 3 (1995): 403–36.

47. Mitsuyuki Hoshiba, "Review of the Nationwide Earthquake Early Warning in Japan during Its First Five Years," in *Earthquake Hazard, Risk and Disasters*, ed. John F. Shroder and Max Wyss (Boston: Academic Press, 2014).

48. Yan Y. Kagan, David D. Jackson, and Robert J. Geller, "Characteristic Earthquake Model, 1884–2011, R.I.P.," *Seismological Research Letters* 83, no. 6 (2012): 951–53.

49. H. Tristram Engelhardt and Arthur L. Caplan, "Patterns of Controversy and Closure: the Interplay of Knowledge, Values, and Political Forces," in *Scientific Controversies: Case Studies in the Resolution and Closure of Disputes in Science and Technology*, ed. H. Tristram Engelhardt and Arthur L. Caplan (Cambridge: Cambridge University Press, 1987), 14–15; Lewenstein, "Science Controversies," 77.

50. Lewenstein, "Science Controversies," 77.

51. For a discussion of "closure through force," see Engelhardt and Caplan, "Patterns of Controversy and Closure," 14; Lewenstein, "Science Controversies," 77.

52. See, for example, Vladimir G. Kossobokov, "Earthquake Prediction: 20 years of Global Experiment," *Natural Hazards* 69, no. 2 (2013): 1155–77; Thomas H. Jordan, "The Prediction Problems of Earthquake System Science," *Seismological Research Letters* 85, no. 4 (2014): 767–69; Matthew E. Pritchard et al., "New Opportunities to Study Earthquake Precursors," *Seismological Research Letters* 91, no. 5 (2020): 2444–47. For a discussion of skepticism toward prediction among scientists in the UK, see Joffe et al., "Stigma in Science."

53. "Watakushi no bōsaisaku," *Asahi shinbun*, August 23, 1977. This article is one of a series in the *Asahi* on "My disaster preparedness plans."

54. "Kōtō chiku bōsai kyoten no Shirahige higashi chiku," *Asahi shinbun*, May 18, 1972.

55. Tsuru Shigeto provides an overview of the incident and the years of legal struggles with the company responsible for the contamination that followed. See Shigeto Tsuru, *The Political Economy of the Environment: The Case of Japan* (London: Athlone Press, 1999), 185–92.

56. Tsuru, *The Political Economy of the Environment*, 192; Muneaki Tajiri, "Rokkakuromu jiken no honshitsu to kadai," *Kōgai kenkyū* 7 (1978): 51–57; "Bōsai kyoten zukuri dai ni dan," *Asahi shinbun*, January 31, 1979. By the time that process was complete, however—it took years—the site-specific problems were no longer the only things preventing the disaster shelter project from moving forward.

57. "Dai 3 no Kōtō bōsai kyoten," *Asahi shinbun*, December 5, 1973; "Ryōgoku bōsai kyoten," *Asahi shinbun*, February 10, 1980; Ehira Teruo, "Tokyo-to ni okeru bōsai toshi zukuri no kiseki to keifu—Kōtō bōsai kyoten o chūshin ni shite," *Kenchiku bōsai* 259 (1999): 26. The footprint of the original six-hub plan is evident in any recent satellite photo of the Kōto delta; just about every large park or green space corresponds to the site of one of the proposed shelters.

58. "Tokyo de daijishin ga okitara," *Asahi shinbun*, November 13, 1975.

59. See "Tokyo de daijishin ga okitara"; "Toshi no jishin kikendo chōsa," *Asahi shinbun*, November 13, 1975, for lists of all locations ranked 3 and 4.

60. "Toshi no jishin kikendo chōsa."

61. Nakabayashi Itsuki, "Tokyo-to shinsai yobō jōrei no kasseika o," *Chiri* 33, no. 6 (1988): 10–11.

62. Critics pointed out these and other shortcomings even before the regulations went into effect. Ward officials, happy to hear the city was going to help them purchase fire extinguishers, discovered that no one in the administration could tell them how to go about getting the funding, or even if there was any. See, for example, "'Shinsai yobō jōrei' no mondai ten," *Asahi shinbun*, October 22, 1971.

63. See chapter 3 of Eiko Maruko Siniawer, *Waste: Consuming Postwar Japan* (Ithaca: Cornell University Press, 2018) for a discussion of Tokyo's "Garbage Wars" of the 1970s.

64. Nakabayashi Itsuki, "Development of Urban Disaster Prevention Systems in Japan - from the Mid-1980s," *Journal of Disaster Research* 1, no. 1 (2006):, 49; Nakabayashi Itsuki, "Tokyo ni okeru jishin kiken no jizen hyōka to jishin ni tsuyoi toshi zukuri," *Toshi kagaku kenkyū* 2 (2008): 76.

65. Nakabayashi Itsuki, "Development of Urban Disaster Prevention Systems in Japan," 49.

66. Nakabayashi Itsuki, "Tokyo ni okeru jishin saigai no yōsō to shinsaitaisaku no tenkai," *Chigaku zasshi* 110, no. 6 (2001): 939.

67. Nakabayashi Itsuki, "Tokyo ni okeru jishin saigai no yōsō to shinsaitaisaku no tenkai," 941.

68. Nakabayashi Itsuki, "Tokyo ni okeru jishin saigai no yōsō to shinsaitaisaku no tenkai," 942.

69. Nakabayashi Itsuki, "Tokyo ni okeru jishin saigai no yōsō to shinsaitaisaku no tenkai," 945.

70. Nakabayashi Itsuki, "Tokyo ni okeru jishin saigai no yōsō to shinsaitaisaku no tenkai," 944.

71. Nakabayashi Itsuki, "Tokyo ni okeru jishin saigai no yōsō to shinsaitaisaku no tenkai," 947.

72. Tokyo-to shinsai taisaku jōrei, February 22, 2000, https://www.reiki.metro.tokyo.lg.jp/reiki/reiki_honbun/g101RG00000173.html.

73. Ishikawa Tomoya, "Shuto chokka jishin kuni sōtei," *Asahi shinbun*, December 20, 2013; Iwata Tomohiro, "Shuto chokka jishin higai sōtei no rosen tenkan," *Aera* (December 30, 2013): 34.

74. Iwata Tomohiro, "Shuto chokka jishin higai sōtei no rosen tenkan."

75. Ishikawa Tomoya, "Shuto chokka jishin kuni sōtei."

76. Ishikawa Tomoya, "Shuto chokka jishin kuni sōtei"; Iwata Tomohiro, "Shuto chokka jishin higai sōtei no rosen tenkan."

77. Nakabayashi Itsuki, "Shuto chokka jishin taisaku no kihon hōkō," 2.

78. Ishikawa Tomoya, "Shuto chokka jishin kuni sōtei."

79. Iwata Tomohiro, "Shuto chokka jishin higai sōtei no rosen tenkan."

80. Nakabayashi Itsuki and Uchida Muneharu, "Intabyū—Shuto chokka jishin no "sōteigai" o sōtei suru," *Tokyojin* 26, no. 7 (2011): 22.

81. If the earthquake occurred around noon in the summer on a day with light winds, which was the council's best-case scenario, the total number of deaths would fall dramatically and fire would be responsible for only one in ten of them. Chūō Bōsai Kaigi Shuto Chokka Jishin Taisaku Kentō Wākinggu Gurūpu, *Shuto chokka jishin no higai sōtei to taisaku ni tsuite (Saishū hōkoku)*, 6, https://www.bousai.go.jp/jishin/syuto/taisaku_wg/pdf/syuto_wg_report.pdf.

82. Tokyo prefecture as a whole takes up around 1800 square kilometers. The twenty-three special wards take up a little over 620.

83. The eight-minute segment from Tokyo MX's September 21, 2011, broadcast of its "Kenshō! Shuto bōsai" series is available at https://www.youtube.com/watch?v=kiH-1XiPgik.

84. Paul S. Boyer, *By the Bomb's Early Light: American Thought and Culture at the Dawn of the Atomic Age* (Chapel Hill: University of North Carolina Press, 1994); Guy Oakes, *The Imaginary War: Civil Defense and American Cold War Culture* (New York: Oxford University Press, 1994); Kenneth D. Rose, *One Nation Underground: The Fallout Shelter in American Culture* (New York: New York University Press, 2001); Spencer R. Weart, *The Rise of Nuclear Fear* (Cambridge, MA: Harvard University Press, 2012).

85. Lee Ben Clarke, *Mission Improbable: Using Fantasy Documents to Tame Disaster* (Chicago: University of Chicago Press, 1999).

86. Clarke, *Mission Improbable*, 41.

87. The 2004 "Act on Special Measures for Promotion of Disaster Management for Trench-type Earthquakes in the Vicinity of the Japan and Chishima Trenches" is an exception to this rule, as it did provide additional funding to communities along parts of Japan's northeastern coast.

88. This was especially evident in the lead-up to the 2020 Summer Olympics. See Alexis Dudden, "An Opportunity for Japan to Change People's Perception," *Asia-Pacific Journal: Japan Focus* 18, no. 5 (2020): 1–4.

89. See, for example, J. Charles Schencking, "1923 Tokyo as a Devastated War and Occupation Zone: The Catastrophe One Confronted in Post Earthquake Japan," *Japanese Studies* 29, no. 1 (2009): 120.

90. "Kōtō ni dai bōsai kōji," *Asahi shinbun*, November 12, 1969. "Watakushi no bōsaisaku (3)," *Asahi shinbun*, August 25, 1977; "Bōsai taisaku wa naze susumanai no ka," *Yomiuri shinbun*, September 1, 1975; "Jishin fuan ni oriau," *Aera* (July 2, 2012): 25.

BIBLIOGRAPHY

Archives and Digital Collections

Caltech Archives and Special Collections, Pasadena, CA.

Full-text Database System for the Minutes of the Diet (Kokkai kaigiroku kensaku shisutemu) (https://kokkai.ndl.go.jp/#/)

National Archives and Records Administration II, College Park, MD.

Magazines and Newspapers

Aera

Asahi shinbun

Mainichi shinbun

Kagaku gijutsuchō nenpō

Nippon Times

Yomiuri shinbun

Works Cited

Abel, Jessamyn. *Dream Super-Express: A Cultural History of the World's First Bullet Train.* Studies of the Weatherhead East Asian Institute, Columbia University. Stanford, CA: Stanford University Press, 2022.

Anderson, Katharine. *Predicting the Weather: Victorians and the Science of Meteorology.* Chicago: University of Chicago Press, 2005.

Andersson, Jenny. "The Great Future Debate and the Struggle for the World." *American Historical Review* 117, no. 5 (2012): 1411–30.

———. "Midwives of the Future: Futurism, Future Studies and the Shaping of the Global Imagination." In *The Struggle for the Long-term in Transnational Science and Politics*, ed. Jenny Andersson, and Eglė Rindzevičiūtė, 16–37. New York: Routledge, Taylor & Francis Group, 2015.

Andō, Masataka. "Possibility of a Major Earthquake in the Tokai District, Japan and its Pre-estimated Seismotectonic Effects." *Tectonophysics* 25 (1975): 69–85.

Anduaga Egaña, Aitor. *Politics, Statistics and Weather Forecasting, 1840–1910: Taming the Weather.* Routledge Studies in the History of Science, Technology and Medicine. London: Routledge, Taylor & Francis Group, 2020.

Asada Toshi. "Chūgoku ni okeru jishin yochi." *Nature* 30, no. 1 (1975): 36–46.

———. "Jishin yochi wa kyōdai kagaku ka?" *Shizen* 33, no. 4 (1978): 48–55.

Ascher, William. "Scientific Information and Uncertainty: Challenges for the Use of Science in Policymaking." *Science and Engineering Ethics* 10, no. 3 (2004): 437–55.

Avenell, Simon. "From Fearsome Pollution to Fukushima: Environmental Activism and the Nuclear Blind Spot in Contemporary Japan." *Environmental History* 17 (2012): 244–76.

Azuma Hideki. "Sengo tōshikeikaku o saikō suru: Takayama Eika no shōgai." *42nd NSRI toshi—kankyō fōramu* 282 (2011): 1–22.

Bartholomew, James R. *The Formation of Science in Japan: Building a Research Tradition*. New Haven, CT: Yale University Press, 1989.

Bates, Alex. *The Culture of the Quake: The Great Kanto Earthquake and Taishō Japan*. Ann Arbor: University of Michigan Center for Japanese Studies, 2015.

Beck, Ulrich. *Risk Society: Towards a New Modernity*. London: Sage Publications, 1992.

Beckert, Jens. *Imagined Futures: Fictional Expectations and Capitalist Dynamics*. Cambridge, MA: Harvard University Press, 2016.

Beroza, Gregory C., and William L. Ellsworth. "Properties of the Seismic Nucleation Phase." *Tectonophysics* 261, no. 1 (1996): 209–27.

Bolt, Bruce A. "Earthquake Studies in the People's Republic of China." *Eos, Transactions, American Geophysical Union* 55, no. 3 (1974): 108–17.

Bolton, Christopher, Komatsu Sakyō, Susan Napier, Tatsumi Takayuki, Kotani Mari, and Otobe Junko. "An Interview with Komatsu Sakyō." *Science Fiction Studies* 29, no. 3 (2002): 323–39.

Borland, Janet. *Earthquake Children: Building Resilience from the Ruins of Tokyo*. Cambridge, MA: Harvard University Asia Center, 2020.

Boudia, Soraya, and Nathalie Jas. "Introduction: Risk and 'Risk Society' in Historical Perspective." *History and Technology* 23, no. 4 (2007): 317–31.

Boyer, Paul S. *By the Bomb's Early Light: American Thought and Culture at the Dawn of the Atomic Age*. Chapel Hill: University of North Carolina Press, 1994.

Chūō Bōsai Kaigi. *Tōkai jishin taisaku taikō*. 2003. https://www.bousai.go.jp/jishin/tokai/pdf/taikou.pdf.

———. *1971-nen 2-gatsu 9-nichi Rosu Anzerusu jishin ni kansuru seifu gōdō chōsadan hōkoku (chūkan hōkoku)*. Tokyo: Chūō Bōsai Kaigi Jimukyoku, 1971.

Chūō Bōsai Kaigi Bōsai Taisaku Suishin Kentō Kaigi Nankai Torafu Kyōdai Jishin Taisaku Kentō Wākingu Gurūpu. *Nankai torafu kyodai jishin no higai sōtei ni tsuite (Dai ichiji hōkoku)*. 2012. https://www.bousai.go.jp/jishin/nankai/taisaku_wg/pdf/20120829_higai.pdf.

———. *Nankai torafu kyōdai jishin taisaku ni tsuite (saishū hōkoku)*. 2013. https://www.bousai.go.jp/jishin/nankai/taisaku_wg/pdf/20130528_honbun.pdf.

Chūō Bōsai Kaigi Shuto Chokka Jishin Taisaku Kentō Wākinggu Gurūpu. *Shuto chokka jishin no higai sōtei to taisaku ni tsuite (Saishū hōkoku)*. 2013. https://www.bousai.go.jp/jishin/syuto/taisaku_wg/pdf/syuto_wg_report.pdf.

Chūō Bōsai Kaigi, Saigai Kyōkun no Keishō ni Kansuru Senmon Chōsakai. *1959 Isewan Taifu hōkokusho*. 2008.

Clancey, Gregory K. *Earthquake Nation: The Cultural Politics of Japanese Seismicity, 1868–1930*. Berkeley: University of California Press, 2006.

Clarke, Lee Ben. *Mission Improbable: Using Fantasy Documents to Tame Disaster*. Chicago: University of Chicago Press, 1999.

Cornell, C. Allin. "Engineering Seismic Risk Analysis." *Bulletin of the Seismological Society of America* 58, no. 5 (1968): 1583–1606.

Davis, Mike. *Ecology of Fear: Los Angeles and the Imagination of Disaster*. 1st ed. New York: Metropolitan Books, 1998.

Dees, Bowen C. *The Allied Occupation and Japan's Economic Miracle: Building the Foundations of Japanese Science and Technology 1945–52*. Surrey, UK: Japan Library, 1997.

Disaster Relief Act (Japan), Provisional Translation, 1947, International Federation of Red Cross and Red Crescent Societies, http://www.ifrc.org/Docs/idrl/502EN.pdf.

Dower, John W. "'Ni' and 'F': Japan's Wartime Atomic Bomb Research." In *Japan in War and Peace: Selected Essays*, 55–100. New York: New Press, 1993.

———. *Embracing Defeat: Japan in the Wake of World War II*. 1st ed. New York: W. W. Norton, 1999.

Dudden, Alexis. "An Opportunity for Japan to Change People's Perception." *Asia-Pacific Journal: Japan Focus* 18, no. 5 (2020): 1–4.

Earthquake Prediction Research Group. *Prediction of Earthquakes: Progress to Date and Plans for Further Development*. Edited by Chuji Tsuboi, Kiyoo Wadati, and Takahiro Hagiwara. Tokyo: Kokusai Bunen Insatsusha, 1962.

Edwards, Paul N. *A Vast Machine: Computer Models, Climate Data, and the Politics of Global Warming*. Cambridge, MA: MIT Press, 2010.

Ehira Teruo. "Tokyo-to ni okeru bōsai toshi zukuri no kiseki to keifu—Kōtō bōsai kyoten o chūshin ni shite." *Kenchiku bōsai* 259 (1999): 24–29.

Enami, Akira. "Shell Structures." In *General Report on the Niigata Earthquake of 1964*, ed. Kawasumi Hiroshi, 283–309. Tokyo: Electrical Engineering College Press, 1968.

Engelhardt, H. Tristram, and Arthur L. Caplan. "Patterns of Controversy and Closure: The Interplay of Knowledge, Values, and Political Forces." In *Scientific Controversies: Case Studies in the Resolution and Closure of Disputes in Science and Technology*, ed. H. Tristram Engelhardt and Arthur L. Caplan, 1–23. Cambridge: Cambridge University Press, 1987.

Fan, Fa-ti. "'Collective Monitoring, Collective Defense': Science, Earthquakes, and Politics in Communist China." *Science in Context* 25, no. 1 (2012): 127–54.

———. "The People's War against Earthquakes: Cultures of Mass Science in Mao's China." In *Cultures Without Culturalism: The Making of Scientific Knowledge*, ed. Karine Chemla and Evelyn Fox Keller, 296–322. Durham, NC: Duke University Press, 2017.

Fedman, David A., and Cary Karacas. "A Cartographic Fade to Black: Mapping the Destruction of Urban Japan during World War II." *Journal of Historical Geography* (2012): 1–23.

Fitch, Thomas J., and Christopher H. Scholz. "Mechanism of Underthrusting in Southwest Japan: A Model of Convergent Plate Interactions." *Journal of Geophysical Research (1896–1977)* 76, no. 29 (1971): 7260–92.

Flüchter, Winfried. "Tokyo before the Next Earthquake: Agglomeration-Related Risks, Town Planning and Disaster Prevention." *Town Planning Review* 74, no. 2 (2003): 213–38.

Frohlich, Cliff. "Kiyoo Wadati and Early Research on Deep Focus Earthquakes: Introduction to Special Section on Deep and Intermediate Focus Earthquakes." *Journal of Geophysical Research* 92, no. B13 (1987): 13777–88.

Gardner, William O. "Narratives of Collapse and Generation: Komatsu Sakyō's Disaster Novels and the Metabolist Movement." *Japan Forum* 26, no. 3 (2014): 306–24.

Geller, Robert J. "Bōsai taisaku to jishin kagaku kenkyū no arikata: risettto no jiki." In *Jishingaku no ima o tō*, ed. Tōhoku Chihō Taiheiyō-oki Jishin Taio Rinji Iinkai, 5–8. Tokyo: Nihon Jishin Gakkai, 2012.

———. "Earthquake Prediction: a Critical Review." *Geophysical Journal International* 131 (1997): 425–50.

———. *Gerā-san Nippon ni mono mōsu*. Tokyo: Tōkyōdō Shuppan, 2018.

———. "Hon shimpojiumu kaisei ni kanshite no hanseiten." In *Nankai torafu jishin rinji jōhō: kagakuteki de-ta ya chiken to katsuyō*, ed. Nihon Jishin Gakkai, 6, 33–34. Tokyo: Nihon Jishin Gakkai, 2020.

———. "Jishin gakusha no shakaiteki kōken." *Nikkei saiensu* (1993): 3.

———. "Jishin yochi saikō." In *Jishin yochi kenkyū shinpojiumu (1994)*, ed. Nihon Gakujutsu Kaigi Jishingaku Kenkyū Renraku Iinkai and Nihon Jishin Gakkai, 131–40. Tokyo: Nihon Jishin Gakkai, 1994.

———. "Machigatta gakusetsu ni tayoru na." In *Jishin Chōsa Kenkyū Suishin Honbu 20-nen no shiryōshū*, ed. Jishin Yochi Sōgō Kenkyū Shinkōkai, 132–37. Tokyo: Monbukagakusho, 2015.

———. *Nihonjin wa shiranai 'Jishin yochi' no shōtai*. Tokyo: Futabasha, 2011.

———. "Problems of Tenure in Japan." *Nature* 345, no. 6274 (1990): 380.

———. "Sakete tōrenai deguchi sakusen." In *"Burupurinto" 50 shūnen jishin kenkyū no ayumi to kongo*, ed. Nihon Jishin Gakkai, 2, 19–21. Tokyo: Nihon Jishin Gakkai, 2013.

———. "Shake-up for Earthquake Prediction." *Nature* 352, no. 6333 (1991): 275–76.

George, Timothy S. *Minamata: Pollution and the Struggle for Democracy in Postwar Japan*. Harvard East Asian Monographs. Cambridge, MA: Harvard University Asia Center, 2001.

Giddens, Anthony. *Modernity and Self-Identity: Self and Society in the Late Modern Age*. Stanford, CA: Stanford University Press, 1991.

Gordon, Andrew. *Labor and Imperial Democracy in Prewar Japan*. Berkeley: University of California Press, 1991.

Hagiwara Takahiro. *Jishingaku hyakunen*. Tokyo: Tokyo Daigaku Shuppankai, 1982.

———. *Jishin yochi to saigai*. Tokyo: Maruzen, 1997.

———. "Naigai ni okeru jishin yochi kenkyū no genjō." In *Jishin yochi kenkyū shinpojiumu: (1967-nen 12-gatsu 1-nichi)*, 208. Tokyo: Nihon Gakujutsu Kaigi, 1968.

Hagiwara Takahiro and Rikitake Tsuneji. "Jishin yochi keikaku hossoku no koro." *Jishin jyānaru* 1 (1986): 11–25.

Hamada, Kazuo, and Robert J. Geller. "Unpredictable Earthquakes (Scientific Correspondence) (With Geller's Reply)." *Nature* 353, no. 71 (1991): 611–12.

Hamblin, Jacob Darwin, and Linda M. Richards. "Beyond the *Lucky Dragon*: Japanese Scientists and Fallout Discourse in the 1950s." *Historia scientiarum*. Second series 25, no. 1 (2015): 36–56.

Hanes, Jeffrey. "From Megalopolis to Megaroporisu." *Journal of Urban History* 19, no. 2 (1993): 56–94.

Harada Katsumasa, and Shiozaki Fumio. *Tokyo Kantō Daishinsai zengo*. Tokyo: Nihon Keizai Hyōronsha, 1997.

Harada Mitsuo. *Omoide no shichijūnen*. Tokyo: Seibundō Shinkōsha, 1966.

Hasegawa Akira. "Jishingaku kenkyūsha—jishingaku comyunitei no shakaiteki yakuwari." In *Jishingaku no ima o tō*, ed. Tōhoku Chihō Taiheiyō-oki Jishin Taio Rinji Iinkai, 18–22. Tokyo: Nihon Jishin gakkai, 2012.

Headquarters for Earthquake Research Promotion. *The Promotion of Earthquake Research: Basic Comprehensive Policy for the Promotion of Earthquake Observation, Measurement, Surveys and Research*. 1999. https://www.jishin.go.jp/main/w_060_f-e.htm.

Headquarters for Earthquake Research Promotion Earthquake Research Committee. *National Seismic Hazard Maps for Japan (2005)*. 2005. https://www.jishin.go.jp/main/chousa/06mar_yosoku-e/NationalSeismicHazardMaps.pdf.

Hearn, Lafcadio. *Gleanings in Buddha Fields*. Boston: Houghton, Mifflin and Company, 1897.

Hein, Carola, Jeffry M. Diefendorf, and Yorifusa Ishida. *Rebuilding Urban Japan After 1945*. New York: Palgrave Macmillan, 2003.

Hewitt, Kenneth. "The Idea of Calamity in a Technocratic Age." In *Interpretations of Calamity from the Viewpoint of Human Ecology*, ed. Kenneth Hewitt, 3–32. Boston: Allen & Unwin, 1983.

Hirata Naoshi. "Sekkyokuteki ni 'Keiken o tsumu' kenkyū o suishin shiyō." In *Jishin yochi kenkyū shinpojiumu (1994)*, 155–57. Tokyo: Nihon Jishin Gakkai, 1994.

Hiroi Osamu. "Niigata jishin to saigai jōhō (1)." *Gekkan shōbō* 9, no. 6 (1987): 26–33.

———. "Niigata jishin to saigai jōhō (2)." *Gekkan shōbō* 9, no. 7 (1987): 44–51.

Hiroyuki Fujiwara, Shinichi Kawai, Shin Aoi, Nobuyuki Morikawa, Shigeki Senna, Kyoko Kobayashi, Toru Ishii, Toshihiko Okumura, and Yuzuru Hayakawa. "National Seismic Hazard Maps of Japan." *Bulletin of the Earthquake Research Institute* 81 (2006): 221–32.

Honkura Yoshimoto. "Tsuitō—Rikitake Tsuneji sensei." *Nihon Jishin Gakkai nyūsureta* 16, no. 4 (2004): 18–19.

Horii, Kenichiro. "Highway Bridges." In *General Report on the Niigata Earthquake of 1964*, ed. Kawasumi Hiroshi, 431–50. Tokyo: Electrical Engineering College Press, 1968.

Horowitz, Andy, and Jacob A. C. Remes. "Introducing Critical Disaster Studies." In *Critical Disaster Studies*, ed. Andy Horowitz and Jacob A. C. Remes, 1–8. Philadelphia: University of Pennsylvania Press, 2021.

Hoshiba, Mitsuyuki. "Review of the Nationwide Earthquake Early Warning in Japan during Its First Five Years." In *Earthquake Hazard, Risk and Disasters*, ed. John F. Shroder and Max Wyss, 505–29. Boston: Academic Press, 2014.

Hough, Susan Elizabeth. *Predicting the Unpredictable: The Tumultuous Science of Earthquake Prediction*. Princeton, NJ: Princeton University Press, 2010.

———. *Richter's Scale: Measure of an Earthquake, Measure of a Man*. Princeton, NJ: Princeton University Press, 2007.

Hudnut, Kenneth W., Anne M. Wein, Dale A. Cox, Keith A. Porter, Laurie A. Johnson, Suzanne C. Perry, Jennifer L. Bruce, and Drew LaPointe. "The HayWired earthquake scenario—We can outsmart disaster." *Fact Sheet* 3016 (2018): 1–6.

Huizenga, John R. *Cold Fusion: The Scientific Fiasco of the Century*. Rochester, NY: University of Rochester Press, 1992.

Ide Satoshi. "Asuperitii—undōgata—jishin yochi." In *Jishingaku no ima o tō*, ed. Tōhoku Chihō Taiheiyō-oki Jishin Taio Rinji Iinkai, 14–17. Tokyo: Nihon Jishin Gakkai, 2012.

Igarashi, Yoshikuni. *Bodies of Memory: Narratives of War in Postwar Japanese Culture, 1945–1970*. Princeton, NJ: Princeton University Press, 2000.

———. "Mothra's Gigantic Egg: Consuming the South Pacific in 1960s Japan." In *In Godzilla's Footsteps: Japanese Pop Culture Icons on the Global Stage*, ed. William M. Tsutsui and Michiko Itō, 83–102. New York: Palgrave Macmillan, 2006.

Imai Makoto. "Saigai taisaku kihonhō ni tsuite (1)." *Jichi kenkyū* 37, no. 12 (1961): 87–101.

Imamura Akitsune. "Daijishin chōsa nikki." *Kagaku chishiki* (October 1923): 12–38.

———. "Daijishin chōsa nikki." *Kagaku chishiki* (November 1923): 76–95.

———. "'Inamura no hi' no oshiekata ni tsuite." *Jishin* 12, no. 8 (1940): 360–74.

———. *Jishin kōwa*. Tokyo: Iwanami Shoten, 1924.

———. *Jishin no kuni*. Tokyo: Bungei Shunjū Shinsha, 1949.

———. *Jishin no seifuku*. Tokyo: Nankōsha, 1926.

———. *Jishingaku*. Tokyo: Dai Nihon Tosho Kabushiki Kaisha, 1905.

———. "Konkai no daishinsai no keika o nobete shōrai ni oyobu." *Kaizō* 5, no. 10 (1923): 131–34.

———. *Namazu no zaregoto*. Tokyo: Sanseidō, 1941.

———. "Nankaidōoki daijishin no nazo." *Jishin* 5, no. 10 (1933): 607–26.

———. "On the Tiltings of the Earth Preceding the Kwanto Earthquake of 1923." *Proceedings of the Imperial Academy* 4, no. 4 (1928): 148–50.

———. "Shigaichi ni okeru jishin no seimei oyobi zaisan ni taisuru songai o keigen suru kanhō." *Taiyō* 11, no. 12 (1905): 162–71.

———. "Sono ato no daijishin chōsa." *Kagaku chishiki* (1924): 2–23.

———. "Supplementary Note on the Result of Precise Levelling in Etigo and Sinano." *Proceedings of the Imperial Academy* 4, no. 3 (1928): 109–11.

Isa Kyōzō. *Daini Kantō daijishin: sono toki dō suru*. Tokyo: Yomiuri shinbunsha, 1975.

Ishibashi Katsuhiko. *Daichi dōran no jidai: jishin gakusha wa keikokusuru*. Tokyo: Iwanami Shoten, 1994.

———. "Daijishin ga 'seikai toshi Tokyo' o chokugeki suru." *Chūō kōron* 102, no. 11 (1987): 124–38.

———. "Daishinhō haishi—jishin kanrenhō ipponka to, hassei yosoku no hojō shudanka o." In *Jishin hassei yosoku to daishinhō oyobi jishin bōsai kenkyū*, ed. Monogurafu 'Jishin Hassei Yosoku to Daishinhō Oyobi Jishin Bōsai Kenkyū' Henshū Iinkai, 5, 29–32. Tokyo: Nihon Jishin Gakkai, 2017.

———. "Genpatsu shinsai—hametsu o sakeru tame ni." *Kagaku* 67, no. 10 (1997): 720–24.

———. "'Jishin yochi keikaku' no kaitai-saihensei: Atarashi 'jishin yochi kenkyū keikaku' to 'jishin saigai keigen' no teian." In *Jishin yochi kenkyū shinpojiumu (1994)*, ed. Nihon Gakujutsu Kaigi Jishingaku Kenkyū Renraku Iinkai and Nihon Jishin Gakkai, 121–30. Tokyo: Nihon Jishin Gakkai, 1994.

———. "Kagakuteki 'jishin yochi' o mezashite." *Kagaku* 65, no. 9 (1995): 573–81.

———. "'Nankai torafu jishin rinji jōhō' taisei e no gimon." In *Nankai torafu jishin rinji jōhō: kagakuteki de-ta ya chiken to katsuyō*, ed. Nihon Jishin Gakkai, 6, 25–28. Tokyo: Nihon Jishin Gakkai, 2020.

———. *Nankai torafu kyodai jishin—rekishi, kagaku, shakai*. Tokyo: Iwanami Shoten, 2014.

———. "Specification of a Soon-to-Occur Seismic Faulting in the Tokai District, Central Japan, Based Upon Seisometrics." In *Earthquake Prediction: an International Review*, ed. David W. Simpson and Paul G. Richards, 297–332. Washington, DC: American Geophysical Union, 1981.

———. "Status of Historical Seismology in Japan." *Annals of Geophysics* 47, no. 2/3 (2004): 339–68.

———. "Surugawan jishinsetsu koshi." *Kagaku* 73, no. 9 (2003): 1057–64.

———. "Tōkai chihō ni yosō sareru daijishin no saikentō—Surugawan jishin no kanosei." *Jishin Yochi Renrakukai kaihō* 17, no. 4–13 (1977): 126–32.

Iwasaki Akira. "Atarashii media no tenkai." *Shisō* 624 (1976): 240–55.

Iwata Tomohiro. "Shuto chokka jishin higai sōtei no rosen tenkan." *Aera* (December 30, 2013): 34.

"Jishin fuan ni oriau," *Aera* (July 2, 2012): 25.

Jishin Yochi Renrakukai Jimukyoku. "Tokutei kansoku chiiki nado no sentei ni itaru keika." *Jishin Yochi Renrakukai kaihō* 3 (1970): 89–91.

Jishin Yochi Renrakukai, ed. *Jishin Yochi Renrakukai 10-nen no ayumi*. Tokyo: Kankōsha Kokudo Chiriin Nihon Sokuryō Kyōkai, 1979.

———. *Jishin Yochi Renrakukai yonjūnen no ayumi*. Tsukuba-shi, Ibaraki-ken: Kokudo Chiriin, 2009.

"Jishin yochi wa hontō ni kanō ka?" *Aera* (January 29, 1996): 26.

Joffe, Helene, Tiziana Rossetto, Caroline Bradley, and Cliodhna O'Connor. "Stigma in Science: The Case of Earthquake Prediction." *Disasters* 42, no. 1 (2018): 81–100.

Jones, Lucile M., Richard Bernknopf, Dale Cox, James Goltz, Kenneth Hudnut, Dennis Mileti, Suzanne Perry, Daniel Ponti, Keith Porter, Michael Reichle, Hope Seligson, Kimberley Shoaf, Jerry Treiman, and Anne Wein. "The ShakeOut Scenario: U.S. Geological Survey Open-File Report 2008–1150 and California Geological Survey Preliminary Report 25." 2008. http://pubs.usgs.gov/of/2008/1150/.

Jordan, Thomas H. "The Prediction Problems of Earthquake System Science." *Seismological Research Letters* 85, no. 4 (2014): 767–69.

Kagan, Yan Y., David D. Jackson, and Robert J. Geller. "Characteristic Earthquake Model, 1884–2011, R.I.P.," *Seismological Research Letters* 83, no. 6 (2012): 951–53.

Kamei, Kojiro. "Fire Following Earthquake in Niigata City." In *General Report on the Niigata Earthquake of 1964*, ed. Kawasumi Hiroshi, 385–401. Tokyo: Electrical Engineering College Press, 1968.

Kaneko Tsutomu. *Ainshutain shokku*. Tokyo: Kawade Shobō Shinsha, 1981.

Kapur, Nick. *Japan At the Crossroads: Conflict and Compromise After Anpo*. Cambridge, MA: Harvard University Press, 2018.

Kasza, Gregory J. "Democracy and the Founding of Japanese Public Radio." *Journal of Asian Studies* 45, no. 4 (1986): 745–67.

Katō Naoyuki. "'Jishin yochi' ni taisuri Nihon gakkai no torikumi." In *"Burupurinto" 50 shūnen jishin kenkyū no ayumi to kongo*, ed. Nihon Jishin Gakkai, 2, 45–48. Tokyo: Nihon Jishin Gakkai, 2013.

Kawabata Nobumasa. "Daikibo jishin taisaku tokubetsu sochihō seitei tōji no naibu jijyō." *Kagaku* 73, no. 9 (2003): 1044–50.

———. "Tōkai jishin bōsai taisei no seiritsu hiwa." *Nihon Saigai Jōhō Gakkai dai 11 kai benkyōkai* (2010): 1–25. http://www.jasdis.gr.jp/_userdata/07benkyokai/011-2.pdf.

Kawana, Sari. "Mad Scientists and Their Prey: Bioethics, Murder, and Fiction in Interwar Japan." *Journal of Japanese Studies* 31, no. 1 (2005): 89–120.

Kawasumi Hiroshi. "A. Imamura's Note on Successful Earthquake Prediction." In *The United States-Japan Conference on Research Related to Earthquake Prediction Problems*, ed. National Science Foundation and Nihon Gakujutsu Shinkōkai, 17–19. Tokyo: Earthquake Research Institute, 1964.

———. "Arasuka jishin no omoide." *Yōbō jihō* 71 (1967): 11–16.

———. "Crustal Movements in the Historical Records." In *The United States-Japan Conference on Research Related to Earthquake Prediction Problems*, ed. National Science Foundation and Nihon Gakujutsu Shinkōkai, 26–27. Tokyo: Earthquake Research Institute, 1964.

———. "Kantō nanbu jishin 69 nen shūki no shōmei to sono hassei no kinpakudo narabi ni taisaku no kinkyūsei to mondai ten." *Chigaku zasshi* 79, no. 3 (1970): 115–38.

———. "Kantō ni daijishin ga yatte kuru." *Nihon oyobi Nihonjin* 1489 (1970): 94–101.

———. "Measures of Earthquake Danger and Expectancy of Maximum Intensity Throughout Japan as Inferred from the Seismic Activity in Historical Times." *Bulletin of the Earthquake Research Institute* 29 (1951): 469–82.

———. "San Furunando (Rosanzerusu) Jishin to sono kunren." *Tsuchi to kiso* 19, no. 8 (1971): 34–40.

———. "Shindo to shindokai." *Jishin* 15, no. 1 (1943): 6–12.

———. "Shindo to shindokai." *Jishin* 15, no. 8 (1943): 1–7.

Kawata Yoshiaki. "Saigai taisaku kihonhō to bōsai kihon keikaku." *Shizen saigai kagaku* 15 (1996): 81–92.

Kazama Norio. "Saigai taisaku kihonhō no seitei: bōsai seisaku netowaku no keisei." *Kinki daigaku hōgaku* 50, no. 1 (2002): 1–82.

Kilb, D., and J. Gomberg. "The Initial Subevent of the 1994 Northridge, California, Earthquake." *Journal of Seismology* 3, no. 4 (1999): 409–20.

Kim Boumsoung. *Meiji Taishō no Nihon no jishingaku: "rōkaru saiensu" o koete*. Tokyo: Tōkyō Daigaku Shuppankai, 2007.

Kimura Kōzō. *Sanriku e nigeru*. Tokyo: Futami Shobō, 1974.

Kisslinger, Carl, Takeshi Mikumo, and Hiroo Kanamori. "U.S.-Japan Quake Prediction Research." *Eos, Transactions, American Geophysical Union* 69, no. 52 (1988): 1672–74.

Kitahara Itoko. *Jishin no shakaishi: Ansei daijishin to minshū*. Kōdansha gakujutsu bunko, Tokyo: Kōdansha, 2000.

———. *Nihon saigaishi*. Tokyo: Yoshikawa Kōbunkan, 2006.

Kitahara Itoko, Matsuura Ritsuko, and Kimura Reo, eds. *Nihon rekishi saigai jiten*. Tokyo: Yoshikawa Kōbunkan, 2012.

Knowles, Scott Gabriel, and Zachary Loeb. "The Voyage of the Paragon: Disaster as Method." In *Critical Disaster Studies*, ed. Jacob A. C. Remes, and Andy Horowitz, 11–31. Philadelphia: University of Pennsylvania Press, 2021.

Kobayashi Tadashi. *Terada Torahiko to jishin yochi*. Tokyo: Tokyo Tosho, 2003.

Koiwai Tadamichi. "Ryōiki no kakudai motomete kagaku kisha yonjūnen." *Shuzai nōto*. 2014. https://www.jnpc.or.jp/journal/interviews/29889/.

Kōketsu Kazuki. "Jishin no kagaku no yukue wa? Higashi Nihon Daishinsai kara ikagetsu, jishin gakusha wa nani o omou ka." *Tokushū: Kenkyūsha intabyū dai 8-kai* (2011): 1–6.

Koketsu, Kazuki, Yusuke Yokota, Naoki Nishimura, Yuji Yagi, Shin'ichi Miyazaki, Kenji Satake, Yushiro Fujii, Hiroe Miyake, Shin'ichi Sakai, Yoshiko Yamanaka, and Tomomi Okada. "A Unified Source Model for the 2011 Tohoku Earthquake." *Earth and Planetary Science Letters* 310, no. 3–4 (2011): 480–87.

Kokudochō, ed. *Bōsai hakusho, Shōwa 56 nenban*. Tōkyō: Ōkurashō Insatsukyoku, 1981.

Kokudochō, ed. *Bōsai hakusho, Shōwa 58 nenban*. Tōkyō: Ōkurashō Insatsukyoku, 1983.

Komatsu Sakyō. *Nihon chinbotsu*. Tokyo: Kobunsha, 1973.

Koshizawa Akira. *Tokyo toshi keikaku no isan: bōsai, fukkō, Orinpikku*. Tokyo: Kabushiki Kaisha Chikuma Shobō, 2014.

Kossobokov, Vladimir G. "Earthquake Prediction: 20 years of Global Experiment." *Natural Hazards* 69, no. 2 (2013): 1155–77.

Kumagai Yoshio. "Jishinkasai no tokuchō to gensai ni mukete." *Yobō jihō* 230 (2007): 36–41.

Kumagai Yoshio and Nojima Yoshiteru. "Urbanization and Disaster Mitigation in Tokyo." In *Crucibles of Hazard: Mega-cities and Disasters in Transition*, ed. James K. Mitchell, 56–91. Tokyo: United Nations University Press, 1999.

Kurosaki, Akira. "Japanese Scientists' Critique of Nuclear Deterrence Theory and Its Influence on Pugwash, 1954–1964." *Journal of Cold War Studies* 20, no. 1 (2018): 101–39.

Kurosawa Tairiku. *'Jishin yochi' no gensō*. Tokyo: Shinchōsha, 2014.

Kuwano Shigeru, dir. *Jishin yochi e no michi: Matsushiro jishin o megutte*. Tokyo: Iwanami Productions, 1967.

Langer, Jessica. "Three Versions of Komatsu Sakyō's *Nihon chinbotsu* (Japan Sinks)." *Science Fiction Film and Television* 2, no. 1 (2009): 45–57.

Levy, Jonathan. *Freaks of Fortune: The Emerging World of Capitalism and Risk in America*. Cambridge, MA: Harvard University Press, 2012.

———. "Radical Uncertainty." *Critical Quarterly* 62, no. 1 (2020): 15–28.

Lewenstein, Bruce V. "From Fax to Facts: Communication in the Cold Fusion Saga." *Social Studies of Science* 25, no. 3 (1995): 403–36.

———. "Science Controversies: Can the Science of Science Communication Provide Management Guidance or Only Analysis?" In *The Oxford Handbook of the Science of Science Communication*, ed. Kathleen Hall Jamieson, Dan M. Kahan, and Dietram A. Scheufele, 73–78. Oxford: Oxford University Press, 2017.

Low, Morris. "Japan: The Political Economy of Japanese Science: Nakasone, Physicists, and the State." In *Scientists and the State: Domestic Structures and the International Context*, ed. Etel Solingen, 93–125. Ann Arbor: University of Michigan Press, 1994.

———. *Science and the Building of a New Japan*. New York: Palgrave Macmillan, 2005.

Mack, Edward Thomas. *Manufacturing Modern Japanese Literature: Publishing, Prizes, and the Ascription of Literary Value*. Durham, NC: Duke University Press, 2010.

Marshall, Amy Bliss. *Magazines and the Making of Mass Culture in Japan*. Toronto: University of Toronto Press, 2019.

Marzocchi, Warner, and J. Douglas Zechar. "Earthquake Forecasting and Earthquake Prediction: Different Approaches for Obtaining the Best Model." *Seismological Research Letters* 82, no. 3 (2011): 442–48.

Matsuzawa Toru. "M9 o sōtei suru tame ni nani ga kaketeita no ka? Kongo dō sureba yoi no ka?" In *Jishingaku no ima o tō*, ed. Tōhoku Chihō Taiheiyō-oki Jishin Taio Rinji Iinkai, 9–13. Tokyo: Nihon Jisshin Gakkai, 2012.

"Meteorological Service Act." Japanese Law Translation Database System. https://www.japaneselawtranslation.go.jp/en/laws/view/1968/en

Mizuide Kōki. "'Bōsai no hi' o meguru saigai no kioku." *Masu komyunikeshun kenkyū* 88 (2016): 157–75.

———. *'Saigo' no kiokushi: Medeiya ni miru Kantō daishinsai—Isewan taifū*. Tokyo: Jinbunshoin, 2019.

Mizuno, Hiromi. *Science for the Empire: Scientific Nationalism in Modern Japan*. Stanford, CA: Stanford University Press, 2009.

Mizuno Hiroo. "Jishin yochi wa shakai no mukete dono yo ni hanashikakarareta kita ka—Kantō jishin kara Tōkai jishin made." *Kagakushi kenkyū* 43 (2004): 239–42.

Mochizuki Toshio and Nakabayashi Itsuki, eds. *Anzen anshin no toshizukuri: anzen de anshinshite sumeru iezukuri machizukuri no tameni*. Tokyo: Tokyo Toritsu Daigaku Shuppankai, 2001.

Mogi Kiyō. "Fujūbun na taisei minaosu kikai ni." *Asahi shinbun*. March 29, 1996.

———. *Jishin yochi o kangaeru*. Tokyo: Iwanami Shoten, 1998.

———. "Two Grave Issues Concerning the Expected Tokai Earthquake." *Earth Planets Space* 56 (2004): li–lxvi.

Mohun, Arwen P. "Constructing the History of Risk. Foundations, Tools, and Reasons Why." *Historical Social Research / Historische Sozialforschung* 41, no. 1 (2016): 30–47.

Molella, Arthur P., and Scott Gabriel Knowles. *World's Fairs in the Cold War: Science, Technology, and the Culture of Progress*. Pittsburgh, PA: University of Pittsburgh Press, 2019.

Moret, Leuren. "Japan's Deadly Game of Nuclear Roulette." *The Asia-Pacific Journal: Japan Focus* 10, no. 54 (2012).

Morimoto Ryōhei. "Medo tatanu jishin yochi." *Sankei shinbun*. August 31, 1994.

Mulargia, Francesco, Philip B. Stark, and Robert J. Geller. "Why is Probabilistic Seismic Hazard Analysis (PSHA) Still Used?" *Physics of the Earth and Planetary Interiors* 264 (2017): 63–75.

Muneaki Tajiri. "Rokka-kuromu jiken no honshitsu to kadai." *Kōgai kenkyū* 7 (1978): 51–57.

Murakami Suminao. "Dai kibō kaihatsu no jitsurei to shisutemu." *Kenchiku zasshi* 85, no. 1028 (1970): 669–73.

———. "Goaisatsu (Bōsai Toshi Keikaku Kenkyūjo)." Laboratory of Urban Safety Planning. http://www.lusp.co.jp/corp1_3.html.

———. "Kōtō chiku bōsai saikaihatsu keikaku ni tsuite." *Toshi keikaku* 60 (1969): 145–52.

———. "Toshi bōsai to keikaku." *Kenchiku zashi* 82, no. 987 (1967): 680–86.

Musha Kinkichi. *Jishin namazu*. Tokyo: Akashi Shoten, 1995.

Naikakufu Daijin Kanbō Seifu Kōhōshitsu. *Daitoshi jūmin no bōsai ishiki ni kansuru yoron chōsa*. 1971. https://survey.gov-online.go.jp/s46/S46-06-46-01.html.

———. *Shōbō gyōsei (dai ichibu) ni kansuru yoron chōsa*. 1952. https://survey.gov-online.go.jp/s26/S26-04-26-01_chosahyo.html.

———. *Shōbō—jishin ni kansuru yoron chōsa*. 1978, 1–19. https://survey.gov-online.go.jp/s53/S53-10-53-10.html.

———. *Shōbō—jishin ni kansuru yoron chōsa*. 1980. https://survey.gov-online.go.jp/s55/S55-09-55-12.html.

———. *Shōbō ni kansuru yoron chōsa*. 1976. https://survey.gov-online.go.jp/s51/S51-07-51-05.html.

Nakabayashi Itsuki. "Development of Urban Disaster Prevention Systems in Japan - from the Mid-1980s." *Journal of Disaster Research* 1, no. 1 (2006): 46–71.

———. "Shuto chokka jishin taisaku no kihon hōkō." *Jishin jyānaru* no. 57 (2014): 1–13.

———. "Tokyo ni okeru jishin kiken no jizen hyōka to jishin ni tsuyoi toshi zukuri." *Toshi kagaku kenkyū* 2 (2008): 69–81.

———. "Tokyo ni okeru jishin saigai no yōsō to shinsaitaisaku no tenkai." *Chigaku zasshi* 110, no. 6 (2001): 938–49.

———. "Tokyo-to shinsai yobō jōrei no kasseika o." *Chiri* 33, no. 6 (1988): 8–11.

Nakabayashi Itsuki and Uchida Muneharu. "Intabyū—Shuto chokka jishin no "sōteigai" o sōtei suru." *Tokyojin* 26, no. 7 (2011): 18–25.

Nakayama, Shigeru. "The American Occupation and the Science Council of Japan." In *The Orientation of Science and Technology: A Japanese View*, Volume 3, ed. Shigeru Nakayama, 222–37. Folkestone: Brill, 2009.

Nakayama, Shigeru, Kunio Gotō, and Hitoshi Yoshioka. *A Social History of Science and Technology in Contemporary Japan, Volume 4: Transformation Period 1970–1979.* Japanese Society Series. Melbourne, Australia: Trans Pacific Press, 2001.

Nankai Torafuzoi no Daikibō Jishin no Yosaku Kanōsei ni Kansuru Chōsabukai. *Nankai torafuzoi no daikibō jishin no yosaku kanōsei ni tsuite.* 2013. 1–11. https://www.bousai.go.jp/jishin/nankai/taisaku_wg/pdf/20130528_houkoku_s3.pdf.

Napier, Susan J. "Panic Sites: The Japanese Imagination of Disaster from Godzilla to Akira." *Journal of Japanese Studies* 19, no. 2 (1993): 327–51.

———. "When Godzilla Speaks." In *In Godzilla's Footsteps: Japanese Pop Culture Icons on the Global Stage*, ed. William M. Tsutsui and Michiko Itō, 9–19. New York: Palgrave Macmillan, 2006.

Narita Ryūichi. "Kantō daishinsai no metahisutori no tame ni." *Shisō* 866 (1996): 61–90.

National Science Foundation and Nihon Gakujutsu Shinkōkai, eds. *The United States-Japan Conference on Research Related to Earthquake Prediction Problems.* Tokyo: Earthquake Research Institute, 1964.

Newren, Edward F. *Japanese Science Films: A Descriptive and Evaluative Catalog of 16mm Motion Pictures, 8mm Cartridges, and Video Tapes.* Bethesda, MD: American Science Films Association, 1973.

NHK Shakaibu. *Jishin ni sonaeru: anata no bōsai taisaku.* Tokyo: Nihon Hōsō Shuppan Kyōkai, 1971.

Nigg, Joanne M. "Predicting Earthquakes: Science, Pseudoscience, and Public Policy Paradox." In *Prediction: Science, Decision Making, and the Future of Nature*, ed. Daniel R. Sarewitz, Roger A. Pielke, and Radford Byerly, 135–56. Washington, DC: Island Press, 2000.

Nihon Gakujutsu Kaigi Jishingaku Kenkyū Renraku Iinkai and Nihon Jishin Gakkai, eds. *Jishin yochi kenkyū shinpojiumu (1994).* Tokyo: Nihon Jishin Gakkai, 1994.

Nishimura Kiyoshi, dir. *Tokyo Daijishin: Magunichūdo 8.1.* Yomiuri Terebi,1980.

Normile, Denis. "Japan Holds Firm to Shaky Science." *Science* 264 (1994): 1656–58.

Nosengo, Nicola. "Scientists Face Trial Over Earthquake Deaths." *Nature*. May 26, 2011. https://doi.org/10.1038/news.2011.325.

O'Bryan, Scott. *The Growth Idea: Purpose and Prosperity in Postwar Japan*. Honolulu: University of Hawaii Press, 2009.

Oakes, Guy. *The Imaginary War: Civil Defense and American Cold War Culture*. New York: Oxford University Press, 1994.

Ogawa Yujiro. "Toshi bōsai no genjō to hyōka." *Sōgō toshi kenkyū* 47 (1992): 91–109.

Ohsaki, Yorihiko. "Building Damage and Soil Condition." In *General Report on the Niigata Earthquake of 1964*, ed. Kawasumi Hiroshi, 354–83. Tokyo: Electrical Engineering College Press, 1968.

Oike Kazuo, Shichi Ryūichi, and Asada Toshi. "Chūgoku ni okeru jishin yochi (Tokubetsu hōkoku)." *Zisin: Journal of the Seismological Society of Japan* 28, no. 1 (1975): 75–94.

Ōkubo Maki. "Tatakau jishin gakusha, mono iu kinzoku zairyō gakusha." *Ronza*, September 2012.

Oliver, Jack. "Earthquake Prediction." *Science* 144, no. 3624 (1964): 1364–65.

Ōmori Fusakichi. "Jishin no fukuin ni tsukite." *Shinsai Yobō Chōsakai hōkoku* 68 (1910): 21–29.

———. "Jishin no toki bunpu." *Shinsai yobō chōsakai hōkoku* 57 (1907): 38–42.

———. "Jishin to tenki to no kankei ni tsukite." *Tōyō gakugei zasshi* 303 (1906): 320–25.

———. "Notes on the Secondary Causes of Earthquakes." *Bulletin of the Imperial Earthquake Investigation Committee* 2, no. 2 (1908): 101–35.

———. "Tokyo ni okurubeki shōrai no jishin ni tsukite." *Shinsai Yobō Chōsakai hōkoku* 57 (1907): 2–17.

———. "Tokyo to daijishin no fusetsu." *Taiyō* 12, no. 4 (1906): 173–76.

Ōmori Kenjirō, dir. *Jishin rettō*. Tōei Eiga, 1980.

Onaga, Lisa. "Measuring the Particular: The Meanings of Low-Dose Radiation Experiments in Post-1954 Japan." *positions: asia critique* 26, no. 2 (2018): 265–304.

———. "Reconstructing the Linear No-Threshold Model in Japan: A Historical Perspective on the Technics of Evaluating Radiation Exposure." *Technology and Culture* 58 (2017): 194–204.

Oreskes, Naomi. "Why Predict? Historical Perspectives on Prediction in Earth Science." In *Prediction: Science, Decision Making, and the Future of Nature*, ed. Daniel R. Sarewitz, Roger A. Pielke, and Radford Byerly, 23–40. Washington, DC: Island Press, 2000.

Paci-Green, Rebekah, Natalie Boles, Tyler Black, Karissa Smith, and Munkh-Erdene, Enkhbayar. "Cascadia Rising: Cascadia Subduction Zone Catastrophic Earthquake and Tsunami Functional Exercise Scenario Document." *Environmental Studies Faculty and Staff Publications* 28 (2015). https://cedar.wwu.edu/envs_facpubs/28.

Panel on Earthquake Prediction of the Committee on Seismology, Assembly of Mathematical and Physical Sciences, National Research Council. *Predicting Earthquakes: A Scientific and Technical Evaluation—with Implications for Society*. Washington, DC: National Academy of Sciences, 1976.

Panel on the Public Policy Implications of Earthquake Prediction of the Advisory Committee on Emergency Planning, Commission on Sociotechnical Systems, National Research Council. *Earthquake Prediction and Public Policy*. Washington, DC: National Academy of Sciences, 1975.

Partner, Simon. *Assembled in Japan: Electrical Goods and the Making of the Japanese Consumer*. Berkeley: University of California Press, 1999.

Pietruska, Jamie L. *Looking Forward: Prediction and Uncertainty in Modern America*. University of Chicago Press, 2017.

Press, Frank. "The 10-Year Program for Earthquake Prediction Research in the U.S." In *Proceedings of the Second United States–Japan Conference on Research Related to Earthquake Prediction*, ed. Robert Page, 8–12. Palisades, NY: Lamont Geological Observatory, 1966.

Pritchard, Matthew E., Richard M. Allen, Thorsten W. Becker, Mark D. Behn, Emily E. Brodsky, Roland Bürgmann, Cindy Ebinger, Jeff T. Freymueller, Matt Gerstenberger, Bruce Haines, Yoshihiro Kaneko, Steve D. Jacobsen, Nate Lindsey, Jeff J. McGuire, Morgan Page, Sergio Ruiz, Maya Tolstoy, Laura Wallace, William R. Walter, William Wilcock, and Harold Vincent. "New Opportunities to Study Earthquake Precursors." *Seismological Research Letters* 91, no. 5 (2020): 2444–47.

Reitherman, Robert K. *Earthquakes and Engineers: An International History*. Reston, VA: American Society of Civil Engineers, 2012.

Rikitake Tsuneji. "Daijishin wa tsune ni yochi sarete iru." *Chūō kōron* 88, no. 9 (1973): 176–82.

———. *Earthquake Forecasting and Warning*. Tokyo: Center for Academic Publications Japan, 1982.

———. "Focal processes and the prediction of earthquakes: Probability of earthquake occurrence as estimated from crustal strain." *Tectonophysics* 23, no. 3 (1974): 299–312.

———. "Jishin yochi wa yumemonogatari de wa nai." *Kagaku Asahi* 38, no. 11 (1978): 72–76.

———. "Nenji keikaku no shinten jōkyō." In *Jishin yochi kenkyū shinpojiumu: (1967-nen 12-gatsu 1-nichi)*, 9–11. Tokyo: Nihon Gakujutsu Kaigi, 1968.

Ritsema, Jeroen, Thorne Lay, and Hiroo Kanamori. "The 2011 Tohoku Earthquake." *Elements* 8 (2012): 183–88.

Rose, Kenneth D. *One Nation Underground: The Fallout Shelter in American Culture*. New York: New York University Press, 2001.

Sabin, Paul. *The Bet: Paul Ehrlich, Julian Simon, and Our Gamble over Earth's Future*. New Haven, CT: Yale University Press, 2013.

Saegusa, Asako. "Japan to Try to Understand Quakes, Not Predict Them." *Nature* 397 (1999): 284.

———. "Japan's Quake Strategy Urged to Switch to Long-term Forecasting." *Nature* 393, no. 6682 (1998): 202.

Sagiya, Takeshi, Hiroo Kanamori, Yuji Yagi, Masumi Yamada, and Jim Mori. "Rebuilding Seismology." *Nature* 472 (2011): 146–48.

Samuels, Richard J. *3.11: Disaster and Change in Japan*. Ithaca, NY: Cornell University Press, 2013.

Satō Takumi. *"Kingu" no jidai: kokumin taishū zasshi no kōkyōsei*. Tokyo: Iwanami Shoten, 2002.

Schencking, J. Charles. "1923 Tokyo as a Devastated War and Occupation Zone: The Catastrophe One Confronted in Post Earthquake Japan." *Japanese Studies* 29, no. 1 (2009): 111–29.

———. *The Great Kantō Earthquake and the Chimera of National Reconstruction in Japan*. New York: Columbia University Press, 2013.

Scholz, C. H. "Kono fukeiki na zenchō." *Chūō kōron* 88, no. 9 (1973): 169–75.

Scholz, Christopher H., Lynn R. Sykes, and Yash P. Aggarwal. "Earthquake Prediction: A Physical Basis." *Science* 181, no. 4102 (1973): 803–10.

Sekino Masaru, ed. *Kōtokuin kokuhō dōzō Amida nyorai zazō shūri kōji hōkokusho*. Kamakura: Kōtokuin, 1961.

Shimamura Hideki. "Daishinhō no gensō." *Ekonomisuto* 82, no. 67 (2004): 33.

———. *'Jishin yochi' wa usodarake*. Tokyo: Kōdansha, 2008.

———. *Kōnin "jishin yochi" o utagau*. Tokyo: Kashiwa Shobo, 2004.

Shinsai Yobō Chōsakai. *Dainihon jishin shiryō*. Tokyo: Maruzen, 1904.

Shizuoka Shinbunsha, ed. *Chinmoku no Surugawan*. Shizuoka-shi: Shizuoka Shinbunsha, 2018.

Shizuoka-ken Kikikanribu Kiki Jōhōka. *Nankai torafu jishin (Tōkai jishin) ni tsuite no kenmin ishiki chōsa*. Jishin taisaku shiryō, vol. 290, 2017.

Shizuoka-ken Kikikanribu Kiki Jōhōka. *Reiwa gannen-do Nankai torafu jishin ni tsuite no kenmin ishiki chōsa*. Jishin taisaku shiryō, vol. 306, 2020.

Shizuoka-ken Kikikanribu Kiki Jōhōshitsu. *Heisei 21-nendo Tōkai jishin ni tsuite no kenmin ishiki chōsa*. Jishin taisaku shiryō, vol. 242, 2010.

Shōbō Shingikai. *Tokyo chihō (Kantō chihō minamibu) ni okeru daishinsai taisaku ni kansuru tōshin*. 1970, 69–165. https://www.fdma.go.jp/singi_kento/singi/council/.

Siniawer, Eiko Maruko. *Waste: Consuming Postwar Japan*. Ithaca, NY: Cornell University Press, 2018.

Smith, Kerry. "Earthquake Prediction in Occupied Japan." *Historical Social Research* 40, no. 2 (2015): 105–33.

Smits, Gregory. "Probabilistic Earthquake Hazard Maps." In *Cartographic Japan: a History in Maps*, ed. Kären Wigen, Fumiko Sugimoto, and Cary Karacas, 226–28. Chicago: University of Chicago Press, 2016.

———. *Seismic Japan: The Long History and Continuing Legacy of the Ansei Edo Earthquake*. Honolulu: University of Hawaii Press, 2013.

———. *When the Earth Roars: Lessons from the History of Earthquakes in Japan*. Lanham, MD: Rowman & Littlefield, 2014.

Sokuchigaku Shingikai. "Dai niji jishin yochi keikaku (Showa 44-nen—48-nen)." *Bunjutsusoku* 9 (1968): 1–9.

———. "Dai sanji jishin yochi keikaku no ichibu minaoshi ni tsuite." *Gakujutsu geppō* 28, no. 7 (1975): 490–493.

Sokuchigaku Shingikai and Jishin Kazan Bukai. *Jishin yochi keikaku no jisshi jōkyo nado no rebyū ni tsuite (Hōkoku)*. Tokyo: Sokuchigaku Shingikai, 1997.

Sorensen, André. *The Making of Urban Japan: Cities and Planning From Edo to the Twenty-First Century*. Nissan Institute/Routledge Japanese Studies Series. London: Routledge, 2002.

Spiegelhalter, D. J. *The Art of Statistics: How to Learn from Data*. First US ed. New York: Basic Books, 2019.

Stallings, Robert A. *Promoting Risk: Constructing the Earthquake Threat*. New York: De Gruyter, 1995.

Stein, Seth, Robert J. Geller, and Mian Liu. "Why Earthquake Hazard Maps Often Fail and What to Do About It." *Tectonophysics* 562–563 (2012): 1–25.

Steinberg, Theodore. *Acts of God: The Unnatural History of Natural Disaster in America*. New York: Oxford University Press, 2000.

Suzuki, Yasumoto. "Kiyoo Wadati and the Path to the Discovery of the Intermediate-deep Earthquake Zone." *Episodes* 24, no. 2 (2001): 118–24.

Swinbanks, David. "Critics Attack Hasty Review of Quake Prediction Efforts." *Nature* 374, no. 6525 (1995): 754.

———. "Earthquake Prediction 'Likely to Grow' After New Review." *Nature* 374, no. 6519 (1995): 205.

———. "Japanese Science Agency Loses Iron Lady." *Nature* 376, no. 6541 (1995): 542.

———. "Kobe Disaster Divides Earthquake Researchers." *Nature* 373 (1995): 373.

———. "Quake Panel Admits Prediction is 'Difficult.'" *Nature* 388, no. 6637 (1997): 4.

———. "Trying to Shake Japan's Faith in Forecasts." *Nature* 356 (1992): 464–465.

Takemura Masayuki. *Kantō Daishinsai: dai Tokyo-ken no yure o shiru*. Tokyo: Kajima Shuppankai, 2003.

———. *Mizō no daisaigai to jishingaku: Kantō Daishinsai*. Tokyo: Kokonshoin, 2009.

Takeuchi Hitoshi. "'Nihon Dajishin' ni sonaeru." *Keizai hyōron* 25, no. 2 (1976): 84–95.

———. *Jishin no kagaku*. Tokyo: Nihon Hoso Shuppan Kyokai, 1973.

———. "Jishin yochi wa jishin taisaku ni arazu." *Keizai hyōron* 26, no. 2 (1977): 122–33.

———. "Jishin yochi yori ido no setsubi o." *Kagaku Asahi* 38, no. 9 (1978): 81–86.

Taubes, Gary. *Bad Science: The Short Life and Weird Times of Cold Fusion*. 1st ed. New York: Random House, 1993.

Terada Torahiko. "Jishin zakkan." In *Taishō Daishin-kasai shi*, 51–54. Tokyo: Kaizōsha, 1924.

———. "Luminous Phenomena Accompanying Destructive Sea-waves (Tunami)." *Proceedings of the Imperial Academy* 9, no. 8 (1933): 367–69.

———. "On a Measure of Uncertainty Regarding the Prediction of Earthquake Based on Statistics." *Proceedings of the Imperial Academy* 9, no. 6 (1933): 255–57.

Tierney, Kathleen J. "From the Margins to the Mainstream? Disaster Research at the Crossroads." *Annual Review of Sociology* 33 (2007): 503–25.

"Tōkai jishin keisei sengen hatsurei wa naze konnan ka," *Aera* (April 1, 1996): 18.

Tokyo Daigaku Hyakunenshi Henshū Iinkai, ed. *Tōkyō Daigaku hyakunenshi, bukyokushi (3)*. Tokyo: Tokyo Daigaku Shuppankai, 1987.

Tokyo Daigaku Shinbun Kenkyūjo, ed. *Jishin yochi to shakaiteki hannō*. Tokyo: Tokyo Daigaku Shuppankai, 1979.

Tokyo Shōbōchō. *Tokyo-to no daishinkasai higai no kentō: taisaku ni taisuru shiryō, dai ippō*. Tokyo: Tokyo Shōbōchō, 1961.

———. *Tokyo-to no daishin kasai higai no kentō: taisaku ni taisuru shiryō, dai nippō*. Tokyo: Tokyo Shōbōcho, 1967.

Tokyo Shōbōcho Shinsai Taisaku Honbu. *Anata no jishin taisaku—Tokyo no shinsai*. Tokyo: Tokyo Shōbōcho Shinsai Taisaku Honbu, 1969.

"Tokyo-to shinsai yobō jōrei (zenbun)." *Zaisei keizai kōhō*, no. 1434 (November 8, 1971): 9–10.

Tomari Jirō. "Beijin gakusha no ronbun ni "yureru" Jishingakkai." *Aera* (December 3, 1991): 66.

———. *Nihon no jishin yochi kenkyū 130-nenshi: Meijiki kara Higashi Nihon Daishinsai made*. Shohan. ed. Tokyo: Tokyo Daigaku Shuppankai, 2015.

———. *Purēto tekutonikusu no kyozetsu to juyō: sengo Nihon no chikyū kagakushi*. Shohan ed. Tokyo: Tokyo Daigaku Shuppankai, 2008.

Tsuboi Chūji. "Ōmori Fusakichi to Nihon no jishingaku." *Chūō kōron* 80, no. 7 (1965): 383–88.

Tsuchida Hiroshige. *Kindai Nihon no "kokumin bōkū" taisei*. Shohan ed. Chiba-shi: Kanda Gaigo Daigaku Shuppankyoku, 2010.

———. "Nijū seiki Nihon no bōsai." *Shigaku zasshi* 127, no. 6 (2018): 6–19.

Tsukui Susumu. *Daisaigai to hō*. Tokyo: Iwanami Shoten, 2012.

Tsuru, Shigeto. *The Political Economy of the Environment: The Case of Japan*. London: Athlone Press, 1999.

Tsutsui, William M. *Godzilla on My Mind: Fifty Years of the King of Monsters*. New York: Palgrave Macmillan, 2004.

———. "Oh No, There Goes Tokyo: Recreational Apocalypse and the City in Postwar Japanese Popular Culture." In *Noir Urbanisms: Dystopic Images of the Modern City*, ed. Gyan Prakash, 104–26. Princeton, NJ: Princeton University Press, 2010.

Tsutsui, William M., and Michiko Itō, eds. *In Godzilla's Footsteps: Japanese Pop Culture Icons on the Global Stage*. New York: Palgrave Macmillan, 2006.

Uchihashi Katsuto and Ishibashi Katsuhiko. "Taidan: 'Mō hitotsu no Nihon' koso ga kongenteki na shinsai taisaku da." *Sekai* 721 (2003): 194–203.

Upham, Frank K. "Unplaced Persons and Movements for Place." In *Postwar Japan as History*, ed. Andrew Gordon, 325–46. Berkeley: University of California Press, 1992.

Usami Tatsuo. "Earthquake Studies and the Earthquake Prediction System in Japan." *Journal of Disaster Research* 1, no. 3 (2006): 416–33.

———. "Great Earthquakes of the Past." In *Earthquake Prediction Techniques: Their Application in Japan*, ed. Asada Toshi, 11–29. Tokyo: University of Tokyo Press, 1982.

———. "Study of Historical Earthquakes in Japan." *Bulletin of the Earthquake Research Institute* 54 (1979): 399–439.

———. "Tokyo ga shindo V ijō no jishin ni osowareru kakuritsu." *Bulletin of the Earthquake Research Institute* 48 (1970): 331–40.

Utsu Tokuji. "Hokkaido shūhen ni okeru daijishin no katsudo to Nemuro nanpō-oki jishin ni tsuite." *Jishin Yochi Renrakukai kaiho* 7 (1972): 7–13.

"Wadati Kiyō," *Asahi gurafu* 1423 (1951): 18.

Walker, Brett L. *Toxic Archipelago: A History of Industrial Disease in Japan*. Seattle: University of Washington Press, 2010.

Wang, Kelin, Qi-Fu Chen, Shihong Sun, and Andong Wang. "Predicting the 1975 Haicheng Earthquake." *Bulletin of the Seismological Society of America* 96, no. 3 (2006): 757–95.

Watanabe Ichirō, Aoki Sukehisa, Yagi Tsuruhei, and Ōtani Keiichi, eds. *Bōsai kankei hōrei (Eibun)*. Vol. 1–352, Bōsai Kagaku Gijutsu Kenkyūjo kenkyu shiryō. Tsukuba: Kokuritsu Bōsai Kagaku Gijutsu Senta, 1978.

Weart, Spencer R. *The Rise of Nuclear Fear*. Cambridge, MA: Harvard University Press, 2012.

Weisenfeld, Gennifer S. *Gas Mask Nation: Visualizing Civil Defense in Wartime Japan*. Chicago: University of Chicago Press, 2023.

Weisenfeld, Gennifer S. *Imaging Disaster: Tokyo and the Visual Culture of Japan's Great Earthquake of 1923*. Berkeley: University of California Press, 2012.

Wittner, David G., and Philip C. Brown. *Science, Technology, and Medicine in the Modern Japanese Empire*. Abingdon, VA: Routledge, 2016.

Wood, Harry O., and B. Gutenberg. "Earthquake Prediction." *Science* 82, no. 2123 (1935): 219–20.

Wyss, Max. "Do Probabilistic Seismic Hazard Maps Address the Need of the Population?" In *Geoethics: Ethical Challenges and Case Studies in Earth Sciences*, ed. Max Wyss and Silvia Peppoloni, 239–49. Oxford: Elsevier, 2015.

Yamamoto Issei and Imamura Akitsune. *Hoshi to sora, kazan to jishin*. Nihon jidō bunko. Tokyo: Arusu, 1930.

Yamamoto Keizaburō. "Tōkai jishin to shutoken chokka jishin." *Keidanren geppō* 9 (1980): 67–71.

Yamashita Fumio. *Jishin yochi no senkusha Imamura Akitsune no shōgai*. Tokyo: Seijisha, 1989.

———. *Kakusareta daishinsai: Taiheiyō Sensō shi hiroku*. Sendai-shi: Tōhoku Daigaku Shuppankai, 2009.

Yanagita Kunio. "'69 nenmoku' o yogen sita otoko." *Bungei shunjū* 51, no. 13 (1973): 246–58.

Yasuda Yasoi. "Dai sanji Kantō Daishinsai no risuku shisutema bunseki to seisaku shimyurāshun 'Daisanji Kantō daishinsai no shimyurāshun' kenkyū hōkokou (1)." *Kanto Gakuin Daigaku keizai keiei kenkyūjo nenpō* 34 (2012): 123–46.

"Yochi sareta Surugawan Enshūnada chōōgata jishin no zenbō." *Shūkan Gendai* 11, no. 50 (1969): 138–142.

Yoshida Minoru and Asada Toshi. "Jishin ni taisuru Chūgoku no katsudō." *Kagaku Asahi* 36, no. 10 (1976): 62–65.

Young, Louise. *Beyond the Metropolis: Second Cities and Modern Life in Interwar Japan*. Berkeley: University of California Press, 2013.

Yoyogi Shintarō. "(Gakkai zuiichi ninki gakusha) Imamura Akitsune hakushi." *Taiyō* 30, no. 1 (1924): 12–14.

Zierler, David. "Robert J. Geller (BS '73, PhD '77), Seismologist." *Caltech Heritage Project*. September 29, 2022. https://heritageproject.caltech.edu/interviews-updates/robert-geller.

INDEX

ACKNOWLEDGMENTS

Authors usually hold off till the end of the acknowledgments to mention their families and loved ones but mine have waited so long for this book to be done that the least I could do is thank them first. Their patience and support made all the difference in the world at each step along the way, of which there were so many more than I could have imagined at the start. Naomi was there through it all. My daughter Alyssa's homemade scones, inspiring artwork and sense of humor are just some of the many gifts she shared with me. I'm deeply grateful to them both. My son Ian was in elementary school when I started working on this project; I finished it shortly after he graduated from college. *Predicting Disasters* is dedicated to him.

I'm fortunate to have had wonderful colleagues in the History Department and in the Department of East Asian Studies at Brown. Cynthia Brokaw, Ethan Pollock, and Robert Self went out of their ways as department chairs to support the project at key stages in its development. Cherrie Guerzon, Julissa Bautista, and Mary Beth Bryson in History and Kathryn Spicer and Melina Packer in East Asian Studies were generous and kind each and every time I turned to them for help. Benjamin Hein, Nancy Jacobs, Tara Nummedal, Joan Richards, and Seth Rockman were excellent company during our collective relocation to Arnold Hall. Mike Vorenberg's insights—shared during long walks in the late afternoon—played important roles in the writing process as the book neared completion.

Elsewhere at Brown, Karen Fischer and Jan Tullis allowed me to sit in on lectures in their team-taught introductory geology course; Karen also answered my poorly framed questions about Imamura Akitsune and seismometry with clarity and patience. Bjorn Sandstede did the same when I needed help understanding some of the statistical analysis Kawasumi Hiroshi performed on the way to his sixty-nine-year theory. Keegan Cothern, Amy Marshall, Aaron Stark, and Lillian Tsay were generous with their time and knowledge of modern Japanese history. Janet Blume, Baylor Fox-Kemper, and Joe Meisel were (and are) excellent people to be in a band with.

I'm also grateful to the many colleagues and panelists whose questions and suggestions at conferences, workshops, and invited talks helped move the manuscript and its arguments forward. David Bello, Janet Borland, Daniel Burton-Rose, Greg Clancey, Fa-ti Fan, Nathan Hopson, Julia Mariko Jacoby, Scott Knowles, Zhao Lu, Jacob Remes, Jordan Sand, and Charles Schencking were all helpful interlocutors, as were two anonymous readers for the University of Pennsylvania Press. Audra Wolfe also read the manuscript and shared her thoughts on how it might be improved.

So many archivists and librarians assisted in this project that I'm sure this list is incomplete, but my thanks to everyone who helped at the Caltech Archives and Special Collections, the Earthquake Research Institute at the University of Tokyo, the Harvard-Yenching Library, Keio University, the National Diet Library, the Stanford University Libraries, the Tokyo Metropolitan Library, the UC Berkeley Library, the University of Delaware Library, and Yale University Library. At Brown, Toshiyuki Minami, Li Wang, and everyone involved in the interlibrary loan process performed miracles on a regular basis.

The Cogut Institute for the Humanities at Brown, the International Consortium for Research in the Humanities (IKGF) at the Friedrich-Alexander-Universität Erlangen-Nuremberg, and the Max Planck Institute for the History of Science in Berlin (Department III, Artefacts, Action, and Knowledge) were generous sources of support and collegiality. Michael Lackner and Petra Hahm at IKGF and Dagmar Schäfer at Max Planck went out of their way to be helpful. The project also received support from the Northeast Asia Council of the Association for Asian Studies.

In Japan, Tomari Jirō and Tsumura Kenshirō were generous with their time; I'm especially grateful to Professor Tsumura and Yokoyama Hiromichi for their help identifying the individuals in Figure 12. I also can't thank Bob Geller enough for his assistance, both in Japan and in many emailed exchanges since. His quick answers to questions both large and small were always helpful; I'm also grateful to him for introducing me to Tomari Jirō and Professor Tsumura.

Erin Greb transformed a lengthy list of events, dates, and locations into three wonderful maps; I hope readers enjoy them as much as I do. At the Press, Bob Lockhart's advice and editorial guidance were indispensable.

www.ingramcontent.com/pod-product-compliance
Lightning Source LLC
LaVergne TN
LVHW102011210926
848483LV00015B/454/J
9781512825374